Selected Titles in This Series

86 **Stanley N. Burris,** Number theoretic density and logical limit laws, 2001

85 **V. A. Kozlov, V. G. Maz′ya, and J. Rossmann,** Spectral problems associated with corner singularities of solutions to elliptic equations, 2001

84 **László Fuchs and Luigi Salce,** Modules over non-Noetherian domains, 2001

83 **Sigurdur Helgason,** Groups and geometric analysis: Integral geometry, invariant differential operators, and spherical functions, 2000

82 **Goro Shimura,** Arithmeticity in the theory of automorphic forms, 2000

81 **Michael E. Taylor,** Tools for PDE: Pseudodifferential operators, paradifferential operators, and layer potentials, 2000

80 **Lindsay N. Childs,** Taming wild extensions: Hopf algebras and local Galois module theory, 2000

79 **Joseph A. Cima and William T. Ross,** The backward shift on the Hardy space, 2000

78 **Boris A. Kupershmidt,** KP or mKP: Noncommutative mathematics of Lagrangian, Hamiltonian, and integrable systems, 2000

77 **Fumio Hiai and Dénes Petz,** The semicircle law, free random variables and entropy, 2000

76 **Frederick P. Gardiner and Nikola Lakic,** Quasiconformal Teichmüller theory, 2000

75 **Greg Hjorth,** Classification and orbit equivalence relations, 2000

74 **Daniel W. Stroock,** An introduction to the analysis of paths on a Riemannian manifold, 2000

73 **John Locker,** Spectral theory of non-self-adjoint two-point differential operators, 2000

72 **Gerald Teschl,** Jacobi operators and completely integrable nonlinear lattices, 1999

71 **Lajos Pukánszky,** Characters of connected Lie groups, 1999

70 **Carmen Chicone and Yuri Latushkin,** Evolution semigroups in dynamical systems and differential equations, 1999

69 **C. T. C. Wall (A. A. Ranicki, Editor),** Surgery on compact manifolds, second edition, 1999

68 **David A. Cox and Sheldon Katz,** Mirror symmetry and algebraic geometry, 1999

67 **A. Borel and N. Wallach,** Continuous cohomology, discrete subgroups, and representations of reductive groups, second edition, 2000

66 **Yu. Ilyashenko and Weigu Li,** Nonlocal bifurcations, 1999

65 **Carl Faith,** Rings and things and a fine array of twentieth century associative algebra, 1999

64 **Rene A. Carmona and Boris Rozovskii, Editors,** Stochastic partial differential equations: Six perspectives, 1999

63 **Mark Hovey,** Model categories, 1999

62 **Vladimir I. Bogachev,** Gaussian measures, 1998

61 **W. Norrie Everitt and Lawrence Markus,** Boundary value problems and symplectic algebra for ordinary differential and quasi-differential operators, 1999

60 **Iain Raeburn and Dana P. Williams,** Morita equivalence and continuous-trace C^*-algebras, 1998

59 **Paul Howard and Jean E. Rubin,** Consequences of the axiom of choice, 1998

58 **Pavel I. Etingof, Igor B. Frenkel, and Alexander A. Kirillov, Jr.,** Lectures on representation theory and Knizhnik-Zamolodchikov equations, 1998

57 **Marc Levine,** Mixed motives, 1998

56 **Leonid I. Korogodski and Yan S. Soibelman,** Algebras of functions on quantum groups: Part I, 1998

55 **J. Scott Carter and Masahico Saito,** Knotted surfaces and their diagrams, 1998

(Continued in the back of this publication)

Number Theoretic Density and Logical Limit Laws

Mathematical
Surveys
and
Monographs

Volume 86

Number Theoretic Density and Logical Limit Laws

Stanley N. Burris

American Mathematical Society

2000 *Mathematics Subject Classification.* Primary 03C13, 05A15, 05A16, 05C30, 11M41, 11M45, 11N45, 11N80, 11P82, 11U09.

This project has been supported by the Natural Sciences and Engineering Research Council of Canada.

ABSTRACT. This book shows how a study of generating series (power series in the additive case and Dirichlet series in the multiplicative case), combined with structure theorems for the finite models of a sentence, lead to general and powerful results on limit laws, including 0–1 laws. Furthermore, a wealth of evidence is collected to support the thesis that local results in the additive case lift to global results in the multiplicative case.

Library of Congress Cataloging-in-Publication Data

Burris, Stanley.
Number theoretic density and logical limit laws / Stanley N. Burris.
p. cm. — (Mathematical surveys and monographs ; v. 86)
Includes bibliographical references and index.
ISBN 0-8218-2666-2 (alk. paper)
1. Generating functions. 2. Limit theorems (Probability theory) I. Title. II. Mathematical surveys and monographs ; no. 86.
QA164.8 .B87 2000
511′.62—dc21 00-048512

∞ The paper used in this book is acid-free and falls within the guidelines established to ensure permanence and durability.
Visit the AMS home page at URL: http://www.ams.org/

10 9 8 7 6 5 4 3 2 1 06 05 04 03 02 01

Contents

Preface

In the late 1980s Kevin Compton published papers showing how to apply an analysis of the popular partition identities to obtain monadic second-order limit laws, including 0–1 laws, for numerous classes of graphs, posets, etc. I was fascinated by this work. When he gave a colloquium talk in the early 1990s in Waterloo on his recent work on a limit law for Abelian groups, the temptation to learn more about this subject was irresistible. This engaged a favorite area of my own research, algebraic structures.

Compton's papers can be somewhat opaque for specialists in combinatorics and number theory, as well as for specialists in logic, because of the intimate way he has woven these subjects together. After seeing the books [**29**], [**30**] of John Knopfmacher on abstract analytic number theory, it seemed worthwhile to separate Compton's treatment into two parts, one on density in number systems, and the other on the application of number theoretic density results to obtain logical limit laws. Each of these parts is of interest in its own right.

The reader will find a leisurely and detailed exposition of Compton's investigations, and closely related work of others, including recent contributions of Jason Bell, Edward Bender, Peter Cameron, Paweł Idziak, Arnold Knopfmacher, John Knopfmacher, Andrew Odlyzko, Bruce Richmond, András Sárközy, Cameron Stewart, Richard Warlimont, Alan Woods, and the author. The presentation is from the perspective of abstract number systems, in the spirit of John Knopfmacher's work in abstract analytic number theory.

This book has been used as an undergraduate special topics reading text at the University of Waterloo. Part 1, on Additive Number Systems, is completely accessible to an advanced undergraduate student. All chapters preceding Chapter 6 are devoted to number theoretic density, requiring only the usual undergraduate background in analysis, especially in power series, and an exposure to abstract mathematics. The well known ratio test plays a central role. The section on asymptotics, at the end of Chapter 5, uses basic complex analysis, including the Cauchy integral formula. Chapter 6 covers the logical aspects for Part 1. This chapter is self-contained so that one can work through it without prior exposure to logic. It features one of the most delightful tools of logic, the Ehrenfeucht-Fraissé games. (Having had a first course in logic, so that one is comfortable with first-order languages and structures, will no doubt make the chapter more rapid reading.)

Part 2, on Multiplicative Number Systems, offers the challenge and reward of becoming reasonably comfortable with Dirichlet series. The parallels with Part 1 show Dirichlet series as a natural companion of power series. Having worked through Part 1, one will be able to predict many of the results to be proved—the local density results of Part 1 seem, as if by magic, to reappear as global density results in Part 2. There is surely some deep connection between power series and Dirichlet series that we have not yet understood. The last chapter introduces the reader to the Feferman-Vaught Theorem, a favorite tool to analyze direct products, and Skolem's analysis of first-order sentences about Boolean algebras.

The reader will find all the material needed to thoroughly understand the method of Compton for proving logical limit laws. Above all, I think one will be delighted to see so many interesting tools from elementary mathematics pull together to help answer the question "What is the probability that a randomly chosen structure has a given property?"

Thanks go to Paweł Idziak for his contributions to the study of limit laws when I was first starting to work in the area, to Andrew Odlyzko for helping me understand what was going on with the Dirichlet series, to András Sárközy for helping develop the general multiplicative theory of limit laws during a visit to Waterloo, to Cameron Stewart for discussions of the ratio test, to Dejan Delic for help with proofreading an early draft of the book, to Bruce Richmond for helping me to locate and understand several relevant results from asymptotics, to John Knopfmacher for challenging me to take a harder look at what was going on with logical limit laws, to Richard Warlimont for a remarkable amount of beautifully handwritten correspondence regarding ways to improve the presentation, to Jason Bell for keeping me informed of his recent research in this area, and to Karen Yeats, a second year mathematics undergraduate at U. Waterloo who eagerly read the entire manuscript during the summer of 2000, giving me detailed feedback on how the text comes across to an undergraduate. Kevin Compton is, of course, the ultimate inspiration for this work, through his publications and his elegant lectures over the years.

Edward Dunne, Christine Thivierge and Elaine Becker of the AMS Book Program did everything possible to make the transition from manuscript to book a pleasant and trouble-free experience; and Barbara Beeton of the AMS Technical Support group was (as always!) able to solve all of my Tex related problems with the document.

Finally, I want to thank the Natural Sciences and Engineering Research Council of Canada for their long standing support of my investigations in universal algebra, logic, and computation, the support that has made it possible to write this book.

For errata and updates see www.thoralf.uwaterloo.ca.

Stanley Burris
Waterloo, 2000

Overview

When studying a subject with a bewildering variety of specimens one hopes to discover simple structural patterns. This happened in the study of finite relational structures with the discovery of 0–1 laws in the mid 1970s.

A finite relational structure $\mathbf{S} = (S, R_1, \ldots, R_n)$ is a set S with a list of relations $R_1, \ldots, R_n$ on S. The most popular relations R are binary, that is, $R \subseteq S \times S$. But $R \subseteq S^k$, for any $k \geq 1$, is also permitted. For the purpose of this overview it suffices to consider relational structures $\mathbf{S} = (S, R)$ with a single binary relation. One can think of R as a directed edge relation on the set of vertices S and draw a picture, for example:

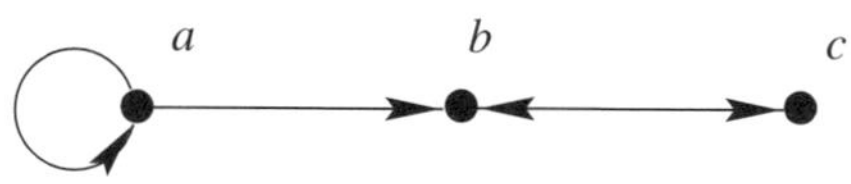

In this picture one sees that aRa, aRb, bRc, cRb. Such structures are *directed graphs.*

As specializations of directed graphs one has several well known classes. If the relation R is irreflexive (not aRa, for any $a \in S$) and symmetric ($aRb \Rightarrow bRa$) one has a *graph.* If it is reflexive, symmetric and transitive, then one has an *equivalence relation.* And if it is reflexive, antisymmetric and transitive, then one has a *poset.* Finite (directed) graphs exhibit incredible diversity and have provided a rich source of examples and problems, from Euler's analysis of the Seven Bridges of Königsberg problem to the modern study of complexity in computer science. Indeed, the interest in finite structures has blossomed in tandem with the growth of theoretical computer science.

One of the fundamental questions is 'What can be said about a randomly chosen structure?' Consider a property $\mathcal{P}$ and a class $\mathcal{K}$ of finite structures. One can ask 'What is the probability that a finite structure chosen randomly from $\mathcal{K}$ satisfies $\mathcal{P}$?' The simplest and most natural definition of this probability is to let p_n be the proportion of structures in $\mathcal{K}$ of size n that have the property $\mathcal{P}$, and then to let the probability that $\mathcal{P}$ holds for a randomly chosen structure from $\mathcal{K}$ be the limit of p_n as n goes to infinity (whenever this limit exists). This probability is called *the asymptotic density* of the collection of structures in $\mathcal{K}$ that satisfy φ.

There is a problem with this definition if there are infinitely many n such that $\mathcal{K}$ has no structures of size n, for then p_n is infinitely often undefined.

This is handled by considering only those p_n that are defined. With this understanding it turns out that the general theory is a minor variation of the theory where one assumes that the p_n are eventually well-defined. So, for the purpose of this overview, it will be assumed that the classes $\mathcal{K}$ have structures of size n for all n greater than some N.

Glebskij, Logan, Liogonkij and Talanov (1969), and independently Fagin (1976), considered properties defined by first-order sentences. For example, the sentence $\forall x \exists y (xRy)$ says, in the case of finite graphs, that there are no isolated vertices; and the sentence $\forall x \forall y \big[(xRy) \rightarrow \exists z(xRz \wedge zRy)\big]$ says one can interpolate a vertex between adjacent vertices. For $\mathcal{K}$ the class of finite directed graphs, or the class of finite graphs, they showed, for any first-order sentence φ, that the probability of φ holding is either 0 or 1. Hence [directed] graphs have a *first-order 0–1 law.* If the probability of a property holding in a class $\mathcal{K}$ is 1 then the property is *almost certainly true* in $\mathcal{K}$.

The method used to prove these results does not generalize readily. First the result is proved for *labeled structures*, using the set of vertices $\{1, \ldots, n\}$ for n-element structures. The following two directed graphs are identified when counting up to isomorphism, but are considered distinct when counting labeled structures:

When counting up to isomorphism one is said to be counting *unlabeled* structures.

To prove a 0–1 law for labeled structures, the original method is to understand the structures in $\mathcal{K}$ well enough to propose a basis Φ for the almost certainly true sentences. Then one proves that each member of Φ is indeed almost certainly true. Finally, to take a labeled 0–1 law back to the unlabeled case one needs to know that the property of being *rigid*[1] is almost certainly true in $\mathcal{K}$. Finding a basis Φ, and proving that rigidity is almost certainly true, are both serious obstacles to extending the applications of this method. Compton was able to carry this out for *posets*, but there are precious few other examples.

In the 1980s an alternate approach to proving 0–1 laws was developed by Compton. By considering *adequate* classes $\mathcal{K}$ of finite structures, that is, classes closed under disjoint union and components, he was able to show that the single condition $a(n-1)/a(n) \rightarrow 1$ is sufficient for a first-order 0–1 law. Here $a(n)$ counts the total number of structures of size n in $\mathcal{K}$ (counting up to isomorphism). This method does not apply to the original examples of graphs and directed graphs as these classes grow so rapidly that one has $a(n-1)/a(n) \rightarrow 0$. However, it is a beautiful technique that does apply to a truly wide range of slowly growing classes such as equivalence relations, permutations, linear forests, etc.

[1] A structure is rigid if the only automorphism is the identity map.

Furthermore, Compton's technique yields a 0–1 law for *monadic second-order logic.* This logic extends the well known first-order logic by allowing quantification over subsets—it is able to express far more properties than first-order logic. For example, in first-order logic one cannot say that a graph is *connected*, but in monadic second-order logic this is expressible by saying that 'if the domain is partitioned into two sets then it is always possible to find two vertices, one from each partition set, with an edge between them':

$$\forall U \forall V \Big[\Big(\exists x(Ux) \wedge \exists y(Vy) \wedge \forall x\big(Ux \leftrightarrow \neg\, Vx\big) \Big) \\ \rightarrow \exists x \exists y \,\big(Ux \wedge Vy \wedge xRy\big) \Big].$$

A considerable portion of the work in Compton's treatment is devoted to studying *Dirichlet density.* (This density is the limiting value of a quotient of two generating series.) He also uses the *fundamental identity*

$$\sum_{n \geq 0} a(n)x^n = \prod_{n \geq 1} \big(1 - x^n\big)^{-p(n)}$$

that relates $a(n)$ to $p(n)$, where $p(n)$ is the number of *components* of size n (counting up to isomorphism). On the surface, Dirichlet density appears to have little to do with the probability that a randomly chosen structure has a property, but under a natural hypothesis one can show that if the probability exists then so does the Dirichlet density, and they are equal. Thus Dirichlet density extends probability (defined as asymptotic density), and the main goal is to find theorems for a converse result that guarantees that if the Dirichlet density exists then it is the asymptotic density. Such converses are called Tauberian Theorems. For the development of the very large part of Compton's theory just described in this paragraph, one does not need the details of the relations involved in the individual structures, but rather just the additive number system associated with $\mathcal{K}$.

Consider the example of the class $\mathcal{K}$ of finite graphs. If one defines the sum of two finite graphs to be their disjoint union then, by identifying isomorphic graphs, one ends up with an additive number system $\mathcal{A}$. The graphs are the 'numbers' in this abstract system. The graph on the empty set gives the zero element of $\mathcal{A}$. The only information that is needed for the asymptotic work with additive number systems is the addition table for the elements and the 'size' (or *norm*) of a graph (defined as the number of vertices). The indecomposable elements, the nonzero elements that cannot be written as the sum of two nonzero elements, are precisely the connected graphs. Clearly every nonzero element of $\mathcal{A}$ can be expressed uniquely as a sum of indecomposable elements since every finite graph with a nonempty set of vertices is uniquely expressible as a disjoint union of connected graphs. The essential features of an additive number system are just the addition operation and the size function. The adjective 'additive' is derived from the fact that the size function $\|\mathsf{a}\|$ is additive, that is, $\|\mathsf{a}+\mathsf{b}\| = \|\mathsf{a}\| + \|\mathsf{b}\|$.

To simplify the presentation, additive number systems are interpolated between adequate classes of structures and the fundamental identities:

Adequate Classes of Structures	—	Additive Number Systems	—	Fundamental Identities

This interpolation allows one to separate the logical aspects from the number-theoretic aspects. Every adequate class gives rise to a (unique) additive number system, and every additive number system can be derived from (many) adequate classes; and each additive number system gives rise to a (unique) fundamental identity that essentially defines the number system.

Given an additive number system $\mathcal{A}$, subsets called *partition sets* play a crucial role in this work. A partition set has the form $\gamma_1\mathsf{P}_1 + \cdots + \gamma_k\mathsf{P}_k$, where $\mathsf{P}_1, \ldots, \mathsf{P}_k$ is a partition of the set P of indecomposable elements of $\mathcal{A}$, and where each of the γ_i is in one of the three forms $(\geq m_i)$, m_i, $(\leq m_i)$. The elements of such a partition set are the members of $\mathcal{A}$ which can be written as a sum of γ_1 members of P_1 plus ... plus γ_k members of P_k. (Repeats are allowed when counting elements.) Thus $1\mathsf{P}$ is just P, and $(\geq 0)\mathsf{P}$ is the entire set of 'numbers' in $\mathcal{A}$. And if $\mathsf{P}_1, \mathsf{P}_2, \mathsf{P}_3$ is a partition of the set P of indecomposable elements of $\mathcal{A}$ then the partition set $(\leq 5)\mathsf{P}_1 + (\geq 4)\mathsf{P}_2 + 2\mathsf{P}_3$ is the set of elements which can be expressed as the sum of at most 5 elements from P_1 plus the sum of at least 4 elements from P_2 plus the sum of exactly 2 elements from P_3.

Now, if $\mathcal{K}$ is an adequate class of structures and $\mathcal{A}$ the corresponding additive number system, then the members of $\mathcal{K}$ that satisfy a given monadic second-order sentence can be described, when viewed in $\mathcal{A}$, as a disjoint union of finitely many partition sets. Thus, if one can show that every partition set of $\mathcal{A}$ has density 0 or 1 then $\mathcal{K}$ has a monadic second-order 0–1 law.

The method of Compton, to establish a 0–1 law for $\mathcal{K}$, is indeed to show that all partition sets of $\mathcal{A}$ have asymptotic density 0 or 1. The necessary and sufficient condition for this to hold is that $a(n-1)/a(n) \to 1$. To obtain interesting conditions that guarantee $a(n-1)/a(n) \to 1$, one turns to the fundamental identity. This ties in with the established work in asymptotic additive number theory, a subject often described by referring to the famous results of Hardy and Ramanujan on the number of partitions of an integer. The applications presented in Chapter 4 are based on an analysis of Bateman and Erdös of additive number systems which have at most one indecomposable of each size.

In a subsequent paper Compton turned to more general limit laws for logic, where one only requires that each sentence φ have a probability of holding (the probability need not be 0 or 1). Again, additive number systems are used to shift to the study of conditions that guarantee that every partition set has an asymptotic density. A rather delicate analysis is needed to establish the main result of Compton. One of the striking corollaries is that if $a(n)$ is asymptotic to $C\beta^n$ then $\mathcal{K}$ has a limit law. Further applications come from the asymptotics of Knopfmacher, Knopfmacher, and Warlimont. By applying the Cauchy integral formula

$$a(n) \;=\; \frac{1}{2\pi i}\int_C \frac{\mathbf{A}(z)}{z^{n+1}}dz$$

to $\mathbf{A}(z) = \sum a(k)z^k$, they are able to find asymptotics for $a(n)$ when $p(n) = C\beta^n + \mathrm{o}(\eta^n)$, where $0 < \eta < \beta$. This gives numerous applications of Compton's theorem, for example, to two-colored linear forests.

In the late 1980s Compton was considering the application of his ideas to the study of randomly chosen members of classes $\mathcal{K}$ of finite algebraic structures—the class of finite Abelian groups is an excellent example. Here the key operation is *direct product*, not disjoint union. The analog of a component (of a relational structure) is, in this setting, a *(directly) indecomposable* structure—that is, a structure which has at least two elements and is not isomorphic to a direct product of smaller structures. The indecomposable finite Abelian groups are precisely the $\mathbf{Z}_{p^n}$, where $\mathbf{Z}_{p^n}$ is the group of integers modulo p^n, for p a prime number. As is well known, every nontrivial finite Abelian group can be uniquely expressed as a product of indecomposable Abelian groups. This is the *unique factorization property* for finite Abelian groups. However, unique factorization does not hold for many important classes of algebraic structures, for example, semigroups.

In the context of finite algebraic structures let us say that $\mathcal{K}$ is *adequate* if it is closed under direct product, direct factors, and it has the unique factorization property. Let $a(n)$ count the number of structures in $\mathcal{K}$ of size n, and let $p(n)$ count the number of indecomposables in $\mathcal{K}$ of size n. (All counting is done up to isomorphism.) From the fact that $\mathcal{K}$ is adequate one has the fundamental identity

$$\sum_{n\geq 1} a(n)n^{-s} \;=\; \prod_{n\geq 2}\left(1-n^{-s}\right)^{-p(n)}.$$

What is the probability that a randomly chosen structure from $\mathcal{K}$ satisfies a given property $\mathcal{P}$? If one defines p_n as before, namely the proportion of structures in $\mathcal{K}$ of size n that have the property $\mathcal{P}$, then very little can be said. However, if one uses the *global* count, that is, let P_n be the proportion of structures in $\mathcal{K}$ of size *at most* n that have the property $\mathcal{P}$, and if one lets the probability be the limit of P_n as n goes to infinity (provided this exists), then the results for relational structures have remarkable parallels in the algebraic setting.

In Part 2 multiplicative number systems are interpolated between adequate classes and fundamental identities, giving an exact analog of the use of additive number systems in Part 1. This is used, as in Part 1, to give a separation of the number theoretic and logical aspects.

Adequate Classes of Structures	—	Multiplicative Number Systems	—	Fundamental Identities

In the case of Abelian groups one obtains the associated number system by letting the 'numbers' be the finite groups **G**, and letting multiplication be direct product. These number systems are said to be multiplicative because the size function is multiplicative, that is, $\|\mathsf{a} \cdot \mathsf{b}\| = \|\mathsf{a}\| \cdot \|\mathsf{b}\|$.

The multiplicative number system analog of Compton's theorem (on general limit laws for relational structures) has, as a corollary, the fact that if $A(x)$ is asymptotic to cx^α, where $A(x) = \sum_{n\leq x} a(n)$, then $\mathcal{K}$ has a first-order limit law. This applies to the example of Abelian groups, and one can show, for example, that the probability of a finite Abelian group having an element of order 2 is $1 - \prod_{n\geq 1}\left(1 - 2^{-n}\right) \approx 0.71$.

Many of the interesting examples, like Abelian groups, have limit laws, but not 0–1 laws. Further examples are again found by turning to complex analysis, using Perron's integral formula, to generalize Oppenheim's asymptotics for the number of ways to factor the numbers less than or equal to n. From this analysis one concludes that if $p(n) = Cn^\alpha + O(n^\beta)$ with $C > 0$, $\alpha \geq 0$, and $\beta < \alpha$, then $\mathcal{K}$ has a first-order limit law. Thus the class of finite lattices that decompose into a product of chains has a first-order limit law.

Chapter 6, the last chapter of Part 1, covers the logic results for adequate classes (with respect to disjoint union) of purely relational structures; and Chapter 12, the last chapter of Part 2, covers the logic results for adequate classes (with respect to direct product) of structures.

These two chapters can be briefly summarized as follows. Given an adequate class $\mathcal{K}$ and a sentence φ, let $\mathcal{K}_\varphi$ be the class of members of $\mathcal{K}$ that satisfy φ. The goal is to prove that $\mathcal{K}_\varphi$ is a disjoint union of finitely many partition classes. Then the number theoretic results on asymptotic density of partition sets can be used.

Three tools are needed to prove these logic results: the Ehrenfeucht-Fraissé games in the additive case; and the Feferman-Vaught Theorem, with Skolem's analysis of sentences about finite Boolean algebras, in the multiplicative case. These tools are fully developed in Chapters 6 and 12.

Notation Guide

The most basic convention used in this text is that **power series** and **Dirichlet series** are designated by upper case boldface letters, with the corresponding lower case italic letters used for the coefficients, and the corresponding upper case italic leters used for the partial sums of the coefficients.

Power series (Part 1)	Dirichlet series (Part 2)	Sums of coefficients
$\mathbf{A}(x) = \sum_n a(n)x^n$	$\mathbf{A}(x) = \sum_n a(n)n^{-x}$	$A(x) = \sum_{n\le x} a(n)$
$\vdots$	$\vdots$	$\vdots$
$\mathbf{S}(x) = \sum_n s(n)x^n$	$\mathbf{S}(x) = \sum_n s(n)n^{-x}$	$S(x) = \sum_{n\le x} s(n)$
$\vdots$	$\vdots$	$\vdots$

For work with abstract number systems the following notational conventions are used:

$\mathcal{A}, \mathcal{B}, \ldots$	number systems
$\mathcal{A} = (\mathsf{A}, \mathsf{P}, +, 0, \|\ \|)$	an additive number system
$\mathcal{A} = (\mathsf{A}, \mathsf{P}, \cdot, 1, \|\ \|)$	a multiplicative number system
$\mathsf{a}, \mathsf{b}, \ldots, \mathsf{p}, \ldots$	elements of the number system
A	set of elements of the number system
P	set of indecomposable elements
$\mathsf{B}, \mathsf{C}, \ldots$	subsets of A
$b(n)$	local count function for B
$B(x) = \sum_{n\le x} b(n)$	global count function for B
$\mathbf{B}(x) = \sum_n b(n)x^n$	generating series for B (additive case)
$\mathbf{B}(x) = \sum_n b(n)n^{-x}$	generating series for B (multiplicative case)

(continued)

ρ	radius of convergence of $\mathbf{A}(x)$ (Part 1)
α^-	abscissa of convergence of $\mathbf{A}(x)$ (Part 2)
α	abscissa of absolute convergence of $\mathbf{A}(x)$ (Part 2)
$\delta(\mathsf{B}) = \lim_{\substack{n\to 0 \\ a(n)\neq 0}} b(n)/a(n)$	local asymptotic density of B
$\Delta(\mathsf{B}) = \lim_{n\to\infty} B(n)/A(n)$	global asymptotic density of B
$\partial(\mathsf{B}) = \lim_{x\to\rho} \mathbf{B}(x)/\mathbf{A}(x)$	Dirichlet density of B (additive case)
$\partial(\mathsf{B}) = \lim_{x\to\alpha} \mathbf{B}(x)/\mathbf{A}(x)$	Dirichlet density of B (multiplicative case)
$f(x) \in \mathcal{PS}_c$	$f(x)$ has a power series expansion that converges absolutely at c
$f(x) \in \mathcal{DS}_c$	$f(x)$ has a Dirichlet series expansion that converges absolutely at c
$\mathsf{supp}\, f(n) = \{n : f(n) > 0\}$	support of $f(n)$
$\mathbf{S}\vert_m(x) = \sum_{k\leq m} s(k)x^k$	mth truncation of the power series $\mathbf{S}(x)$
$\mathbf{S}\vert_m(x) = \sum_{k\leq m} s(k)k^{-x}$	mth truncation of the Dirichlet series $\mathbf{S}(x)$
$0\mathsf{B} = \{0\}$	(for additive systems: Part 1)
$\mathsf{B}^0 = \{1\}$	(for multiplicative systems: Part 2)
$m\mathsf{B} = \{\mathsf{a}_1 + \cdots + \mathsf{a}_m : \mathsf{a}_i \in \mathsf{B}\}$	(for additive systems: Part 1)
$\mathsf{B}^m = \{\mathsf{a}_1 \cdots \mathsf{a}_m : \mathsf{a}_i \in \mathsf{B}\}$	(for multiplicative systems: Part 2)
$(\leq m)\mathsf{B} = 0\mathsf{B} \cup \cdots \cup m\mathsf{B}$	(for additive systems: Part 1)
$\mathsf{B}^{\leq m} = \mathsf{B}^0 \cup \cdots \cup \mathsf{B}^m$	(for multiplicative systems: Part 2)
$(\geq m)\mathsf{B} = \bigcup_{n\geq m} n\mathsf{B}$	(for additive systems: Part 1)
$\mathsf{B}^{\geq m} = \bigcup_{n\geq m} \mathsf{B}^n$	(for multiplicative systems: Part 2)

for $\mathsf{Q} \subseteq \mathsf{P}$

$\mathsf{A}_\mathsf{Q} = \omega\mathsf{Q} = \bigcup_{n\geq 0} n\mathsf{Q}$	(for additive systems: Part 1)
$\mathsf{A}_\mathsf{Q} = \mathsf{Q}^\omega = \bigcup_{n\geq 0} \mathsf{Q}^n$	(for multiplicative systems: Part 2)
$\partial_\mathsf{Q}(\gamma\mathsf{Q}) = \lim_{x\to\rho} \frac{[\gamma\mathbf{Q}](x)}{\mathbf{A}_\mathsf{Q}(x)}$	relativized Dirichlet density, additive case
$\partial_\mathsf{Q}(\mathsf{Q}^\gamma) = \lim_{x\to\alpha} \frac{[\mathbf{Q}^\gamma](x)}{\mathbf{A}_\mathsf{Q}(x)}$	relativized Dirichlet density, multiplicative case

Part 1

Additive Number Systems

CHAPTER 1

Background from Analysis

This chapter reviews basic facts about power series and infinite products that are needed in Part 1. The reader will no doubt be familiar with most of this material and can skip over it. However, note the meaning of $\mathcal{PS}_c$ in Definition 1.16.

The set of complex numbers, and significant subsets, are designated as follows:

symbol	meaning
$\mathbb{C}$	set of complex numbers
$\mathbb{R}$	set of real numbers
$\mathbb{Q}$	set of rational numbers
$\mathbb{Z}$	set of integers
$\mathbb{N}$	set of natural numbers $\{0, 1, 2, \dots\}$
$\mathbb{P}$	set of positive integers $\{1, 2, \dots\}$

A sequence, series, or function, can have $-\infty$ or ∞ as a limit. But the word 'converges' is only used when the limit is finite. If $\lim_n f(n) = \infty$, or $\sum_n s(n) = \infty$, the function, respectively series, is said to *diverge to* ∞.

At times it will be convenient to allow functions to take values in the *extended reals* $[-\infty, \infty]$, and to extend the domain of functions like e^x to the extended reals. The usual partially defined arithmetic[1] of infinity is:

$$\begin{aligned}
\infty + r &= r + \infty = \infty, && \text{for } r \in (-\infty, \infty] \\
\infty \cdot r &= r \cdot \infty = \infty, && \text{for } r \in (0, \infty] \\
\infty / r &= \infty, && \text{for } r \in (0, \infty) \\
r/\infty &= 0, && \text{for } r \in (-\infty, \infty).
\end{aligned}$$

If $f(x)$ is a function such that $\lim\limits_{x\to\infty} f(x) = \infty$ then the expression $f(\infty) = \infty$ is permitted, etc. For example,

$$\begin{aligned}
e^{\infty} &= \infty \\
e^{-\infty} &= 0.
\end{aligned}$$

[1]Note that division by 0 is not defined since it does not have the 'continuity' property. For example, if $1/0$ were assigned a value in the extended reals then

$$\lim_{x\to 0} \frac{1}{x} = \frac{1}{0}$$

would be impossible since the limit from the left is $-\infty$, and from the right ∞.

This is used in obvious ways, for example,

$$e^{f(x)} = g(x), \qquad \text{for } x \geq 0,$$

means that, for each $x \in [0, \infty)$, either both $f(x)$ and $g(x)$ are finite and the equation holds, or $f(x) = g(x) = \infty$, or $f(x) = -\infty$ and $g(x) = 0$. Another example of acceptable usage of the extended reals is:

$$\sum_{n \geq 0} x^n = \begin{cases} \dfrac{1}{1-x} & \text{if } 0 \leq x < 1 \\ \infty & \text{if } x \geq 1. \end{cases}$$

1.1. Series and infinite products

For real valued (or complex valued) functions f, g, the operations of scalar multiplication, addition, multiplication, and composition are defined pointwise, that is:

Definition 1.1. Let λ be a constant, and f, g functions. Then

$$\begin{aligned} (\lambda f)(x) &= \lambda \cdot f(x) \\ (f+g)(x) &= f(x) + g(x) \\ (f \cdot g)(x) &= f(x) \cdot g(x) \\ (f \circ g)(x) &= f\big(g(x)\big), \end{aligned}$$

whenever the right sides are defined.

1.1.1. Power series basics.

Definition 1.2. A series $\mathbf{S}(x)$ of the form $\sum_{n \geq 0} s(n)x^n$ is called a *power series*. $s(n)$ is the *coefficient* of x^n in $\mathbf{S}(x)$, also written as $\big[x^n\big]\mathbf{S}(x)$.

The convention used in Part 1 is that power series are designated by upper case boldface letters, with the corresponding lower case italic letters used for the coefficients, and the corresponding upper case italic letters used for partial sums of the coefficients.

Power series	Partial sums of coefficients
$\mathbf{A}(x) = \sum_n a(n)x^n$	$A(x) = \sum_{n \leq x} a(n)$
$\vdots$	$\vdots$
$\mathbf{S}(x) = \sum_n s(n)x^n$	$S(x) = \sum_{n \leq x} s(n)$
$\vdots$	$\vdots$

The logical limit laws of Part 1 are based on making a connection between logic and power series.

Definition 1.3. The *Cauchy product* $\mathbf{S}(x) * \mathbf{T}(x)$ of two power series $\mathbf{S}(x)$ and $\mathbf{T}(x)$ is defined by

$$\mathbf{S}(x) * \mathbf{T}(x) \;=\; \sum_{n\geq 0}\Big(\sum_{i+j=n} s(i)\cdot t(j)\Big)x^n.$$

The Cauchy product is associative, and, for finitely many power series $\mathbf{S}_k(x) = \sum_n s_k(n)x^n$, $1 \leq k \leq m$, the Cauchy product is given by:

$$\begin{aligned}\prod^{*}_{1\leq k\leq m} \mathbf{S}_k(x) \;&=\; \mathbf{S}_1(x) * \cdots * \mathbf{S}_m(x)\\ &=\; \sum_n\Big(\sum_{n_1+\cdots+n_m=n} s_1(n_1)\cdots s_m(n_m)\Big)x^n.\end{aligned}$$

Proposition 1.4. *Let $\mathbf{S}(x)$ and $\mathbf{T}(x)$ be power series with real coefficients, and let $c > 0$.*

(a) *There is a $\rho \in [0,\infty]$ such that $\mathbf{S}(x)$ converges absolutely for $|x| < \rho$, and diverges for $|x| > \rho$. ρ is called the* radius of convergence *of $\mathbf{S}(x)$ and will usually be designated as ρ_S.*

(b) *The radius of convergence of $\mathbf{S}(x)$ is*

$$\rho_s \;=\; 1\Big/\limsup_{n\to\infty}|s(n)|^{1/n}.$$

(c) *If all $s(n) \geq 0$ then $\mathbf{S}(x)$ defines a function on $[0,\rho_s)$ that is nonnegative and nondecreasing.*

(d) *If $\mathbf{S}(x)$ converges absolutely at c then $\mathbf{S}(x)$ converges uniformly on $[0,c]$.*

(e) *If $\mathbf{S}(x)$ converges absolutely at c then $\mathbf{S}(x)$ is continuous on $[0,c]$.*

(f) *If $\mathbf{S}(x)$ and $\mathbf{T}(x)$ converge absolutely at c then*

$$\mathbf{S}(x)\cdot\mathbf{T}(x) \;=\; \mathbf{S}(x) * \mathbf{T}(x)$$

holds on $[0,c]$.

(g) *One can differentiate a power series term by term, namely*

$$\frac{d^k}{dx^k}\mathbf{S}(x) \;=\; \sum n(n-1)\cdots(n-k+1)s(n)x^{n-k}, \qquad \textit{for } |x| < \rho_S.$$

PROOF. These are standard results in a first course in real analysis. For example, (d) follows from the Weierstrass M–test since $|s(n)x^n| \leq |s(n)c^n|$, for $x \in [0,c]$. Then (e) follows from uniform convergence. □

1.1.2. Double series. A double series $\sum_i\sum_j s_{ij}$ is *absolutely convergent* if, for each i, the series $\sum_j |s_{ij}|$ is convergent, and the series $\sum_i(\sum_j |s_{ij}|)$ is convergent. This implies that the series $\sum_i\big|\sum_j s_{ij}\big|$, $\sum_j s_{ij}$, and $\sum_i\sum_j s_{ij}$ are convergent. One can interchange the order of summation in a double series that is absolutely convergent.

Lemma 1.5. *If $\sum_i \sum_j s_{ij}$ is absolutely convergent then*

$$\sum_i \Big(\sum_j s_{ij}\Big) = \sum_j \Big(\sum_i s_{ij}\Big),$$

and all sums mentioned are absolutely convergent.

In the case that the s_{ij} are nonnegative, the summation signs can always be interchanged—either both sides of the above equation will converge to the same value or both will diverge to ∞.

1.1.3. Infinite Cauchy products. Underlying the success of Part 1 are the power series that can be expressed as products of geometric power series. In this section the Cauchy product of infinitely many geometric power series is defined.

Definition 1.6. Given $g : \mathbb{N} \to \mathbb{P}$ with the property $g(n) \to \infty$, let

$$\begin{aligned} \mathbf{S}_n^g(x) &= 1 + x^{g(n)} + x^{2g(n)} + \cdots + x^{kg(n)} + \cdots \\ \mathbf{T}_n^g(x) &= \mathop{\boxed{*}\!\!\prod}_{j \le n} \mathbf{S}_j^g(x) = \Big(\sum_{k \ge 0} x^{kg(0)}\Big) * \cdots * \Big(\sum_{k \ge 0} x^{kg(n)}\Big). \end{aligned}$$

$\mathbf{S}_n^g(x)$ is a geometric power series with increment factor $x^{g(n)}$, and $\mathbf{T}_n^g(x)$ is the product of the first $n+1$ of these geometric series. A remarkable property of the coefficients of the $\mathbf{T}_n^g(x)$ is that they *stabilize*. This means, for any fixed m, that the sequence of coefficients of x^m,

$$\big[x^m\big]\mathbf{T}_0^g(x), \quad \big[x^m\big]\mathbf{T}_1^g(x), \quad \ldots, \quad \big[x^m\big]\mathbf{T}_n^g(x), \quad \ldots,$$

eventually becomes constant. The reason is simple—if $g(k) > m$ then the power series

$$\mathbf{S}_k^g(x) = 1 + x^{g(k)} + x^{2g(k)} + \cdots$$

does not contribute to the coefficient of x^m in $\mathbf{T}_n^g(x)$.

Definition 1.7. Given $g : \mathbb{N} \to \mathbb{P}$ with $g(n) \to \infty$, the *infinite Cauchy product* $\mathop{\boxed{*}\!\!\prod}_{n \ge 0} \mathbf{S}_n^g(x)$ is the power series $\mathbf{T}^g(x)$ whose coefficients are given by

$$\big[x^m\big]\mathbf{T}^g(x) = \lim_{n \to \infty} \big[x^m\big]\mathbf{T}_n^g(x),$$

that is, the mth coefficient of the infinite Cauchy product is the stable value of the mth coefficients of the $\mathbf{T}_n^g(x)$.

Clearly this defines a power series with nonnegative integer coefficients. However, work is needed to show that there is an analytic connection between this power series and $\lim_{n \to \infty} \mathbf{T}_n^g(x)$.

1.1.4. Convergence of infinite products.

Definition 1.8. Let a sequence $s(n)$ be given. Define the infinite product $\prod_n s(n)$ by

$$\prod_{n\geq 0} s(n) \;=\; \lim_{n\to\infty}\big(s(0)\cdots s(n)\big),$$

provided the limit exists.[2]

Infinite products with $s(n) \geq 1$, for $n \geq 0$, will be of primary interest. This means one can write $s(n) = 1 + r(n)$ with $r(n) \geq 0$. ($s(n) = r(n) = \infty$ is permitted.)

Lemma 1.9. *Given a sequence $r(n)$ of nonnegative extended reals, the infinite product*

$$\prod_{n\geq 0}\big(1 + r(n)\big)$$

always exists, and it is either a real number ≥ 1, or ∞. Furthermore, the value is independent of rearrangements of the sequence $r(n)$.

PROOF. The existence of the infinite product follows from noting that the sequence of finite products

$$\big(1 + r(1)\big)\cdots\big(1 + r(n)\big)$$

has terms ≥ 1, and is nondecreasing.

Suppose that $r'(n)$ is a rearrangement of $r(n)$. Then, for any k, there is an ℓ such that the terms $r(0), \ldots, r(k)$ appear in the list $r'(0), \ldots, r'(\ell)$, so

$$\prod_{n\leq k}\big(1 + r(n)\big) \;\leq\; \prod_{n\leq \ell}\big(1 + r'(n)\big) \;\leq\; \prod_{n\geq 0}\big(1 + r'(n)\big).$$

Thus

$$\prod_{n\geq 0}\big(1 + r(n)\big) \;\leq\; \prod_{n\geq 0}\big(1 + r'(n)\big).$$

By symmetry, the two infinite products must be equal. □

(If any of the $r(n) = \infty$ then the infinite product is ∞.)

Definition 1.10. Given a power series $\mathbf{R}(x)$, define the *mth truncation* $\mathbf{R}|_m(x)$ of $\mathbf{R}(x)$ to be the polynomial

$$\mathbf{R}|_m(x) \;=\; r(0) \,+\, r(1)x \,+\cdots+\, r(m)x^m.$$

Lemma 1.11. *If $\mathbf{R}(x)$ has nonnegative coefficients then, for $x \geq 0$,*

$$\lim_{m\to\infty}\mathbf{R}|_m(x) \;=\; \mathbf{R}(x).$$

[2]It is usual to assume $s(0)\cdots s(n) \not\to 0$.

PROOF. This just uses the definition of the value of a power series $\mathbf{R}(x)$ at a real number x, namely

$$\mathbf{R}(x) \;=\; \lim_{m\to\infty} \mathbf{R}|_m(x),$$

provided the right side exists. When the coefficients are nonnegative, the limit exists. □

For $g : \mathbb{N} \to \mathbb{P}$, and, for $x \geq 0$, the $\mathbf{S}_n^g(x)$ take values in $[1,\infty]$, so the infinite product $\prod_{n\geq 0} \mathbf{S}_n^g(x)$ exists, possibly with the value ∞. Since $g(n) \to \infty$, this infinite product has the same value as the infinite Cauchy product.

Lemma 1.12. *If $g : \mathbb{N} \to \mathbb{P}$ with $g(n) \to \infty$ then, for $x \geq 0$,*

$$\mathbf{T}^g(x) \;=\; \mathop{\prod\nolimits^{*}}_{n\geq 0} \mathbf{S}_n^g(x) \;=\; \prod_{n\geq 0} \mathbf{S}_n^g(x).$$

PROOF. Given m, choose N such that $n \geq N$ implies $g(n) > m$. Then, making use of the fact that the Cauchy product of finitely many polynomials is the product of the polynomials,

$$\begin{aligned}
\Big(\mathop{\prod\nolimits^{*}}_{n\geq 0} \mathbf{S}_n^g\Big)\Big|_m(x) \;&\leq\; \mathop{\prod\nolimits^{*}}_{n\leq N} \mathbf{S}_n^g|_m(x) \;=\; \prod_{n\leq N} \mathbf{S}_n^g|_m(x) \\
&\leq\; \prod_{n\leq N} \mathbf{S}_n^g(x) \;\leq\; \prod_{n\geq 0} \mathbf{S}_n^g(x),
\end{aligned}$$

so

$$\mathop{\prod\nolimits^{*}}_{n\geq 0} \mathbf{S}_n^g(x) \;\leq\; \prod_{n\geq 0} \mathbf{S}_n^g(x).$$

For the other inequality,

$$\prod_{n\leq M} \mathbf{S}_n^g\Big|_m(x) \;=\; \mathop{\prod\nolimits^{*}}_{n\leq M} \mathbf{S}_n^g\Big|_m(x) \;\leq\; \mathop{\prod\nolimits^{*}}_{n\geq 0} \mathbf{S}_n^g(x),$$

so

$$\prod_{n\geq 0} \mathbf{S}_n^g(x) \;\leq\; \mathop{\prod\nolimits^{*}}_{n\geq 0} \mathbf{S}_n^g(x).$$

□

The content of this lemma can be expressed as 'One can take the infinite Cauchy product of the $\mathbf{S}_n^g$ and then evaluate at x, or one can evaluate each $\mathbf{S}_n^g$ at x, and then take the infinite product of their values. The results are the same.'

This gives a key tool when working with fundamental identities.

Proposition 1.13. *If $g : \mathbb{N} \to \mathbb{P}$ with $g(n) \to \infty$ then, for $0 \leq x < 1$,*

$$\mathop{\prod\nolimits^{*}}_{n\geq 0} \sum_{k\geq 0} x^{kg(n)} \;=\; \prod_{n\geq 0} \sum_{k\geq 0} x^{kg(n)} \;=\; \prod_{n\geq 0} \big(1 - x^{g(n)}\big)^{-1}.$$

PROOF. This follows immediately from Lemma 1.12, noting that, for $0 \le x < 1$,

$$\mathbf{S}_n^g(x) \;=\; \sum_{k\ge 0} x^{kg(n)} \;=\; \big(1 - x^{g(n)}\big)^{-1}.$$

□

A classic test for when an infinite product converges (to a finite value) is the following.

Lemma 1.14. *For $r(n) \ge 0$,*

$$\prod_n \big(1 + r(n)\big) \;\; \textit{converges iff} \;\; \sum_n r(n) \;\; \textit{converges.}$$

PROOF. As

$$\sum_{n\le N} r(n) \;\le\; \prod_{n\le N} \big(1 + r(n)\big) \;\le\; \prod_{n\le N} e^{r(n)} \;=\; \exp\Big(\sum_{n\le N} r(n)\Big),$$

the convergence of the product guarantees the convergence of the sum, and vice versa. □

Lemma 1.15. *Let $f : \mathbb{P} \to \mathbb{N}$ be given. Then, for $x \in [0,1)$,*

$$\prod_{n\ge 1} (1 - x^n)^{-f(n)} \;\; \textit{converges iff} \;\; \sum_{n\ge 1} f(n)x^n \;\; \textit{converges.}$$

PROOF. Since

$$(1 - x^n)^{-f(n)} \;=\; \Big(1 + \frac{x^n}{1 - x^n}\Big)^{f(n)},$$

Lemma 1.14 shows that $\prod_n (1-x^n)^{-f(n)}$ converges iff $\sum_n \dfrac{f(n)x^n}{1-x^n}$ converges. For $x \in [0,1)$, and, for sufficiently large n,

$$x^n \;\le\; \frac{x^n}{1 - x^n} \;\le\; 2x^n,$$

so, by the comparison test, $\sum_n \dfrac{f(n)x^n}{1-x^n}$ converges iff $\sum_n f(n)x^n$ converges.

□

1.2. Power series expansions

The following definition of a power series expansion of a function is tailored to the later work on asymptotic density.

Definition 1.16. A power series $\mathbf{S}(x)$ is a *power series expansion* of a function $f(x)$ if, for some $c > 0$, one has $f(x) = \mathbf{S}(x)$ on $[0,c)$. Then one says that $f(x)$ has a *power series expansion* $\mathbf{S}(x)$ *on* $[0,c)$. The expression $f(x) \in \mathcal{PS}_c$ means that $f(x)$ has a power series expansion on $[0,c)$ that converges absolutely at c.

If $f(x)$ has a power series expansion $\mathbf{S}_1(x)$ on $[0, c_1)$, where $c_1 > 0$, and a power series expansion $\mathbf{S}_2(x)$ on $[0, c_2)$, where $c_2 > 0$, then the two power series are identical. Thus if $f(x)$ has *a* power series expansion then one can speak of *the* power series expansion of $f(x)$.

Any reference to $\mathcal{PS}_c$ automatically assumes that $c > 0$. For $\mathbf{S}(x)$ a power series, note that $\mathbf{S}(x) \in \mathcal{PS}_c$ holds iff $\mathbf{S}(c)$ converges absolutely.

Lemma 1.17. *Suppose* $f(x) \in \mathcal{PS}_c$ *has the power series expansion* $\mathbf{S}(x)$. *Then*

$$\lim_{x \to c^-} f(x) = \mathbf{S}(c).$$

PROOF. $f(x) \in \mathcal{PS}_c$ means that $\mathbf{S}(x)$ converges absolutely at c, so $\mathbf{S}(x)$ is continuous on $[0, c]$. Then

$$\lim_{x \to c^-} \mathbf{S}(x) = \mathbf{S}(c).$$

As $f(x) = \mathbf{S}(x)$ on $[0, c)$, the lemma follows. □

The next elementary lemma says that $\mathcal{PS}_c$ is an algebra of functions, in view of the obvious fact that constant functions are in $\mathcal{PS}_c$.

Lemma 1.18. *If* $f, g \in \mathcal{PS}_c$ *and* $\lambda \in \mathbb{R}$ *then*

$$\begin{aligned} \lambda f &\in \mathcal{PS}_c \\ f + g &\in \mathcal{PS}_c \\ f \cdot g &\in \mathcal{PS}_c. \end{aligned}$$

The next lemma shows how to find the power series expansions corresponding to these operations.

Lemma 1.19. *Suppose* $\mathbf{S}(x)$, $\mathbf{T}(x) \in \mathcal{PS}_c$, *and* $\lambda \in \mathbb{R}$. *Then the following hold, for* $0 \le x \le c$:

$$\begin{aligned} \lambda \mathbf{S}(x) &= \sum \big(\lambda s(n)\big) x^n \\ \mathbf{S}(x) + \mathbf{T}(x) &= \sum \big(s(n) + t(n)\big) x^n \\ \mathbf{S}(x) \cdot \mathbf{T}(x) &= \mathbf{S}(x) * \mathbf{T}(x). \end{aligned}$$

Furthermore, if the power series $\mathbf{S}(x)$ *and* $\mathbf{T}(x)$ *have nonnegative coefficients, and* $\lambda > 0$, *then the above hold for any* $x \ge 0$.

PROOF. The above results are standard except, perhaps, for the last statement. In the case that the coefficients are nonnegative, the first two equations are straightforward to verify, for $x \ge 0$. For the third, let

$$\mathbf{R}(x) = \mathbf{S}(x) * \mathbf{T}(x).$$

Then, for $x \ge 0$,

$$\mathbf{S}|_M(x) \cdot \mathbf{T}|_M(x) \le \mathbf{R}|_{2M}(x) \le \mathbf{S}|_{2M}(x) \cdot \mathbf{T}|_{2M}(x)$$

leads to the desired conclusion. □

Lemma 1.20. *Suppose $f_k(x) \in \mathcal{PS}_c$, say $f_k(x)$ has the power series expansion $\sum_n s_k(n)x^n$, and suppose that the double series $\sum_k \sum_n s_k(n)c^n$ converges absolutely. Then $\sum_k f_k(x) \in \mathcal{PS}_c$, and the power series expansion of $\sum_k f_k(x)$ is $\sum_n \big(\sum_k s_k(n)\big)x^n$.*

PROOF. For $x \in [0, c]$, the comparison test shows that $\sum_k \sum_n s_k(n)x^n$, a double series, converges absolutely. It follows that

$$\sum_k f_k(x) \;=\; \sum_k \sum_n s_k(n)x^n,$$

with both sides convergent, for $x \in [0, c)$. By Lemma 1.5, for $x \in [0, c]$,

$$\begin{aligned}\sum_k \sum_n s_k(n)x^n &= \sum_n \sum_k s_k(n)x^n \\ &= \sum_n \Big(\sum_k s_k(n)\Big)x^n,\end{aligned}$$

and all sums mentioned are absolutely convergent. Thus $\sum_k f_k(x) \in \mathcal{PS}_c$. Furthermore, the power series expansion of $\sum_k f_k(x)$ is as described. □

1.2.1. Exponentiation. Exponentiation plays a big role in understanding asymptotic density.

Lemma 1.21. *If $f(x) \in \mathcal{PS}_c$ then $e^{f(x)} \in \mathcal{PS}_c$.*

PROOF. Suppose $f(x) \in \mathcal{PS}_c$. Then there is a power series $\mathbf{S}(x) = \sum_n s(n)x^n$ such that $f(x) = \mathbf{S}(x)$ on $[0, c)$, and $\sum_n |s(n)|c^n$ converges. Thus, for $x \in [0, c)$,

$$\begin{aligned}e^{f(x)} &= \sum_k \frac{1}{k!}\mathbf{S}(x)^k \\ &= \sum_k \frac{1}{k!} \sum_n \sum_{n_1+\cdots+n_k=n} s(n_1)\cdots s(n_k)x^n \\ &= \sum_k \sum_n \Big(\frac{1}{k!} \sum_{n_1+\cdots+n_k=n} s(n_1)\cdots s(n_k)\Big)x^n. \qquad (1.21.1)\end{aligned}$$

Since, by Lemma 1.18, $\mathbf{S}(x)^k/k! \in \mathcal{PS}_c$, it follows that if the doubly infinite series (1.21.1) is absolutely convergent at $x = c$ then, by Lemma 1.20, one has $e^{f(x)} \in \mathcal{PS}_c$.

To show (1.21.1) is absolutely convergent at $x = c$ consider the power series $\mathbf{T}(x) = \sum_n |s(n)|x^n$. As noted above, this series is (absolutely) convergent at $x = c$. Thus

$$\begin{aligned}e^{\mathbf{T}(c)} &= \sum_k \frac{1}{k!}\mathbf{T}(c)^k \\ &= \sum_k \frac{1}{k!} \sum_n \sum_{n_1+\cdots+n_k=n} |s(n_1)|\cdots|s(n_k)|c^n\end{aligned}$$

$$= \sum_k \sum_n \Big(\frac{1}{k!} \sum_{n_1+\cdots+n_k=n} |s(n_1)| \cdots |s(n_k)| \Big) c^n. \tag{1.21.2}$$

Thus the double series (1.21.2) is convergent (to $e^{\mathbf{T}(c)}$). As

$$\Big| \sum_{n_1+\cdots+n_k=n} s(n_1)\cdots s(n_k) \Big| \le \sum_{n_1+\cdots+n_k=n} |s(n_1)|\cdots|s(n_k)|,$$

the comparison test shows that (1.21.1) is indeed absolutely convergent at $x = c$. This finishes the proof. □

This section concludes with a look at the radius of convergence of the power series expansion of $e^{\mathbf{S}(x)}$, given that $\mathbf{S}(x)$ is a power series with radius of convergence $\rho_s > 0$. For $0 < c < \rho_s$, $\mathbf{S}(x) \in \mathcal{PS}_c$, so, by Lemma 1.21, $e^{\mathbf{S}(x)} \in \mathcal{PS}_c$. This means that the radius of convergence $\widehat{\rho}$ of the power series expansion of $e^{\mathbf{S}(x)}$ is at least ρ_s. However, it can be much more, for example, let $\mathbf{S}(x) = \sum -x^m/m$. Then $\rho_s = 1$ but $\widehat{\rho} = \infty$ (as $\mathbf{S}(x) = \log(1-x)$ on $[0,1)$). There is an interesting setting where $\widehat{\rho} = \rho_s$, and this will be important to find asymptotic formulas (see Proposition 3.46).

Lemma 1.22. *Suppose $\mathbf{S}(x)$ is a power series with nonnegative coefficients and radius of convergence $\rho_s > 0$. Then $e^{\mathbf{S}(x)}$ has a power series expansion with nonnegative coefficients and radius of convergence ρ_s.*

Proof. From the above, $\rho_s \le \widehat{\rho}$. Let $e^{\mathbf{S}(x)}$ have the power series expansion $\mathbf{T}(x)$. As the $s(n)$ are nonnegative it follows that the $t(n)$ are nonnegative, since $t(n)$ is the coefficient of x^n in the power series expansion of

$$\sum_j \frac{1}{j!} \Big(\sum_n s(n)x^n \Big)^j = 1 + \big(s(0) + \cdots + s(n)x^n + \cdots \big) + \cdots .$$

Indeed $t(n) \ge s(n)$, for all n, and since $\sum_n s(n)x^n$ diverges when $x > \rho_s$, it follows that $\sum_n t(n)x^n$ also diverges when $x > \rho_s$. Thus $\rho_s \ge \widehat{\rho}$, so $\rho_s = \widehat{\rho}$. □

1.3. Big O, little o, and $\sim$ notation

To compare real-valued functions $f(n)$ and $g(n)$ at large values of their arguments, the following concepts are used, provided $g(n)$ is eventually positive:

Notation	Meaning
$f(n) = \mathrm{O}\big(g(n)\big)$	$(\exists C)(\exists N)\big[n \ge N \Rightarrow \lvert f(n)\rvert \le Cg(n)\big]$
$f(n) = \mathrm{o}\big(g(n)\big)$	$\lim_{n\to\infty} \frac{f(n)}{g(n)} = 0$
$f(n) \sim g(n)$	$\lim_{n\to\infty} \frac{f(n)}{g(n)} = 1.$

These notations are read as follows:

Notation	Reading
$f(n) = \mathrm{O}(g(n))$	$f(n)$ **is big O of** $g(n)$
$f(n) = \mathrm{o}(g(n))$	$f(n)$ **is little o of** $g(n)$
$f(n) \sim g(n)$	$f(n)$ **is asymptotic to** $g(n)$.

These concepts apply to functions $f(n)$ and $g(n)$ that are *eventually* defined on $\mathbb{N}$, that is, each is defined for suitably large n.

For functions $f(x)$ and $g(x)$ that are eventually defined on the reals, with $g(x)$ eventually positive, these concepts are described as follows:

Notation	Meaning
$f(x) = \mathrm{O}(g(x))$	$(\exists C)(\exists N)\left[x \ge N \Rightarrow \|f(x)\| \le Cg(x)\right]$
$f(x) = \mathrm{o}(g(x))$	$\lim_{x\to\infty} \frac{f(x)}{g(x)} = 0$
$f(x) \sim g(x)$	$\lim_{x\to\infty} \frac{f(x)}{g(x)} = 1.$

The readings of these three notations are the same as those just given for sequences.

1.4. The radius of convergence and RT_ρ

The ratio test plays an important role in Part 1.

Proposition 1.23. *Let* $\mathbf{S}(x)$ *be a power series whose coefficients are eventually nonzero. If*

$$\lim_{n\to\infty} \left|\frac{s(n-1)}{s(n)}\right| = \rho$$

then ρ *is the radius of convergence of* $\mathbf{S}(x)$.

It will be important in the study of asymptotic density to be able to find the radius of convergence by the ratio test.

Definition 1.24. Given $\rho \ge 0$, a real-valued function $f(n)$ that is eventually defined on $\mathbb{N}$, and eventually positive, is in RT_ρ if

$$\lim_{n\to\infty} \frac{f(n-1)}{f(n)} = \rho.$$

A power series $\mathbf{S}(x) = \sum_n s(n)x^n$ is in RT_ρ if the sequence of coefficients $s(n)$ is in RT_ρ.

For $0 < \rho < \infty$, the simplest example of a sequence $s(n) \in \mathsf{RT}_\rho$ is surely $s(n) = 1/\rho^n$. This example occurs in a back and forth mapping between RT_1 and RT_ρ, namely $f(n) \in \mathsf{RT}_1$ iff $f(n)/\rho^n \in \mathsf{RT}_\rho$.

Proposition 1.25. *Suppose* $f(n)$ *is eventually defined on* $\mathbb{N}$, *and eventually positive. Then, for* ρ *a positive real,*

$$f(n) \in \mathsf{RT}_\rho \quad \textit{iff} \quad f(n)\rho^n \in \mathsf{RT}_1.$$

PROOF.

$$\begin{aligned} f(n) \in \mathsf{RT}_\rho \quad &\Leftrightarrow \quad \lim_{n\to\infty} \frac{f(n-1)}{f(n)} = \rho \\ &\Leftrightarrow \quad \lim_{n\to\infty} \frac{f(n-1)\rho^{n-1}}{f(n)\rho^n} = 1 \\ &\Leftrightarrow \quad f(n)\rho^n \in \mathsf{RT}_1. \end{aligned}$$

□

Three of the simplest examples from RT_1 are:

$$\begin{array}{rclll} f(n) &=& c & \text{for} & c > 0 \\ f(n) &=& n^c & \text{for} & c \in \mathbb{R} \\ f(n) &=& d^{n^c} & \text{for} & c < 1 < d. \end{array}$$

A substantial collection of $f(n)$ in RT_ρ can be constructed from these examples and the following proposition.

Proposition 1.26.

(a) *If* $f(n) \sim g(n) \in \mathsf{RT}_\rho$ *then* $f(n) \in \mathsf{RT}_\rho$.

(b) *For* $0 < \rho, \sigma < \infty$, *if* $f(n) \in \mathsf{RT}_\rho$ *and* $g(n) \in \mathsf{RT}_\sigma$ *then one has* $1/f(n) \in \mathsf{RT}_{1/\rho}$, *and* $f(n) \cdot g(n) \in \mathsf{RT}_{\rho\sigma}$.

(c) RT_1 *is closed under multiplication, division, and asymptotic equality.*

PROOF. (a) Suppose $f(n) \sim g(n) \in \mathsf{RT}_\rho$. Then $f(n)$ must be eventually positive, and

$$\frac{f(n-1)}{f(n)} = \frac{f(n-1)/g(n-1)}{f(n)/g(n)} \cdot \frac{g(n-1)}{g(n)} \to \rho \qquad \text{as } n \to \infty,$$

so $f \in \mathsf{RT}_\rho$.

(b) Suppose $f(n) \in \mathsf{RT}_\rho$ and $g(n) \in \mathsf{RT}_\sigma$. Then

$$\frac{1/f(n-1)}{1/f(n)} = \frac{f(n)}{f(n-1)} \to 1/\rho \qquad \text{as } n \to \infty,$$

so $1/f(n) \in \mathsf{RT}_{1/\rho}$, and

$$\frac{f(n-1)g(n-1)}{f(n)g(n)} = \frac{f(n-1)}{f(n)} \cdot \frac{g(n-1)}{g(n)} \to \rho\sigma \qquad \text{as } n \to \infty,$$

so $f(n) \cdot g(n) \in \mathsf{RT}_{\rho\sigma}$.

(c) This follows from parts (a) and (b). □

For example, with $a, c, \rho > 0$ and $b \in \mathbb{R}$, if $f(n) \sim an^b e^{c\sqrt{n}}/\rho^n$ then it is easy to see that $f(n) \in \mathsf{RT}_\rho$. Such an example occurs in the asymptotics of §5.5.

The following lemma gives constraints on the growth of members of RT_ρ.

Lemma 1.27. *Suppose* $f(n) \in \mathsf{RT}_\rho$ *with* $0 < \rho < \infty$. *Then, for* ε *such that* $0 < \varepsilon < \rho^{-1}$, *there are* $C, D, N > 0$ *such that*

(1.27.1) $$C(\rho^{-1} - \varepsilon)^n < f(n) < D(\rho^{-1} + \varepsilon)^n, \qquad \text{for } n \geq N.$$

PROOF. Choose N such that $f(n) > 0$, for $n \geq N$, and

$$\left| \frac{f(n)}{f(n-1)} - \rho^{-1} \right| < \varepsilon.$$

Then, for $n \geq N$,

$$\rho^{-1} - \varepsilon < \frac{f(n)}{f(n-1)} < \rho^{-1} + \varepsilon,$$

so

$$\left(\rho^{-1} - \varepsilon\right)^{n-N} < \frac{f(n)}{f(N)} < \left(\rho^{-1} + \varepsilon\right)^{n-N}.$$

Let $C = f(N) \cdot (\rho^{-1} - \varepsilon)^{-N}$ and $D = f(N) \cdot (\rho^{-1} + \varepsilon)^{-N}$. □

A smaller ρ leads to faster growing sequences.

Corollary 1.28. *If $f(n) \in \mathsf{RT}_\sigma$ and $g(n) \in \mathsf{RT}_\tau$ with $0 < \sigma < \tau < \infty$ then $g(n) = o\big(f(n)\big)$.*

PROOF. For $0 < \varepsilon < \min\{\tau^{-1}, (\sigma^{-1} - \tau^{-1})/2\}$, choose $C, D, N > 0$ such that

$$n \geq N \quad \Rightarrow \quad \begin{cases} f(n), g(n) > 0 \\ f(n) > C(\sigma^{-1} - \varepsilon)^n \\ g(n) < D(\tau^{-1} + \varepsilon)^n. \end{cases}$$

Then, for $n \geq N$,

$$0 < \frac{g(n)}{f(n)} < \frac{D}{C}\left(\frac{\tau^{-1} + \varepsilon}{\sigma^{-1} - \varepsilon}\right)^n \to 0 \qquad \text{as } n \to \infty,$$

so $g(n) = o\big(f(n)\big)$. □

1.5. Growth rate of coefficients

Definition 1.29. For $f : \mathbb{N} \to \mathbb{R}$, let ρ_f be the radius of convergence of $\sum f(n)x^n$.

Definition 1.30. Given a function $f : \mathbb{N} \to \mathbb{R}$,

(a) f is *exponentially bounded* if

$$f(n) = O(\beta^n), \qquad \text{for some } \beta > 1;$$

(b) f is *subexponential* if

$$f(n) = O(\beta^n), \qquad \text{for every } \beta > 1.$$

Lemma 1.31. *Let $f : \mathbb{N} \to \mathbb{R}$ be given. Then*
(a) $\rho_f > 0$ *iff* f *is exponentially bounded, and*
(b) $\rho_f \geq 1$ *iff* f *is subexponential.*
(c) *For the case that* $f : \mathbb{N} \to \mathbb{N}$, $\rho_f > 1 \Leftrightarrow \rho_f = \infty \Leftrightarrow \sum_n f(n) < \infty$.

PROOF. For (a),

$$\begin{aligned}
\rho_f > 0 \quad &\text{iff} \quad \limsup_{n\to\infty} |f(n)|^{1/n} < \infty \\
&\text{iff} \quad (\exists \beta > 1)(\exists M)\Big(n \geq M \Rightarrow |f(n)|^{1/n} < \beta\Big) \\
&\text{iff} \quad (\exists \beta > 1)(\exists M)\Big(n \geq M \Rightarrow |f(n)| < \beta^n\Big) \\
&\text{iff} \quad (\exists \beta > 1)\Big(f(n) = \mathrm{O}(\beta^n)\Big) \\
&\text{iff} \quad f \text{ is exponentially bounded.}
\end{aligned}$$

For (b),

$$\begin{aligned}
\rho_f \geq 1 \quad &\text{iff} \quad \limsup_{n\to\infty} |f(n)|^{1/n} \leq 1 \\
&\text{iff} \quad (\forall \beta > 1)\Big(\limsup_{n\to\infty} |f(n)|^{1/n} < \beta\Big) \\
&\text{iff} \quad (\forall \beta > 1)(\exists M)\Big(n \geq M \Rightarrow |f(n)|^{1/n} < \beta\Big) \\
&\text{iff} \quad (\forall \beta > 1)\Big(f(n) = o\big(\beta^n\big)\Big) \\
&\text{iff} \quad f \text{ is subexponential.}
\end{aligned}$$

For (c), observe that $\rho_f > 1$ iff $\limsup_{n\to\infty} f(n)^{1/n} < 1$, and the latter can happen iff f is eventually 0, since the values of $f(n)$ are nonnegative integers. Now f is eventually 0 is equivalent to $\sum f(n) < \infty$, again because the range of f is in the nonnegative integers. □

CHAPTER 2

Counting Functions and Fundamental Identities

Although this chapter starts off with an abstract definition of additive number systems, the reader will soon see examples of the very concrete systems that are of interest, systems that are derived from well known classes of relational structures such as equivalence relations and linear forests.

2.1. Defining additive number systems

Additive number systems are generalizations of the familiar system of nonnegative integers $\mathbb{N} = \{0, 1, \dots\}$ under addition. The definition of such systems requires the notion of a countable free commutative monoid.

2.1.1. Commutative monoids.

Definition 2.1. $\mathcal{M} = (M, +, 0)$ is a *commutative monoid* if $+$ is a commutative and associative operation on M, and $0 \in M$ is an identity element for the operation $+$, that is, the following identities hold in $\mathcal{M}$:

$$\begin{array}{rcll} x + y &=& y + x & \text{for } x, y \in M \\ x + (y + z) &=& (x + y) + z & \text{for } x, y, z \in M \\ x + 0 &=& x & \text{for } x \in M. \end{array}$$

There are many examples of commutative monoids to be found in the complex plane—just take any set M of complex numbers that is closed under addition and has 0 in it. For example, $(\mathbb{C}, +, 0)$, $(\mathbb{R}, +, 0)$, $(\mathbb{Q}, +, 0)$, $(\mathbb{Z}, +, 0)$, and $(\mathbb{N}, +, 0)$ are all commutative monoids. However, among these examples, only the last one is found in additive number systems because it is the only one that is countable and free.

2.1.2. Free commutative monoids.

Definition 2.2. The *indecomposable* elements of a commutative monoid $\mathcal{M}$ are the nonzero elements that cannot be written as a sum of two nonzero elements.

The set of indecomposable elements of a commutative monoid will be denoted by P. Here are some examples of commutative monoids and their indecomposable elements:

commutative monoid	indecomposable elements
$(\mathbb{C}, +, 0)$	none
$(\mathbb{R}, +, 0)$	none
$(\mathbb{Q}, +, 0)$	none
$(\mathbb{Z}, +, 0)$	none
$(\mathbb{N}, +, 0)$	1
$(\{m + in : m, n \in \mathbb{N}\}, +, 0)$	$1, i$

Only the last two examples have $P \neq \varnothing$. They also have a very important property regarding P, namely every element of the commutative monoid can be uniquely expressed as a sum of elements from P.[1]

Definition 2.3. A commutative monoid $\mathcal{M}$ is *free* if every element of $\mathcal{M}$ can be *uniquely* expressed, up to commutativity and associativity, as a sum of indecomposable elements. If $\mathcal{M}$ is free then P is called a set of *free generators* of $\mathcal{M}$, $\mathcal{M}$ is said to be *freely generated* by P, and $|P|$, the cardinality of P, is called the *rank* of $\mathcal{M}$.

Only the last two of the examples above are free commutative monoids, namely $(\mathbb{N}, +, 0)$ and $(\{m + in : m, n \in \mathbb{N}\}, +, 0)$. The first has rank 1 and the second has rank 2. (The last example is the set of Gaussian integers in the first quadrant.)

Proposition 2.4. *A free commutative monoid is determined, up to isomorphism, by its rank.*

PROOF. Suppose $\mathcal{M}$ is freely generated by P, $\mathcal{M}'$ is freely generated by P', and $|P| = |P'|$. Let $\gamma : P \to P'$ be a bijection. Then one can extend γ to $\beta : M \to M'$ by $\beta(0) = 0$ and

$$\beta(p_1 + \cdots + p_n) = \gamma(p_1) + \cdots + \gamma(p_n).$$

This gives an isomorphism between $\mathcal{M}$ and $\mathcal{M}'$. □

For *countable* free commutative monoids, the possible ranks are the natural numbers and countably infinite, that is, $0, 1, \ldots, \aleph_0$.

Natural examples of free commutative monoids are the monoids $\sum_I \mathcal{N}$, where $\mathcal{N} = (\mathbb{N}, +, 0)$, and the elements of $\sum_I \mathcal{N}$ are the maps $f : I \to \mathbb{N}$ which have *finite support*, that is, the set $\{i \in I : f(i) \neq 0\}$ is finite. The operation $+$ is defined on $\sum_I \mathcal{N}$ coordinatewise, that is, $(f_1 + f_2)(i) = f_1(i) + f_2(i)$. The free generators of $\sum_I \mathcal{N}$ are the functions p_i, for $i \in I$, defined by

$$p_i(j) = \begin{cases} 1 & \text{if } j = i \\ 0 & \text{otherwise.} \end{cases}$$

Up to isomorphism, the countable free commutative monoids are as given in Figure 1.

[1]The convention used here is that 0 *is the sum of the empty set of indecomposables.*

rank	monoid
0	$(\{0\},+,0)$
1	$\mathcal{N}$
2	$\mathcal{N}+\mathcal{N}$
$\vdots$	$\vdots$
$\aleph_0$	$\sum_{\aleph_0}\mathcal{N}$

FIGURE 1. Countable Free Commutative Monoids

2.1.3. A useful notation. Let $\mathcal{M} = (M,+,0)$ be a free commutative monoid. For $a \in M$ and $n \in \mathbb{N}$ let na be the sum of n copies of a, with $0a$ the zero element of $\mathcal{M}$. Then any element of M can be expressed in the form

$$\sum n_p p,$$

where p runs over the members of P, and the n_p are nonnegative integers. This convention requires that all but finitely many of the n_p be 0, and that

$$\sum n_p p \;=\; \begin{cases} 0 & \text{if all } n_p = 0 \\ \displaystyle\sum_{n_p \neq 0} n_p p & \text{otherwise.} \end{cases}$$

From the unique decomposition property it follows that each element of M has a unique expression in the form $\sum n_p p$. n_p is called the *coefficient* of p in the decomposition of the element into indecomposables.

2.1.4. Additive number systems. An additive number system is a countable free commutative monoid with a size function called an additive norm.

Definition 2.5. An *additive norm* $\| \ \|$ on a commutative monoid $\mathcal{M}$ is a mapping from M to the nonnegative integers such that

- $\|a\| = 0 \;$ iff $a = 0$
- $\|a+b\| = \|a\| + \|b\|$.

For example, $\|n\| = 34n$ defines an additive norm on $\mathcal{N}$, and $\|m+in\| = 4m + 7n$ defines an additive norm on the Gaussian integers in the first quadrant.

Lemma 2.6. *If $\mathcal{M}$ is a free commutative monoid, freely generated by P, then any additive norm on $\mathcal{M}$ is completely determined by its restriction to the set P. Furthermore, any mapping $\| \ \|$ from P to the set $\mathbb{P}$ of positive integers extends (uniquely) to an additive norm on $\mathcal{M}$ via*

$$\Big\|\sum n_p p\Big\| \;=\; \sum n_p \|p\|. \tag{2.6.1}$$

PROOF. Given a norm $\| \ \|$ on a free commutative monoid $\mathcal{M}$, with P as its set of indecomposables, for any $m \in M$ one has m expressible as $\sum n_p p$. Then

$$\|m\| = \sum n_p \|p\|.$$

Thus the values of the norm are determined by its values on P. (This part of the proof only requires that every element be expressible as a sum of indecomposables, but not necessarily uniquely.)

Given any mapping $\| \ \|$ from P to $\mathbb{P}$, it is routine to verify that the map on M defined by (2.6.1) is indeed a norm for $\mathcal{M}$. (This part of the proof does require that every element be uniquely expressible as a sum of indecomposables, in order to be sure that one has indeed defined a mapping on M.) □

Definition 2.7. An *additive number system*

$$\mathcal{A} = (\mathsf{A}, \mathsf{P}, +, 0, \| \ \|)$$

consists of a countable free commutative monoid $(\mathsf{A}, +, 0)$, with P being the set of indecomposable elements, and with $\| \ \|$ an additive norm that satisfies, for every $n \in \mathbb{N}$,

$$\{\mathsf{a} \in \mathsf{A} : \|\mathsf{a}\| = n\} \text{ is finite,}$$

that is, the norm is a finite-to-one mapping. The *rank* of $\mathcal{A}$ is $|\mathsf{P}|$.

Throughout Part 1 the letter $\mathcal{A}$ will denote an additive number system whose rank is positive, that is, $\mathsf{P} \neq \varnothing$.

Remark 2.8. Lower case sans serif letters a, b, ... , p, ... , are used for the 'numbers' of an additive number system $\mathcal{A}$, the upper case sans serif letter A is the set of all such 'numbers', and other upper case sans serif letters B, C, etc., denote subsets of A, with P being the set of indecomposables. The choice of the letter A is motivated by the study of *abstract* number systems, both additive and multiplicative, and the letter P reminds one of *primes*, the indecomposables of the usual multiplicative number system.

Definition 2.9. An additive number system $\mathcal{B} = (\mathsf{B}, \mathsf{Q}, +, 0, \| \ \|)$ is a *subsystem* of $\mathcal{A} = (\mathsf{A}, \mathsf{P}, +, 0, \| \ \|)$ if $\mathsf{Q} \subseteq \mathsf{P}$, and if the operation $+$, the constant 0, and the norm $\| \ \|$ of $\mathcal{B}$ agree with the corresponding items of $\mathcal{A}$. This is expressed by $\mathcal{B} \leq \mathcal{A}$, and $\mathcal{B}$ is the *subsystem generated by* Q. To indicate that $\mathcal{B}$ is a *proper* subsystem, that is, $\mathsf{B} \neq \mathsf{A}$, the expression $\mathcal{B} < \mathcal{A}$ is used. A subsystem $\mathcal{B}$ is *trivial* if $\mathsf{Q} = \varnothing$.

The concept of two additive number systems being 'essentially the same' is captured in the next definition. (Some readers might prefer the phrase 'norm preserving isomorphism' instead of the single word 'isomorphism'.)

Definition 2.10. Let $\mathcal{A}_i = (\mathsf{A}_i, \mathsf{P}_i, +_i, 0_i, \| \ \|_i)$, where $i = 1, 2$, be two additive number systems. $\mathcal{A}_1$ is *isomorphic to* $\mathcal{A}_2$, written $\mathcal{A}_1 \cong \mathcal{A}_2$, if there is a bijection $\mu : \mathsf{A}_1 \to \mathsf{A}_2$ that preserves addition, zero, and norm, that is,

$$\begin{aligned} \mu(\mathsf{a} +_1 \mathsf{b}) &= \mu(\mathsf{a}) +_2 \mu(\mathsf{b}) \qquad \text{for } \mathsf{a}, \mathsf{b} \in \mathsf{A}_1 \\ \mu(\mathsf{0}_1) &= \mathsf{0}_2 \\ \|\mu(\mathsf{a})\|_2 &= \|\mathsf{a}\|_1 \qquad \text{for } \mathsf{a} \in \mathsf{A}_1. \end{aligned}$$

Such a mapping μ is called an *isomorphism*.

Some of the detailed notational distinction between two additive number systems used in the above definition is, as is common practice in algebra, sacrificed for the clarity that comes with ambiguous notation. The $+_i$, $\mathsf{0}_i$, and $\|\ \|_i$ are replaced by $+$, $\mathsf{0}$, and $\|\ \|$.

As is usual with the study of isomorphism, inverses of isomorphisms are isomorphisms, and compositions of isomorphisms are isomorphisms. Thus, since every additive number system is isomorphic to itself, isomorphism is an equivalence relation on the class of additive number systems. Note that the indecomposables were not mentioned in the definition of isomorphism. This is because the above definition of isomorphism ensures that indecomposables map to indecomposables.

Lemma 2.11. *If $\mu : \mathcal{A}_1 \to \mathcal{A}_2$ is an isomorphism then $\mu(\mathsf{P}_1) = \mathsf{P}_2$.*

PROOF. An element $\mathsf{a} \in \mathsf{A}_1 \setminus \{\mathsf{0}\}$ is indecomposable iff $\mathsf{a} = \mathsf{b} + \mathsf{c}$ implies either $\mathsf{b} = \mathsf{0}$ or $\mathsf{c} = \mathsf{0}$. Now $\mathsf{a} = \mathsf{b} + \mathsf{c}$ iff $\mu(\mathsf{a}) = \mu(\mathsf{b}) + \mu(\mathsf{c})$, $\mathsf{b} = \mathsf{0}$ iff $\mu(\mathsf{b}) = \mathsf{0}$, and $\mathsf{c} = \mathsf{0}$ iff $\mu(\mathsf{c}) = \mathsf{0}$. Thus a is indecomposable iff $\mu(\mathsf{a})$ is indecomposable. □

2.2. Examples of additive number systems

One has the familiar additive number system of nonnegative integers $\mathbb{N}$ with the usual $+$ and 0, and with the norm being the identity map, that is, $\|n\| = n$. This is a system of rank 1. The set of Gaussian integers in the first quadrant with the usual 'norm' from number theory, namely $\|m + in\| = m^2 + n^2$, is not an additive number system because this is not an additive norm. However one can easily create additive norms, such as $\|m + in\| = 4m + 7n$, to form an additive number system of rank 2.

Although there are plenty of such 'ordinary' additive number systems, this study is concerned with additive number systems derived from classes of structures (as studied in logic)—they provide the connection with logical limit laws. Several examples are introduced in this section, to be used throughout Part 1. The idea behind derived number systems is simply to consider structures up to isomorphism, with addition being disjoint union and the norm being the number of elements in the underlying set of the structure. The structure on the empty set gives the zero element of the derived additive number system.

2.2.1. Equivalence relations. By an *equivalence relation* on a finite set S is meant a relational structure[2] $(S, \equiv)$, where $\equiv$ satisfies the reflexive, symmetric and transitive laws:

$$\begin{array}{llll} x \equiv x & & & \text{for all } x \in S \\ x \equiv y & \text{implies} & y \equiv x & \text{for all } x, y \in S \\ x \equiv y \text{ and } y \equiv z & \text{imply} & x \equiv z & \text{for all } x, y, z \in S. \end{array}$$

Such a structure can be pictured as a graph. Figure 2 gives one on 12 elements.[3] The edge relation is the equivalence relation. The maximal connected pieces, called *components* in graph theory, are the *equivalence classes*. (The components in the graph of an equivalence relation are *cliques*.)

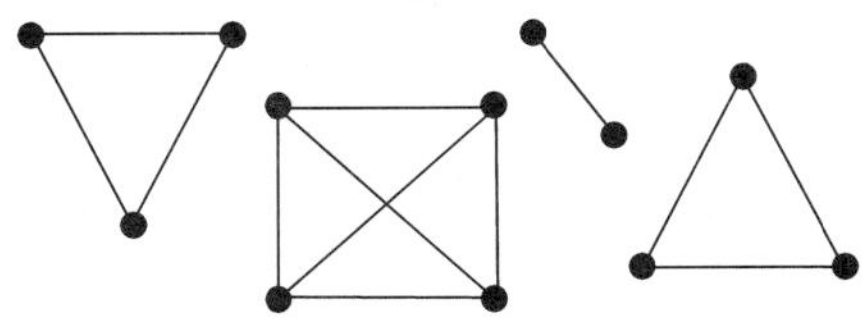

FIGURE 2. An Equivalence Relation

Clearly every finite equivalence relation on a nonempty set can be viewed as a disjoint union of cliques. By considering the structures up to isomorphism one has the additive number system of equivalence relations, where addition is disjoint union. The norm of an equivalence relation $(S, \equiv)$ is just the number of elements in S. The indecomposables are the equivalence relations on nonempty sets with a single equivalence class, that is, their graph is a single clique. Thus, for each $n \in \mathbb{P}$, there is exactly one indecomposable (a clique) of norm n.

2.2.2. Permutations. By a *permutation* on a finite set S is meant a relational structure (S, R) where R is the directed graph of a bijection from S to itself. This means:

- for every $x \in S$, there is a unique $y \in S$ such that xRy;
- for every $y \in S$, there is a unique $x \in S$ such that xRy.

Such a structure can be pictured as a directed graph, as in Figure 3. The components are called *cycles*, and each permutation is a disjoint union of cycles.

By considering the structures up to isomorphism one has the additive number system of permutations, with addition and norm defined as in the

[2]The phrase *equivalence relation* is used to denote both the structure $(S, \equiv)$ as well as the binary relation $\equiv$. This should not cause any confusion. A similar comment applies to the next example on permutations.

[3]To be more precise it is a *reflexive* graph $[\forall x\,(x \equiv x)]$, as graphs are considered (at least by logicians) to be *irreflexive* $[\forall x\,(x \not\equiv x)]$. In the graph picture the loops at each vertex, to indicate that the relation is reflexive, have been omitted.

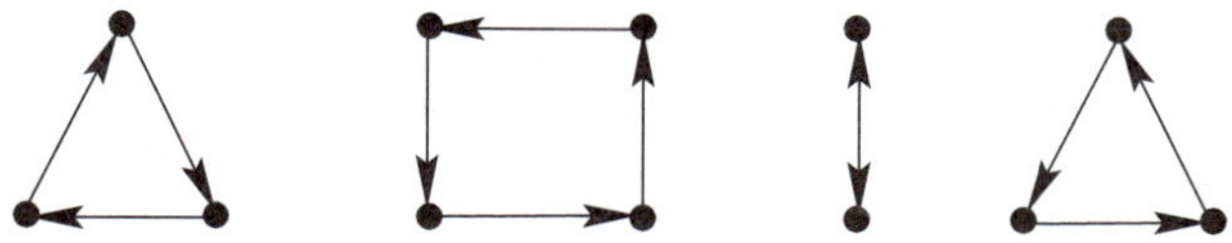

FIGURE 3. A Permutation

previous example. The indecomposables are the cycles, and there is exactly one indecomposable (a cycle) of norm n, for each $n \in \mathbb{P}$.

2.2.3. Linear forests (as posets). By a *linear forest* (as a *poset*) on a finite set S is meant a poset $(S, \leq)$ that is a disjoint union of linearly ordered posets (called *chains*). This class is defined by the properties:

$$\begin{array}{llll} x \leq x & & & \text{for all } x \in S \\ x \leq y \text{ and } y \leq x & \text{imply} & x = y & \text{for all } x, y \in S \\ x \leq y \text{ and } y \leq z & \text{imply} & x \leq z & \text{for all } x, y, z \in S \\ x \leq z \text{ and } y \leq z & \text{imply} & x \leq y \text{ or } y \leq x & \text{for all } x, y, z \in S \\ z \leq x \text{ and } z \leq y & \text{imply} & x \leq y \text{ or } y \leq x & \text{for all } x, y, z \in S. \end{array}$$

Such a structure can be pictured using a Hasse diagram, as in Figure 4. The additive number system of linear forests has exactly one indecomposable (a chain) of norm n, for each $n \in \mathbb{P}$.

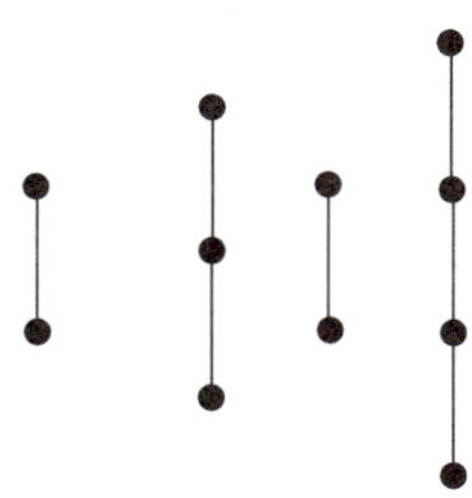

FIGURE 4. A Linear (Poset) Forest

2.2.4. Linear forests (as graphs). By a *linear forest* (as a *graph*) on a finite set S is meant a graph (S, R) whose components are chains, that is, each vertex of the graph is adjacent to at most two other vertices, and there are no cycles. Figure 5 shows an example. The additive number system of linear forests has exactly one indecomposable (a chain) of norm n, for each $n \in \mathbb{P}$.

FIGURE 5. A Linear (Graph) Forest

2.2.5. Two-colored linear forests (as posets). By a *two-colored linear forest* (as a poset) on a finite set S is meant a structure $(S, \leq, B, G)$ such that $(S, \leq)$ is a linear forest, and B and G denote two colors (blue and gold) such that each vertex has exactly one of the two colors. The latter can be expressed by: $Bx \leftrightarrow \neg Gx$, for $x \in S$. See Figure 6 for an example. The additive number system of two-colored linear forests has exactly 2^n indecomposables (two-colored chains) of norm n, for each $n \in \mathbb{P}$.

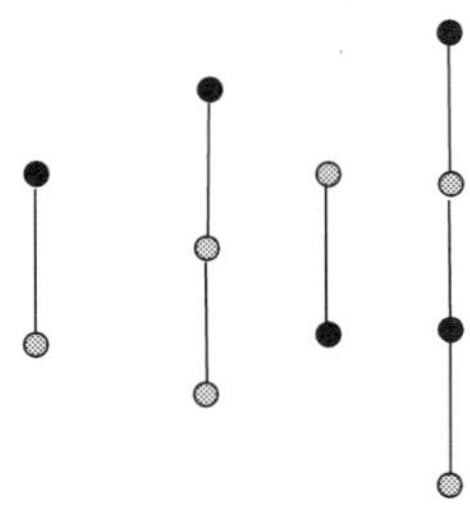

FIGURE 6. A Two-colored Linear (Poset) Forest

2.2.6. Two-colored linear forests (as graphs). By a *two-colored linear forest* (as a graph) on a finite set S is meant a structure (S, R, B, G) such that (S, R) is a linear forest, and B and G denote two colors such that each vertex has exactly one of the two colors. See Figure 7 for an example.

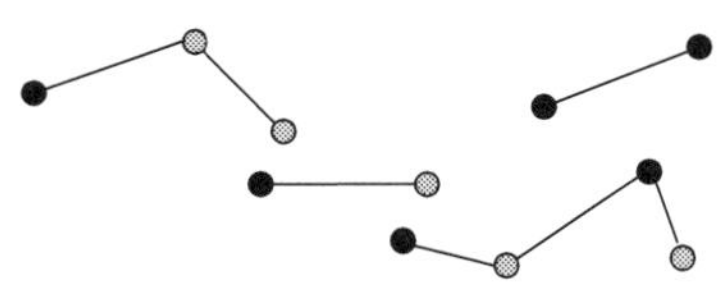

FIGURE 7. A Two-colored Linear (Graph) Forest

The additive number system of two-colored linear forests (as graphs) has a more complicated formula giving the number of indecomposables (two-colored chains) of norm n due to the possibility of symmetry. Indeed, a two-colored chain is either rigid, having only the identity automorphism, or it is symmetric and has one nonidentity automorphism. Some examples are in Figure 8. There are $2^{(n+1)/2}$ symmetric chains of size n (up to isomorphism) when n is odd, and $2^{n/2}$ when n is even, so the total number of chains of size n is

$$\begin{array}{lcll} \frac{1}{2}\left(2^n - 2^{(n+1)/2}\right) + 2^{(n+1)/2} & = & 2^{n-1} - 2^{(n-1)/2} + 2^{(n+1)/2}, & \text{for } n \text{ odd} \\ \frac{1}{2}\left(2^n - 2^{n/2}\right) + 2^{n/2} & = & 2^{n-1} - 2^{(n-2)/2} + 2^{n/2}, & \text{for } n \text{ even.} \end{array}$$

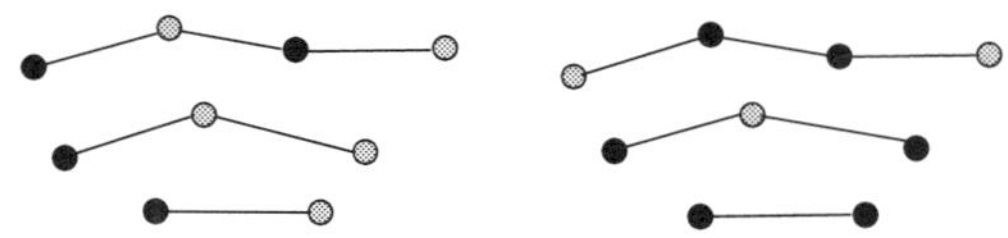

FIGURE 8. Examples of Rigid and Symmetric Chains

2.2.7. Finite Abelian p-groups. One can also obtain additive number systems from certain classes of algebraic structures, provided the sizes of the structures are always a power of some integer. The idea is again to work up to isomorphism, but with addition being direct product, and the norm being the logarithm of the size.

For $\mathbf{G}$ a finite Abelian p-group, define $\|\mathbf{G}\| = n$ if the number of elements in $\mathbf{G}$ is p^n. With addition being direct product, there is a unique indecomposable of norm n, for each $n \in \mathbb{P}$, namely the cyclic p-group $\mathbf{Z}_{p^n}$.

In reviewing the above examples, all but the two-colored examples give rise to essentially the same additive number system, namely one in which there is one indecomposable of each size. Having such a derived additive number system leads to a 0–1 law for these classes. Weaker logical limit laws hold for the two-colored classes.

2.3. Counting functions, fundamental identities

Definition 2.12. Given an additive number system $\mathcal{A} = (\mathsf{A}, \mathsf{P}, +, 0, \|\ \|)$ define, for each set $\mathsf{B} \subseteq \mathsf{A}$, the *(local) counting function*

$$b(n) \quad = \quad \left|\{\mathsf{b} \in \mathsf{B} : \|\mathsf{b}\| = n\}\right|,$$

that is, $b(n)$ gives the number of elements in B with norm n. In some situations, particularly when the set B has a more complicated name than just a single letter, the notation $|\mathsf{B}|_n$ is used to denote $b(n)$. The corresponding power series $\mathbf{B}(x)$, called the *(ordinary) generating series* of B, is defined by:[4]

$$\mathbf{B}(x) \quad = \quad \sum_{n \geq 0} b(n)x^n.$$

$\mathbf{B}(x)$ can also be expressed by

$$\mathbf{B}(x) \quad = \quad \sum_{\mathsf{b} \in \mathsf{B}} x^{\|\mathsf{b}\|}.$$

The two most significant counting functions are $a(n)$ and $p(n)$, associated with A and P, called the *(local) counting functions of* $\mathcal{A}$. Note that $a(0) = 1$ and $p(0) = 0$. So the generating series of A and P are:

$$\mathbf{A}(x) \quad = \quad \sum_{n \geq 0} a(n)x^n$$

[4]Notations similar to $Z_{\mathsf{B}}(x)$, instead of our $\mathbf{B}(x)$, are preferred by some authors.

$$\mathbf{P}(x) = \sum_{n\geq 1} p(n)x^n.$$

An additive number system is completely determined, up to isomorphism, by either of its counting functions $a(n)$ or $p(n)$.

Proposition 2.13. *Let $\mathcal{A}_i$, where $i = 1, 2$, be two additive number systems, and let $a_i(n), p_i(n)$ be their counting functions. Then*

$$\mathcal{A}_1 \cong \mathcal{A}_2 \quad \textit{iff} \quad p_1 = p_2 \quad \textit{iff} \quad a_1 = a_2.$$

PROOF. If $\mu : \mathcal{A}_1 \to \mathcal{A}_2$ is an isomorphism then, by Lemma 2.11, $\mathsf{a} \in \mathsf{A}_1$ is an indecomposable iff $\mu(\mathsf{a})$ is an indecomposable in A_2. As $\|\mathsf{a}\| = \|\mu(\mathsf{a})\|$, it follows that a is an indecomposable of norm n iff $\mu(\mathsf{a})$ is an indecomposable of norm n. Thus $p_1(n) = p_2(n)$.

If $p_1 = p_2$ then let $\mu : \mathsf{P}_1 \to \mathsf{P}_2$ be a bijection that preserves the norm. Now extend this to A_1 by

$$\mu\Big(\sum n_{\mathsf{p}}\mathsf{p}\Big) = \sum n_{\mathsf{p}}\mu(\mathsf{p}).$$

It is easily seen that this defines an isomorphism, so $\mathcal{A}_1 \cong \mathcal{A}_2$.

In any additive number system $\mathcal{A}$ with counting functions $a(n), p(n)$, one can determine $a(n)$ from $p(n)$ since the elements of norm n are uniquely expressible as sums of indecomposables of size at most n. Also, one can recursively find $p(n)$ from $a(n)$ by noting that $p(1) = a(1)$, and, having determined $p(m)$ for $m < n$, compute the number $b(n)$ of elements of norm n that can be expressed as sums of indecomposables of size less than n. Then $p(n) = a(n) - b(n)$. □

Definition 2.14. The *fundamental identity*[5] of $\mathcal{A}$ is

$$\sum_{n\geq 0} a(n)x^n = \prod_{n\geq 1}(1 - x^n)^{-p(n)}. \tag{2.14.1}$$

A detailed discussion of how to understand this, as describing a method to compute the values of $a(n)$ from $p(n)$, is given in §2.3.4, as well as a proof that it holds as an identity between two functions, for $0 \leq x < 1$.

[5]These identities are often referred to as *partition identities* since the original identity studied by Hardy and Ramanujan,

$$\sum_n a(n)x^n = \prod_n (1 - x^n)^{-1},$$

is such that $a(n)$ counts the number of ways to (additively) *partition* a positive integer n, that is, to express n as a sum of positive numbers (up to commutativity and associativity). The name *fundamental identity* is preferred in this study as the word 'partition' is heavily used in the context of partition sets. The corresponding identities in Part 2 are also called 'fundamental identities'.

2.3.1. Examples where $p(n) = 1$. Returning to the additive number systems of equivalence relations, etc., $p(n) = 1$, for $n \geq 1$, holds for all but the two-colored examples. Now, $p(n) = 1$ gives the fundamental identity

$$\sum_{n\geq 0} a(n)x^n = \prod_{n\geq 1}(1-x^n)^{-1},$$

the famous identity studied by Hardy and Ramanujan. In this case $a(n)$ is the number of ways to partition the number n into a sum of positive numbers, so $a(1) = 1$, $a(2) = 2$, $a(3) = 3$, $a(4) = 5$, etc. Later (Chapter 4 + Chapter 6) it will be seen that, simply because $p(n) \leq 1$, for $n \geq 1$, the classes $\mathcal{K}$ of relational structures with this derived number system will have a monadic second-order 0–1 law. This means that if φ is any monadic second-order sentence then the probability that a structure chosen at random from $\mathcal{K}$ will satisfy φ is either 0 or 1.

However, it is important to note that the probability of a given φ being true still depends on the choice of $\mathcal{K}$. For example, the probability that

$$\exists x \exists y \exists z \ \big(x \neq y \text{ and } xRz \text{ and } yRz\big)$$

holds is 1 in equivalence relations (where R is $\equiv$), but 0 in permutations.

Remark 2.15. Another word of caution is in order. Abelian p-groups also have a derived additive number system with $p(n) = 1$, for $n \geq 1$, but since they are a class of algebras closed under direct products, and not a class of relational structures closed under disjoint union, the conclusion is weaker. One can only say that they have a *first-order* 0–1 law. (Indeed, it is proved in [**28**] that they do *not* have a monadic second-order limit law.)

2.3.2. Two-colored linear forests. For the additive number system of two-colored linear forests (as posets), $p(n) = 2^n$, for $n \geq 1$, so the fundamental identity is

$$\sum_{n\geq 0} a(n)x^n = \prod_{n\geq 1}(1-x^n)^{-2^n}.$$

Later it will be seen (Chapter 5 + Chapter 6) that simply because the class of two-colored linear forests (as posets) has this fundamental identity it must have a monadic second-order limit law. This means, for any given monadic second-order sentence φ, that the probability of being true exists, but need not be 0 or 1.

For the additive number system of two-colored linear forests (as graphs),

$$p(n) = \begin{cases} 2^{n-1} - 2^{(n-1)/2} + 2^{(n+1)/2}, & \text{for } n \text{ odd} \\ 2^{n-1} - 2^{(n-2)/2} + 2^{n/2}, & \text{for } n \text{ even,} \end{cases}$$

so the fundamental identity is

$$\sum_{n\geq 0} a(n)x^n = \prod_{n\geq 1}(1-x^{2n-1})^{-2^{2n-2}+2^{n-1}-2^n} \cdot (1-x^{2n})^{-2^{2n-1}+2^{n-1}-2^n}.$$

It will be shown (Chapter 5 + Chapter 6) that, since the class of two-colored linear forests (as graphs) has this fundamental identity, it must have a monadic second-order limit law.

Here is a simple example to illustrate what is meant by saying that two-colored linear forests have a monadic second-order limit law. Consider the question:

> *What is the probability that a randomly chosen finite linear forest has at least one component (a chain) with strictly alternating colors?*

In either of the additive number systems $\mathcal{A} = (\mathsf{A}, \mathsf{P}, +, 0, \|\ \|)$ derived from the two-colored linear forests, let B be the collection of linear forests (up to isomorphism) with the desired property. Then it has a generating series

$$\mathbf{B}(x) \;=\; \sum_{n\geq 0} b(n)x^n,$$

and the question being asked is: What is $\lim\limits_{n\to\infty} \dfrac{b(n)}{a(n)}$, or indeed, does this limit even exist? Later it will be shown, by the methods of Chapter 5, that the probability requested does indeed exist. In the case of linear forests as posets, the probability is

$$\lim_{x\to\frac{1}{2}} \frac{\mathbf{B}(x)}{\mathbf{A}(x)} \;=\; 1 - \Big(\prod_{n=1}^{\infty}(1-2^{-n})\Big)^2 \;\approx\; 0.92.$$

So there is a probability of about 92% that a randomly chosen linear forest has a component with strictly alternating colors. In the case of linear forests as graphs, the probability is

$$\lim_{x\to\frac{1}{2}} \frac{\mathbf{B}(x)}{\mathbf{A}(x)} \;=\; 1 - \prod_{n=1}^{\infty}(1-2^{-2n+1})^2(1-2^{-2n}) \;\approx\; 0.88,$$

or about 88%.

This is a sample of the kind of question to which the methods developed in Part 1 can be applied. The question asked concerns a property that can be expressed in monadic second-order logic[6]—indeed, in §6.2.1 one sees how to express the property of having some component with strictly alternating colors. The methods of Part 1 show that any property about two-colored linear forests that can be expressed in monadic second-order logic will be such that the probability of a randomly chosen finite two-colored linear forest satisfying the property will be proved to exist. Furthermore, the probability will be expressed in a form that often leads to explicit computations such as those above.

---

[6]However, not in the weaker first-order logic in the graph case.

2.3.3. A Cauchy product. As a prelude to understanding the fundamental identity, consider the Cauchy product

$$\prod^{*}_{\mathsf{p}\in\mathsf{P}} \left(1 + x^{\|p\|} + x^{2\|p\|} + \cdots\right).$$

The case that P is infinite will be examined in this section. The discussion when P is finite is essentially the same, but easier, as one is not concerned with the convergence of infinite products.

Given that P is infinite, let p_n be an ordering of the elements of P, and let $g(n) = \|\mathsf{p}_n\|$. Then $g : \mathbb{N} \to \mathbb{P}$, and $g(n) \to \infty$. In the notation of §1.1.3,

$$\mathbf{S}^g_n(x) \;=\; 1 + x^{\|\mathsf{p}_n\|} + x^{2\|\mathsf{p}_n\|} + \cdots,$$

so

$$\prod^{*}_{n\geq 0} \left(1 + x^{\|\mathsf{p}_n\|} + x^{2\|\mathsf{p}_n\|} + \cdots\right)$$

is the infinite Cauchy product $\prod^{*}_{k\geq 1} \mathbf{S}^g_k(x)$ discussed in §1.1.3.

Lemma 2.16. *For $x \geq 0$,*

$$\prod^{*}_{n\geq 0} \left(1 + x^{\|\mathsf{p}_n\|} + x^{2\|\mathsf{p}_n\|} + \cdots\right) \;=\; \prod_{n\geq 0} \left(1 + x^{\|\mathsf{p}_n\|} + x^{2\|\mathsf{p}_n\|} + \cdots\right),$$

and the products do not depend on the ordering of the factors.

PROOF. This follows from Lemmas 1.9 and 1.12. □

Since the ordering of the factors is unimportant, it is preferable, when possible, not to refer to any particular ordering. Then the above equation can be written more simply as

$$\prod^{*}_{\mathsf{p}\in\mathsf{P}} \left(1 + x^{\|\mathsf{p}\|} + x^{2\|\mathsf{p}\|} + \cdots\right) \;=\; \prod_{\mathsf{p}\in\mathsf{P}} \left(1 + x^{\|\mathsf{p}\|} + x^{2\|\mathsf{p}\|} + \cdots\right).$$

Lemma 2.17. *For $x \geq 0$,*

$$\prod_{\mathsf{p}\in\mathsf{P}} \left(1 + x^{\|\mathsf{p}\|} + x^{2\|\mathsf{p}\|} + \cdots\right) \;=\; \prod_{n\geq 1} \left(1 + x^{n} + x^{2n} + \cdots\right)^{p(n)}.$$

PROOF. Since the left side has $p(n)$ factors involving $\mathsf{p} \in \mathsf{P}$ with $\|\mathsf{p}\| = n$, the right side is just a particular arrangement of the left side. □

Proposition 2.18. *For $0 \leq x < 1$,*

$$\begin{aligned}
\prod_{n\geq 1} \left(1 - x^{n}\right)^{-p(n)} &= \prod_{\mathsf{p}\in\mathsf{P}} \left(1 - x^{\|\mathsf{p}\|}\right)^{-1} \\
&= \prod_{\mathsf{p}\in\mathsf{P}} \left(1 + x^{\|\mathsf{p}\|} + x^{2\|\mathsf{p}\|} + \cdots\right) \\
&= \prod^{*}_{\mathsf{p}\in\mathsf{P}} \left(1 + x^{\|\mathsf{p}\|} + x^{2\|\mathsf{p}\|} + \cdots\right).
\end{aligned}$$

PROOF. This follows from Proposition 1.13 and the previous two lemmas. □

2.3.4. $\mathbf{A}(x)$ as a Cauchy product. The Cauchy product described in the last section is actually $\mathbf{A}(x)$. This gives a combinatorial connection between the counting functions $a(n)$ and $p(n)$ that implies the fundamental identity.

Proposition 2.19. *For any additive number system $\mathcal{A}$,*

$$\mathbf{A}(x) \;=\; \mathop{\boxed{*}\!\!\prod}_{\mathsf{p}\in\mathsf{P}} \left(1 + x^{\|\mathsf{p}\|} + x^{2\|\mathsf{p}\|} + \cdots\right).$$

PROOF. The case that P is infinite will be discussed—the case that P is finite requiring a similar but simpler analysis. The equation asserts that $\mathbf{A}(x)$ is the indicated infinite Cauchy product, so the claim is that $a(m)$, the number of elements in A of norm m, is the stable value of the coefficient of x^m in the initial products of the infinite Cauchy product:

$$a(m) \;=\; [x^m] \mathop{\boxed{*}\!\!\prod}_{\substack{\mathsf{p}\in\mathsf{P}\\ \|\mathsf{p}\|\le n}} \left(1 + x^{\|\mathsf{p}\|} + x^{2\|\mathsf{p}\|} + \cdots\right),$$

for any $n \ge m$.

To see that this is true, proceed as follows. Let $\mathsf{p}_0, \mathsf{p}_1, \ldots$ be a listing of the indecomposable elements of A. There is a natural one-to-one correspondence between the elements of A and the sequences (m_i) of nonnegative integers that have only finitely many nonzero entries, namely, (m_i) corresponds to the element a such that m_i is the coefficient of p_i in the decomposition of a into indecomposables. Thus the sequences $\left(x^{m_i\|\mathsf{p}_i\|}\right)$ of monomials, where almost all monomials are 1, are also naturally in a one-to-one correspondence with the elements of A.

Now, $a(m)$ is the number of sequences (m_i) of nonnegative integers such that $m = \sum m_i\|\mathsf{p}_i\|$. This is because each element of A with norm m can be expressed as a sum $\sum m_i\mathsf{p}_i$ with $\sum m_i\|\mathsf{p}_i\| = m$, and distinct sequences (m_i) give rise to distinct elements $\sum m_i\mathsf{p}_i$ of A. Note that $m = \sum m_i\|\mathsf{p}_i\|$ iff $x^m = \prod x^{m_i\|\mathsf{p}_i\|}$.

Thus $a(m)$ is the number of sequences (m_i) of nonnegative integers such that $x^m = \prod x^{m_i\|\mathsf{p}_i\|}$. Now consider the product

$$(2.19.1) \qquad \left(1 + x^{\|\mathsf{p}_0\|} + x^{2\|\mathsf{p}_0\|} + \cdots\right) \cdots \left(1 + x^{\|\mathsf{p}_n\|} + x^{2\|\mathsf{p}_n\|} + \cdots\right),$$

where n is sufficiently large to ensure that all indecomposables of norm at most m appear in the list $\mathsf{p}_0, \ldots, \mathsf{p}_n$. Then each sequence (m_i) with $x^m = \prod x^{m_i\|\mathsf{p}_i\|}$ corresponds to a choice of one monomial from each of the sums $1 + x^{\|\mathsf{p}_i\|} + x^{2\|\mathsf{p}_i\|} + \cdots$ above, and thus $a(m)$ is the coefficient of x^m in the infinite Cauchy product $\boxed{*}\!\!\prod_{\mathsf{p}\in\mathsf{P}} \left(1 + x^{\|\mathsf{p}\|} + x^{2\|\mathsf{p}\|} + \cdots\right)$. □

Combining the results of this section and the previous one gives the following.

Theorem 2.20. *For $x \geq 0$,*

$$\begin{aligned} \mathbf{A}(x) &= \prod^{*}_{\mathsf{p}\in\mathsf{P}} \left(1 + x^{\|\mathsf{p}\|} + x^{2\|\mathsf{p}\|} + \cdots\right) \\ &= \prod_{\mathsf{p}\in\mathsf{P}} \left(1 + x^{\|\mathsf{p}\|} + x^{2\|\mathsf{p}\|} + \cdots\right) \\ &= \prod_{n\geq 1} \left(1 + x^{n} + x^{2n} + \cdots\right)^{p(n)}, \end{aligned}$$

and, for $0 \leq x < 1$, the following also hold:

$$\begin{aligned} \mathbf{A}(x) &= \prod_{\mathsf{p}\in\mathsf{P}} \left(1 - x^{\|\mathsf{p}\|}\right)^{-1} \\ &= \prod_{n\geq 1} \left(1 - x^{n}\right)^{-p(n)}. \end{aligned}$$

Remark 2.21. Requiring that the $a(n), p(n)$ in a fundamental identity be obtained from an additive number system $\mathcal{A}$ is not a restriction on the possible fundamental identities. For suppose $a(n), p(n)$ are two functions with values in $\mathbb{N}$ that satisfy (2.14.1). Take a free commutative monoid of rank $r = \sum_n p(n)$ and endow it with a norm such that the number of generators with norm n will be exactly $p(n)$. For this additive number system $\mathcal{A}$, the associated counting functions are precisely the given $a(n), p(n)$.

2.3.5. The radius of convergence of $\mathcal{A}$.

Definition 2.22. The *radius of convergence ρ of $\mathcal{A}$* is the radius of convergence of the generating series $\mathbf{A}(x)$, that is, $\rho = \rho_a$, the radius of convergence of $\sum_n a(n)x^n$.

Lemma 2.23. $$0 \leq \rho \leq 1.$$

PROOF. Since $\mathcal{A}$ has positive rank (see p. 20) it follows that $a(n) \geq 1$ holds infinitely often, so $\mathbf{A}(1) = \infty$. Thus $\rho \leq 1$; and one always has $\rho \geq 0$. □

Corollary 2.24. *The fundamental identity*

$$\sum_{n\geq 0} a(n)x^n = \prod_{n\geq 1}(1 - x^n)^{-p(n)}$$

of $\mathcal{A}$ is valid for $0 \leq x < 1$. The two sides converge for $0 \leq x < \rho$. For $\rho < x < 1$, the two sides have the value ∞.

PROOF. Just add to Theorem 2.20 the observation that $\mathbf{A}(x)$ converges when $0 \leq x < \rho$, and diverges when $x > \rho$. □

Corollary 2.25.

(a) *For $x \in [0, 1)$,*

$$\mathbf{A}(x) \ \text{converges iff} \ \mathbf{P}(x) \ \text{converges.}$$

(b) $\rho = \min(1, \rho_p)$.
(c) *The following conditions are equivalent:*
- $a(n)$ *is exponentially bounded*
- $\rho > 0$
- $p(n)$ *is exponentially bounded*
- $\rho_p > 0$.

(d) *The following conditions are equivalent:*
- $a(n)$ *is subexponential*
- $\rho = 1$
- $p(n)$ *is subexponential*
- $\rho_p \geq 1$.

PROOF. (a) Let $x \in [0, 1)$. Then

$$\begin{aligned} \mathbf{A}(x) \text{ converges} \quad &\text{iff} \quad \textstyle\prod_n (1 - x^n)^{-p(n)} \text{ converges} \quad && \text{by Corollary 2.24} \\ &\text{iff} \quad \mathbf{P}(x) \text{ converges} && \text{by Lemma 1.15.} \end{aligned}$$

(b) Since $\rho_p \geq \rho$, use (a) and Lemma 2.23.

(c) $a(n)$ is exponentially bounded
iff $\rho > 0$ by Lemma 1.31 (a)
iff $\rho_p > 0$ by part (b)
iff $p(n)$ is exponentially bounded by Lemma 1.31 (a).

(d) $a(n)$ is subexponential
iff $\rho = 1$ by Lemma 1.31 (b) and (b) above
iff $\rho_p \geq 1$ by part (b)
iff $p(n)$ is subexponential by Lemma 1.31 (b).

□

In the examples from §2.2, those with $p(n) = 1$ have $\rho_p = 1$, and hence $\rho = 1$; and the two-colored linear forests have $\rho_p = 1/2$, and hence $\rho = 1/2$, in both the poset and graph cases. $\rho = 1$ is needed to prove logical 0–1 laws, and Chapter 4 is devoted to this case. The case $0 < \rho < 1$ is more difficult and is treated in Chapter 5.

Remark 2.26. One cannot extend part (a) of Corollary 2.25 to include $x = 1$ since $\mathbf{A}(1) = \infty$, but $\mathbf{P}(1)$ could be either finite or infinite, depending on whether P is finite or infinite.

Remark 2.27. If $a(n)$, equivalently $p(n)$, is not exponentially bounded, then there is no analytical information available in the fundamental identity. For example, if $p(n) > 2^{n \log n}$ holds infinitely often, then $\rho = 0$.

Compton uses the terminology *fast growing* to describe additive number systems with $\rho = 0$, and *slowly growing* when $\rho > 0$. There is very little information on density in fast growing systems—the density results in this study are for slowly growing systems.

2.4. Global counts

Definition 2.28. For $f : \mathbb{N} \to \mathbb{N}$, define the *global* (or *cumulative*) count function $F : [0, \infty) \to \mathbb{N}$ by

$$F(x) = \sum_{n \leq x} f(n).$$

For $\mathsf{B} \subseteq \mathsf{A}$, the *global* count function $B(x)$ is

$$B(x) = \sum_{n \leq x} b(n).$$

In some situations, particularly when the set B has a more complicated name than just a single letter, the notation $|\mathsf{B}|_{\leq x}$ is used to denote $B(x)$.

One can think of a global count function $F(x)$ as a function from $\mathbb{N}$ to $\mathbb{N}$ by simply restricting its domain, and consequently one can consider the radius of convergence ρ_F of $\sum F(n)x^n$. $F(n)$ is called the (discrete) global counting function of f.

Lemma 2.29. *For $f : \mathbb{N} \to \mathbb{N}$, with $\sum_n f(n) > 0$,*

$$\rho_F = \min(1, \rho_f).$$

PROOF. The argument is in three cases. (These cases are the only possibilities because $f(n)$ is a nonnegative *integer* valued function.)

Case 1: $\limsup_{n\to\infty} f(n)^{1/n} = 0$.

Then f is eventually 0, so F is eventually a positive constant. This gives $\rho_f = \infty$ and $\rho_F = 1$.

Case 2: $\limsup_{n\to\infty} f(n)^{1/n} = \infty$.

Since $f(n) \leq F(n)$, $\limsup_{n\to\infty} F(n)^{1/n} = \infty$. Then $\rho_f = \rho_F = 0$.

Case 3: $\limsup_{n\to\infty} f(n)^{1/n} = r$, where $1 \leq r < \infty$.

Then, given $\varepsilon > 0$, there is an m such that, for $n \geq m$,

$$f(n)^{1/n} < r + \varepsilon,$$

so, for $n \geq m$,

$$\begin{aligned} F(n) &= f(0) + \cdots + f(m-1) + f(m) + \cdots + f(n) \\ &\leq F(m-1) + n \cdot (r+\varepsilon)^n \\ &\leq \big(1 + F(m-1)\big) \cdot n \cdot (r+\varepsilon)^n \qquad (\text{as } r \geq 1). \end{aligned}$$

Thus

$$F(n)^{1/n} \leq \big(1 + F(m-1)\big)^{1/n} \cdot n^{1/n} \cdot (r+\varepsilon),$$

and this leads to

$$\limsup_{n\to\infty} F(n)^{1/n} \leq r + \varepsilon.$$

Since this holds when $\varepsilon > 0$,

$$\limsup_{n\to\infty} F(n)^{1/n} \;\le\; r.$$

Also, since $f(n) \le F(n)$, for all n,

$$r \;=\; \limsup_{n\to\infty} f(n)^{1/n} \;\le\; \limsup_{n\to\infty} F(n)^{1/n},$$

and thus

$$\limsup_{n\to\infty} f(n)^{1/n} \;=\; \limsup_{n\to\infty} F(n)^{1/n}.$$

In this case $\rho_F = \rho_f \le 1$. □

Lemma 2.30. *For $f : \mathbb{N} \to \mathbb{N}$, with $\sum f(n) > 0$, and for $x \in [0, \rho_F)$,*

$$\sum_{n\ge 0} F(n)x^n \;=\; (1-x)^{-1} \cdot \sum_{n\ge 0} f(n)x^n.$$

Proof. The Cauchy product of the two power series $1 + x + x^2 + \cdots$ and $\sum_n f(n)x^n$ is the series $\sum_n F(n)x^n$, so the above equation holds when $|x| < \min(1, \rho_f) = \rho_F$. □

Lemma 2.31. *With $A(x)$ the global count function of the additive number system $\mathcal{A}$ the following hold:*

(a) *$A(x)$ is nondecreasing,*
(b) $\lim_{x\to\infty} A(x) \;=\; \infty$,
(c) $\rho_A \;=\; \rho$, *and*
(d) *for $x \in [0, \rho)$,*

$$\sum_{n\ge 0} A(n)x^n \;=\; (1-x)^{-1} \cdot \prod_{n\ge 1} (1-x^n)^{-p(n)}.$$

Proof. (a) and (b) are obviously true. (c) follows from Lemmas 2.23 and 2.29. Item (d) follows from Corollary 2.24 and Lemma 2.30. □

Remark 2.32. From item (d) above, one can interpret $A(n)$, the global count function of $\mathcal{A}$, as the local count function of an additive number system obtained by augmenting $\mathcal{A}$ by a single indecomposable p with $\|\mathsf{p}\| = 1$.

2.5. Alternate version of the fundamental identity

The fundamental identity gives a connection between the two count functions $a(n), p(n)$ of an additive number system $\mathcal{A}$. As will be seen, a variation on this identity is useful to study density of partition sets.

Proposition 2.33. *The identity*

$$\mathbf{A}(x) \;=\; \exp\Big(\sum_{m\ge 1} \mathbf{P}(x^m)/m\Big) \tag{2.33.1}$$

holds, for $x \ge 0$.

PROOF. For $x \geq 1$, both sides of (2.33.1) are clearly ∞. For $0 \leq x < 1$,

$$\begin{aligned}(1-x^n)^{-p(n)} &= \exp\Big(\log\big((1-x^n)^{-p(n)}\big)\Big)\\ &= \exp\big(-p(n)\log(1-x^n)\big)\\ &= \exp\Big(p(n)\sum_{m\geq 1} x^{mn}/m\Big),\end{aligned}$$

thus,

$$\prod_{1\leq n\leq M}(1-x^n)^{-p(n)} = \prod_{1\leq n\leq M}\exp\Big(p(n)\sum_{m\geq 1} x^{mn}/m\Big).$$

So, from Corollary 2.24,

$$\begin{aligned}\mathbf{A}(x) &= \prod_{n\geq 1}(1-x^n)^{-p(n)}\\ &= \lim_{M\to\infty}\prod_{1\leq n\leq M}\exp\Big(\sum_{m\geq 1} p(n)x^{mn}/m\Big)\\ &= \lim_{M\to\infty}\exp\Big(\sum_{1\leq n\leq M}\sum_{m\geq 1} p(n)x^{mn}/m\Big)\\ &= \exp\Big(\sum_{n\geq 1}\sum_{m\geq 1} p(n)x^{mn}/m\Big)\\ &= \exp\Big(\sum_{m\geq 1}\sum_{n\geq 1} p(n)x^{mn}/m\Big)\\ &= \exp\Big(\sum_{m\geq 1}\mathbf{P}(x^m)/m\Big).\end{aligned}$$

□

This leads to the fact that $\mathbf{A}(x)$ exponentially dominates $\mathbf{P}(x)$.

Corollary 2.34. *For* $x \geq 0$,

$$e^{\mathbf{P}(x)} \leq \mathbf{A}(x).$$

Definition 2.35. Let $\mathbf{H}(x) = \sum h(n)x^n$ be the power series defined by

$$h(n) = \begin{cases} 0 & \text{for } n = 0\\ \displaystyle\sum_{jm=n}\frac{p(j)}{m} & \text{for } n \geq 1.\end{cases}$$

Thus $h(n) \geq p(n)$, for $n \geq 0$, and $\mathbf{H}(x) = \sum_{n\geq 1} h(n)x^n$.

Proposition 2.36. *The identity*

$$\mathbf{H}(x) = \sum_{m\geq 1}\mathbf{P}(x^m)/m$$

holds when $x \geq 0$.

PROOF. For $x \geq 0$ and $\ell \geq k \geq 0$,

$$\begin{aligned}\sum_{n=1}^{k} h(n)x^n &= \sum_{n=1}^{k}\sum_{jm=n}\frac{p(j)}{m}x^n = \sum_{1\leq jm\leq k}\frac{p(j)}{m}x^{jm}\\ &\leq \sum_{1\leq m\leq k}\sum_{1\leq j\leq \ell}\frac{p(j)}{m}x^{jm} \leq \sum_{1\leq jm\leq k\ell}\frac{p(j)}{m}x^{jm}\\ &= \sum_{n=1}^{k\ell}\sum_{jm=n}\frac{p(j)}{m}x^n = \sum_{n=1}^{k\ell}h(n)x^n \leq \mathbf{H}(x),\end{aligned}$$

so,

$$\sum_{n=1}^{k} h(n)x^n \leq \sum_{1\leq m\leq k}\Big(\sum_{1\leq j\leq \ell} p(j)(x^m)^j\Big)\Big/m \leq \mathbf{H}(x).$$

Taking the limit as $\ell \to \infty$, and then as $k \to \infty$, gives the result. □

With this the alternate version of the fundamental identity of $\mathcal{A}$ becomes $\mathbf{A}(x) = e^{\mathbf{H}(x)}$, and one has the following information from Proposition 2.33.

Proposition 2.37. *The identity*

$$\mathbf{A}(x) = e^{\mathbf{H}(x)} \tag{2.37.1}$$

holds for $x \geq 0$.

Corollary 2.38. *For $x \geq 0$, $\mathbf{P}(x) \leq \mathbf{H}(x) \leq \mathbf{A}(x)$, and*

$$\mathbf{H}(x) < \infty \quad \textit{iff} \quad \mathbf{A}(x) < \infty.$$

Corollary 2.39. *Suppose $\mathbf{A}(x) \in \mathcal{PS}_c$. Then $1/\mathbf{A}(x) \in \mathcal{PS}_c$ and, letting its power series expansion be $[\mathbf{1/A}](x)$,*

$$1/\mathbf{A}(x) = [\mathbf{1/A}](x) \qquad \textit{on } [0,c].$$

PROOF. Since $\mathbf{A}(x) \in \mathcal{PS}_c$, by Corollary 2.38 it follows that $\mathbf{H}(x) \in \mathcal{PS}_c$, so $-\mathbf{H}(x) \in \mathcal{PS}_c$, and then, by Lemma 1.21,

$$1/\mathbf{A}(x) = e^{-\mathbf{H}(x)} \in \mathcal{PS}_c.$$

Let $[\mathbf{1/A}](x)$ be the power series expansion of $1/\mathbf{A}(x)$. Then one has $1/\mathbf{A}(x) = [\mathbf{1/A}](x)$ on $[0,c)$. Since $1/\mathbf{A}(x)$ and $[\mathbf{1/A}](x)$ are left continuous at $x = c$, they must also be equal at $x = c$. □

2.5.1. A formula for $a(n)$. The results of the last section lead to the following explicit formula to calculate $a(n)$ from $p(n)$.

Proposition 2.40. *For $n \geq 1$,*

$$a(n) = \sum_{j=1}^{n}\frac{1}{j!}\sum_{\substack{n_1+\cdots+n_j=n\\ n_i\geq 1}}\ \sum_{a_1b_1=n_1}\cdots\sum_{a_jb_j=n_j}\frac{p(a_1)\cdots p(a_j)}{b_1\cdots b_j}.$$

PROOF.

$$\begin{aligned} a(n) &= [x^n]e^{\mathbf{H}(x)} \\ &= \sum_{j=1}^{n} \frac{1}{j!} \sum_{\substack{n_1+\cdots+n_j=n \\ n_i \geq 1}} h(n_1)\cdots h(n_j) \\ &= \sum_{j=1}^{n} \frac{1}{j!} \sum_{\substack{n_1+\cdots+n_j=n \\ n_i \geq 1}} \Big(\sum_{a_1 b_1 = n_1} \frac{p(a_1)}{b_1} \Big) \cdots \Big(\sum_{a_j b_j = n_j} \frac{p(a_j)}{b_j} \Big). \end{aligned}$$

□

Thus, for example,

$$\begin{aligned} a(1) &= p(1) \\ a(2) &= p(2) + \frac{1}{2}p(1) + \frac{1}{2}p(1)^2 \\ a(3) &= p(3) + \frac{1}{3}p(1) + \frac{1}{2}p(1)^2 + p(1)p(2) + \frac{1}{6}p(1)^3. \end{aligned}$$

2.6. Reduced additive number systems

Let $\mathcal{A} = \big(\mathsf{A}, \mathsf{P}, +, 0, \| \ \|\big)$ be an additive number system with counting functions $a(n), p(n)$ and radius of convergence ρ.

Definition 2.41. For $f : \mathbb{N} \to \mathbb{N}$ define $\mathsf{supp}\, f(n)$, the *support* of f, to be the set of n for which $f(n)$ is positive, that is,

$$\mathsf{supp}\, f(n) \;=\; \{n \in \mathbb{N} : f(n) > 0\}.$$

Lemma 2.42.

$$\gcd\big(\mathsf{supp}\, a(n)\big) \;=\; \gcd\big(\|\mathsf{a}\| : \mathsf{a} \in \mathsf{A}\big) \;=\; \gcd\big(\mathsf{supp}\, p(n)\big) \;=\; \gcd\big(\|\mathsf{p}\| : \mathsf{p} \in \mathsf{P}\big).$$

PROOF. $\mathsf{supp}\, a(n) = \{\|\mathsf{a}\| : \mathsf{a} \in \mathsf{A}\}$ and $\mathsf{supp}\, p(n) = \{\|\mathsf{p}\| : \mathsf{p} \in \mathsf{P}\}$ follow immediately from the definition. Since $\mathsf{supp}\, p(n) \subseteq \mathsf{supp}\, a(n)$ it suffices to show that $\gcd\big(\mathsf{supp}\, p(n)\big) \,|\, \gcd\big(\mathsf{supp}\, a(n)\big)$.

Let $d = \gcd\big(\mathsf{supp}\, p(n)\big)$. Then $k \in \mathsf{supp}\, a(n)$ iff there is an $\mathsf{a} \in \mathsf{A}$ with $\|\mathsf{a}\| = k$. Express a as a sum of indecomposables, say $\mathsf{a} = \sum m_i \mathsf{p}_i$. Then $d|k$ since $k = \|\mathsf{a}\| = \sum m_i \|\mathsf{p}_i\|$, and each $\|\mathsf{p}_i\|$ is divisible by d. Thus $d|\gcd\big(\mathsf{supp}\, a(n)\big)$. □

Definition 2.43. $\mathcal{A}$ is *reduced* if

$$\gcd\big(\mathsf{supp}\, p(n)\big) \;=\; 1.$$

Note that the examples of equivalence relations, etc., that were discussed in §2.2 are reduced—indeed they all have $p(1) > 0$.

2.6.1. The reduced form of $\mathcal{A}$.

Definition 2.44. Let $d = \gcd\big(\mathsf{supp}\, p(n)\big)$. Define $\mathcal{A}^\star$, the *reduced form of* $\mathcal{A}$, by altering the norm of $\mathcal{A}$, namely let

$$\|\mathsf{a}\|^\star \;=\; \|\mathsf{a}\|/d.$$

Let $a^\star(n), p^\star(n)$ be the counting functions of $\mathcal{A}^\star$, and let $\rho^\star$ be the radius of convergence of $\mathcal{A}^\star$.

Lemma 2.45. $\mathcal{A}^\star = (\mathsf{A}, \mathsf{P}, +, \mathsf{0}, \|\ \|^\star)$ *is a reduced additive number system. Furthermore,*

(a) $a^\star(n) \;=\; a(nd)$ *and* $p^\star(n) \;=\; p(nd)$
(b) $\gcd\big(\mathsf{supp}\, a^\star(n)\big) \;=\; \gcd\big(\mathsf{supp}\, p^\star(n)\big) \;=\; 1$
(c) $\rho^\star = \rho^d$.

PROOF. For $\mathsf{a}, \mathsf{b} \in \mathsf{A}$,

$$\begin{aligned} &\|\mathsf{a}\|^\star = 0 \quad \text{iff} \quad \|\mathsf{a}\|/d = 0 \quad \text{iff} \quad \mathsf{a} = \mathsf{0}\\ &\|\mathsf{a}+\mathsf{b}\|^\star \;=\; \|\mathsf{a}+\mathsf{b}\|/d \;=\; \|\mathsf{a}\|/d \;+\; \|\mathsf{b}\|/d \;=\; \|\mathsf{a}\|^\star \;+\; \|\mathsf{b}\|^\star. \end{aligned}$$

Also

$$\begin{aligned} a^\star(n) &= \big|\{\mathsf{a} \in \mathsf{A} : \|\mathsf{a}\|^\star = n\}\big|\\ &= \big|\{\mathsf{a} \in \mathsf{A} : \|\mathsf{a}\|/d = n\}\big|\\ &= \big|\{\mathsf{a} \in \mathsf{A} : \|\mathsf{a}\| = nd\}\big|\\ &= a(nd). \end{aligned}$$

Likewise, $p^\star(n) = p(nd)$.

Next,

$$\begin{aligned} \gcd\big(\mathsf{supp}\, p^\star(n)\big) &= \gcd\big(\{n \in \mathbb{N} : p^\star(n) \neq 0\}\big)\\ &= \gcd\big(\{n \in \mathbb{N} : p(nd) \neq 0\}\big)\\ &= (1/d) \cdot \gcd\big(\{nd : n \in \mathbb{N},\ p(nd) \neq 0\}\big)\\ &= 1. \end{aligned}$$

Finally, $\rho^\star = \rho^d$ since

$$\begin{aligned} \limsup_{n\to\infty} \big(a^\star(n)\big)^{1/n} &= \limsup_{n\to\infty} \big(a(nd)\big)^{1/n}\\ &= \Big(\limsup_{n\to\infty} \big(a(nd)\big)^{1/nd}\Big)^d\\ &= \Big(\limsup_{n\to\infty} \big(a(n)\big)^{1/n}\Big)^d. \end{aligned}$$

□

2.7. Finitely generated number systems

Let $\mathcal{A} = \big(\mathsf{A}, \mathsf{P}, +, 0, \|\ \|\big)$ be an additive number system with counting functions $a(n), p(n)$ and radius of convergence ρ.

The fact that the elements of an additive number system can be constructed by closing a set of indecomposables under addition, and the fact that the norm is additive, leads to the pleasant fact that the norm values $\|\mathsf{a}\|$ eventually take on a regular behavior, namely they eventually look like the multiples of an integer d. This simple fact is nontrivial though, and is based on a careful analysis of the counting functions $a(n)$ of finitely generated additive number systems.

An additive number system is *finitely generated* iff its rank $r = \sum p(n)$ is finite, that is, the set P of indecomposables is finite. The fundamental identity of such a system was investigated by means of partial fractions in $\mathbb{C}(x)$ by Cayley and Sylvester in the 1850's. The general results about $a(n)$, obtained by this method, have been attributed to I. Schur (see Wilf [**56**]) and to G. Polya (see Bateman and Erdös [**4**]). The basic result is that $a(n)$ has polynomial growth on its support. First, a simple lemma about powers of geometric series is proved.

Lemma 2.46. *For* $0 \le x < 1$,

$$\frac{1}{(1-x)^r} = \sum_{n\ge 0} \binom{n+r-1}{r-1} x^n.$$

Proof. For $0 \le x < 1$,

$$\begin{aligned}
\frac{1}{(1-x)^r} &= \frac{1}{(r-1)!} \cdot \frac{d^{r-1}}{dx^{r-1}} \frac{1}{1-x} \\
&= \frac{1}{(r-1)!} \cdot \frac{d^{r-1}}{dx^{r-1}} \sum_{n\ge 0} x^n \\
&= \frac{1}{(r-1)!} \cdot \sum_{n\ge r-1} n\cdot(n-1)\cdots(n-r+2)x^{n-r+1} \\
&= \sum_{n\ge r-1} \binom{n}{r-1} x^{n-r+1} \\
&= \sum_{n\ge 0} \binom{n+r-1}{r-1} x^n.
\end{aligned}$$

□

Lemma 2.47. *Given an integer* $j \ge 0$,

$$\binom{n+j}{j} = \frac{n^j}{j!} + \mathrm{o}\big(n^j\big).$$

PROOF. This follows easily from the fact that $\binom{n+j}{j}$ is a polynomial in n:

$$\binom{n+j}{j} = \frac{n(n-1)\cdots(n-j+1)}{j!}$$
$$= \frac{n^j}{j!} + \text{terms of lower degree in } n,$$

and since any polynomial in n of lower degree than n^j is $o(n^j)$. □

Theorem 2.48. *Suppose $\mathcal{A}$ is a reduced additive number system with finite rank r. Let $\mathsf{supp}\, p(n) = \{d_1, \ldots, d_k\}$ and let $p(d_i) = m_i$. Then*

$$a(n) \sim \frac{1}{(r-1)!\prod {d_i}^{m_i}} \cdot n^{r-1}.$$

PROOF. The fundamental identity of $\mathcal{A}$ is

$$\sum_{n\geq 0} a(n)x^n = \prod_{i=1}^{k}(1-x^{d_i})^{-m_i},$$

and it holds when $0 \leq x < 1$ since $\mathbf{P}(x) = m_1x^{d_1} + \cdots + m_kx^{d_k}$ shows, by Corollary 2.25 (b), that $\rho = 1$. Apply the method of partial fractions to decompose the right side into linear factors. One would be inclined to express this as

$$\prod_{i=1}^{k}(1-x^{d_i})^{-m_i} = \sum_{j,\theta} \frac{C_{j,\theta}}{(x-\theta)^j},$$

where the $C_{j,\theta} \in \mathbb{C}$, the variable θ runs over the roots of unity that appear in the factorization of the various $1 - x^{d_i}$, and $j \leq r$. To take advantage of Lemma 2.46, one can rewrite each of the factors $x - \theta$ as $-\theta(1 - \theta^{-1}x)$. Since the θ^{-1} are also roots of unity, this means we can write the partial fraction decomposition in the form

$$\prod_{i=1}^{k}(1-x^{d_i})^{-m_i} = \sum_{j,\theta} \frac{C_{j,\theta}}{(1-\theta x)^j}.$$

Note that, because $\gcd(\mathsf{supp}\, p(n)) = 1$, all roots except $\theta = 1$ of $\prod_{i=1}^{k}(1-x^{d_i})^{m_i}$ have multiplicity less than r. For suppose θ is a root of all the factors $1 - x^{d_i}$. Then $\theta^{d_i} = 1$, for $1 \leq i \leq k$. Since $\gcd(d_1, \ldots, d_k) = 1$, there are integers $\ell_1, \ldots, \ell_k$ with $\ell_1 d_1 + \cdots + \ell_k d_k = 1$. Then from $\theta^{\ell_i d_i} = 1$ it follows that $\theta = \theta^{\ell_1 d_1 + \cdots + \ell_k d_k} = 1$.

Now, $\theta = 1$ has multiplicity $r = m_1 + \cdots + m_k$. So there are constants C and $C_{j,\theta}$ such that, with $\ell \leq \max(d_1, \ldots, d_k)$ and $0 \leq x < 1$, one has

$$\begin{aligned}
\sum_{n\geq 0} a(n)x^n &= \prod_{i=1}^{k}(1-x^{d_i})^{-m_i} \\
&= \frac{C}{(1-x)^r} + \sum_{\substack{j<r\\ \theta^\ell=1}} \frac{C_{j,\theta}}{(1-\theta x)^j} \qquad (2.48.1) \\
&= C\cdot\sum_{n=0}^{\infty}\binom{n+r-1}{r-1}x^n + \sum_{\substack{j<r\\ \theta^\ell=1}} C_{j,\theta}\cdot\sum_{n=0}^{\infty}\binom{n+j-1}{j-1}(\theta x)^n \\
&= \sum_{n=0}^{\infty}\left[C\cdot\binom{n+r-1}{r-1} + \sum_{\substack{j<r\\ \theta^\ell=1}} C_{j,\theta}\cdot\binom{n+j-1}{j-1}\theta^n\right]x^n.
\end{aligned}$$

Thus

$$\begin{aligned}
a(n) &= C\cdot\binom{n+r-1}{r-1} + \sum_{\substack{j<r\\ \theta^\ell=1}} C_{j,\theta}\cdot\binom{n+j-1}{j-1}\theta^n \\
&= C\cdot\left(\frac{n^{r-1}}{(r-1)!} + o\big(n^{r-1}\big)\right) + o\big(n^{r-1}\big) \\
&= C\cdot\frac{n^{r-1}}{(r-1)!} + o\big(n^{r-1}\big) \\
&\sim C\cdot\frac{n^{r-1}}{(r-1)!}.
\end{aligned}$$

To determine C, multiply

$$\frac{C}{(1-x)^r} + \sum_{\substack{j<r\\ \theta^\ell=1}} \frac{C_{j,\theta}}{(1-\theta x)^j} = \prod_{i=1}^{k}(1-x^{d_i})^{-m_i}$$

by $(1-x)^r$. Then taking the limit as $x \to 1^-$ gives

$$\begin{aligned}
C &= \lim_{x\to 1^-}(1-x)^r\cdot\prod_{i=1}^{k}(1-x^{d_i})^{-m_i} \\
&= \lim_{x\to 1^-}\prod_{i=1}^{k}(1+x+\cdots+x^{d_i-1})^{-m_i} \\
&= \prod_{i=1}^{k} d_i^{-m_i}.
\end{aligned}$$

□

Corollary 2.49 (Hua's Theorem). *Suppose $\mathcal{A}$ has finite rank r with d the gcd of* $\operatorname{supp} p(n)$. *Let* $\operatorname{supp} p(n) = \{d_1, \ldots, d_k\}$ *and let* $p(d_i) = m_i$. *Then*

$$a(nd) \sim \frac{d^r}{(r-1)! \prod {d_i}^{m_i}} \cdot n^{r-1}.$$

In particular, it follows that

(a) $\operatorname{supp} a(n)$ *eventually agrees with* $d\mathbb{N} = \{0, d, 2d, 3d, \ldots\}$, *so* $a(n)$ *is eventually positive on the multiples of* d, *and* 0 *off the multiples of* d,

(b) $\lim_{n\to\infty} \dfrac{a\big((n-1)d\big)}{a(nd)} = 1,$

(c) $\lim_{n\to\infty} \dfrac{a(n)}{A(n)} = 0.$

PROOF. Since the reduced form $\mathcal{A}^\star$ of $\mathcal{A}$ is such that the hypotheses of Theorem 2.48 are satisfied,

$$\begin{aligned} a^\star(n) &\sim \frac{1}{(r-1)! \prod (d_i/d)^{m_i}} \cdot n^{r-1} \\ &= \frac{d^r}{(r-1)! \prod d_i^{m_i}} \cdot n^{r-1}, \end{aligned}$$

and thus

$$a(nd) \sim \frac{d^r}{(r-1)! \prod d_i^{m_i}} \cdot n^{r-1}.$$

(a) Consequently $a(n)$ is eventually positive on multiples of d. $a(n)$ is 0 off the multiples of d since $\operatorname{supp} a(n)$ is a subset of the multiples of d.

(b) From $a(nd) \sim cn^{r-1}$, for some c, it follows that

$$\frac{a\big((n-1)d\big)}{a(nd)} \sim \frac{(n-1)^{r-1}}{n^{r-1}}$$

and thus

$$\lim_{n\to\infty} \frac{a\big((n-1)d\big)}{a(nd)} = 1.$$

(c) By Remark 2.32, $A(n)$ can be thought of as the local count function of an additive number system of rank $r+1$ that is reduced (since it has an indecomposable of norm 1). Thus, by Theorem 2.48, for some constant c, $A(n) \sim cn^r$, so

$$\lim_{n\to\infty} \frac{A(n-1)}{A(n)} = 1.$$

Now, since $a(n) = A(n) - A(n-1)$,

$$\frac{a(n)}{A(n)} = 1 - \frac{A(n-1)}{A(n)},$$

and (c) follows. □

Remark 2.50. Given a finite P with norms $\{d_1, \ldots, d_k\}$, it seems to be quite difficult to determine precisely when $\mathsf{supp}\, a(n)$ starts to behave like an arithmetical sequence. (See Wilf [**56**], p. 97.)

2.8. $a^\star(n)$ is eventually positive

Now the restriction that $\mathcal{A}$ be finitely generated is dropped, and the general result that the support of the norm of $\mathcal{A}$ is eventually an arithmetical sequence, that is, $a^\star(n)$ is eventually positive, is proved.

Definition 2.51. A function $f : \mathbb{N} \to \mathbb{N}$ has *superpolynomial* growth if, for all constants c,

$$\lim_{n\to\infty} \frac{n^c}{f(n)} = 0.$$

Theorem 2.52. *Let* $d = \gcd\left(\mathsf{supp}\, p(n)\right)$. *Then*

(a) $\mathsf{supp}\, a(n)$ *eventually equals the multiples of* d,
(b) $a^\star(n)$ *is eventually positive,*
(c) *if the rank of* $\mathcal{A}$ *is finite then* $a(n)$ *has polynomially bounded growth, that is,* $a(n) = \mathrm{O}(n^c)$, *for some* c, *and*
(d) *if the rank of* $\mathcal{A}$ *is infinite then* $a(n)$ *has superpolynomial growth on its support, that is, for all* c,

$$\lim_{n\to\infty} \frac{n^c}{a^\star(n)} = 0.$$

Proof. Let $\{d_1, \ldots, d_k\}$ be a finite subset of $\mathsf{supp}\, p(n)$ such that d is the gcd of $\{d_1, \ldots, d_k\}$. Then choose one indecomposable p_i of norm d_i, for $1 \le i \le k$. By Corollary 2.49 (a) the elements $\mathsf{b} = m_1\mathsf{p}_1 + \cdots + m_k\mathsf{p}_k$, $m_i \ge 0$, have norms that eventually coincide with the multiples of d. Since all elements of A have norms that are multiples of d, (a) holds. (b) is just a restatement of (a). (c) and (d) then follow from Theorem 2.48. □

In 1956 Bateman and Erdös found necessary and sufficient conditions for $a(n)$ to be eventually strictly increasing, given that $p(n) \le 1$, for all n. This result is in Appendix D.

CHAPTER 3

Density and Partition Sets

Now, having introduced the basic theory of counting functions, the discussion turns to the task of developing methods to determine, for B a subset of an additive number system $\mathcal{A}$, what the probability is that a randomly chosen element of A will belong to B. The focus will be on subsets B called partition sets, probability means asymptotic density, and the key tool for the analysis is Dirichlet density.

3.1. Asymptotic density

Let $\mathcal{A} = \big(\mathsf{A}, \mathsf{P}, +, 0, \|\ \|\big)$ be an additive number system with counting functions $a(n), p(n)$ and radius of convergence ρ.

Definition 3.1. For $\mathsf{B} \subseteq \mathsf{A}$, the (*local*) *asymptotic density* $\delta(\mathsf{B})$ of B is as follows, provided the indicated limit exists:

$$\delta(\mathsf{B}) \quad = \quad \lim_{\substack{n\to\infty \\ a(n)\neq 0}} \frac{b(n)}{a(n)}.$$

This notion of the density of B captures the idea of *the probability that a randomly chosen element from* A *is in* B. In the example of two-colored linear forests discussed in §2.3.2, the problem is to determine the asymptotic density of the collection B of finite two-colored linear forests with at least one component having strictly alternating colors.

The following lemma says that the collection of subsets of A that have local asymptotic density is closed under complement as well as union of disjoint members.

Lemma 3.2. *Suppose* B, C *are subsets of* A *such that* $\delta(\mathsf{B})$ *and* $\delta(\mathsf{C})$ *exist. Then*

(a) $\delta(\mathsf{A} \smallsetminus \mathsf{B}) \ = \ 1 - \delta(\mathsf{B})$
(b) $\mathsf{B} \cap \mathsf{C} = \varnothing \quad \Rightarrow \quad \delta(\mathsf{B} \cup \mathsf{C}) \ = \ \delta(\mathsf{B}) + \delta(\mathsf{C})$.

PROOF. From $\big|\mathsf{A} \smallsetminus \mathsf{B}\big|_n \ = \ a(n) - b(n)$ one has

$$\frac{\big|\mathsf{A} \smallsetminus \mathsf{B}\big|_n}{a(n)} \ = \ 1 - \frac{b(n)}{a(n)},$$

so taking the limit as $n \to \infty$ gives (a).

If B and C are disjoint then $\left|\mathsf{B}\cup\mathsf{C}\right|_n = b(n)+c(n)$, so

$$\frac{\left|\mathsf{B}\cup\mathsf{C}\right|_n}{a(n)} = \frac{b(n)}{a(n)} + \frac{c(n)}{a(n)}.$$

Now take the limit as $n\to\infty$ to derive (b). □

Lemma 3.3. *$\mathcal{A}$ and $\mathcal{A}^\star$ have the same local density function, that is, any subset B of A has local asymptotic density in $\mathcal{A}$ iff it has local asymptotic density in $\mathcal{A}^\star$, and when these densities exist they are the same.*

PROOF. Let $d=\gcd(\mathsf{supp}\,p(n))$. Since the support of $a(n)$ is eventually the multiples of d,

$$\delta(\mathsf{B}) = \lim_{n\to\infty}\frac{b(nd)}{a(nd)}.$$

Now use the fact that $b^\star(n) = b(nd)$, $a^\star(n) = a(nd)$, where $a^\star(n), b^\star(n)$ count in $\mathcal{A}^\star$. □

As seen in Theorem 2.52, $\mathcal{A}$ being reduced guarantees that $a(n)$ is eventually positive. Thus, for $\mathcal{A}$ reduced, there is a slightly simpler definition of density:

$$\delta(\mathsf{B}) = \lim_{n\to\infty}\frac{b(n)}{a(n)}.$$

Corresponding to the global count function there is *global asymptotic density.*

Definition 3.4. For $\mathsf{B}\subseteq\mathsf{A}$, the *global asymptotic density* $\Delta(\mathsf{B})$ is as follows, whenever the indicated limit exists:

$$\Delta(\mathsf{B}) = \lim_{n\to\infty}\frac{B(n)}{A(n)}.$$

The diagrams in Figure 1 illustrate the notion of local asymptotic density as the limit of the ratio of the lengths of the line segments measuring $b(n)$ and $a(n)$, and global asymptotic density as the limit of the ratio of the areas measuring $B(n)$ and $A(n)$.

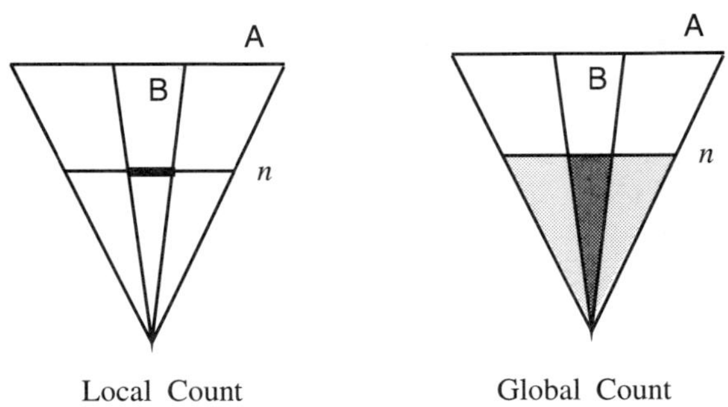

FIGURE 1

The main interest in additive number systems is local asymptotic density, so, in Part 1, the unmodified phrase 'asymptotic density' takes the default meaning of 'local asymptotic density'.

The collection of subsets of A that have global asymptotic density is closed under complement as well as union of disjoint members.

Lemma 3.5. *Suppose* B, C *are subsets of* A *such that* $\Delta(\mathsf{B})$ *and* $\Delta(\mathsf{C})$ *exist. Then*

(a) $\Delta(\mathsf{A} \smallsetminus \mathsf{B}) \ = \ 1 - \Delta(\mathsf{B})$

(b) $\mathsf{B} \cap \mathsf{C} = \varnothing \quad \Rightarrow \quad \Delta(\mathsf{B} \cup \mathsf{C}) \ = \ \Delta(\mathsf{B}) + \Delta(\mathsf{C})$.

PROOF. From $\big|\mathsf{A} \smallsetminus \mathsf{B}\big|_{(\leq n)} \ = \ A(n) - B(n)$ one has

$$\frac{\big|\mathsf{A} \smallsetminus \mathsf{B}\big|_{(\leq n)}}{A(n)} \ = \ 1 - \frac{B(n)}{A(n)},$$

so taking the limit as $n \to \infty$ gives (a).

If B and C are disjoint then $\big|\mathsf{B} \cup \mathsf{C}\big|_{(\leq n)} \ = \ B(n) + C(n)$, so

$$\frac{\big|\mathsf{B} \cup \mathsf{C}\big|_{(\leq n)}}{A(n)} \ = \ \frac{B(n)}{A(n)} + \frac{C(n)}{A(n)}.$$

Now take the limit as $n \to \infty$ to derive (b). □

As one might expect, the global asymptotic density $\Delta(\mathsf{B})$ exists whenever $\delta(\mathsf{B})$ exists.

Proposition 3.6 ($\delta \subseteq \Delta$). *Suppose* B *is a subset of* A. *If the local asymptotic density* $\delta(\mathsf{B})$ *exists then the global asymptotic density* $\Delta(\mathsf{B})$ *also exists, and* $\Delta(\mathsf{B}) = \delta(\mathsf{B})$.

PROOF. This follows from noting that if a_n, b_n are two sequences of nonnegative integers with the a_n eventually positive, then $b_n/a_n \to c$ implies $B_n/A_n \to c$, where $A_n = a_0 + \cdots + a_n$ and $B_n = b_0 + \cdots + b_n$. For then

$$\begin{aligned}
\left|\frac{B_n}{A_n} - c\right| \ &= \ \big|B_n - cA_n\big|/A_n \\
&= \ \Big|\sum_{k \leq n}(b_k - ca_k)\Big|\Big/A_n \\
&\leq \ \underbrace{\Big|\sum_{k \leq M}(b_k - ca_k)\Big|\Big/A_n}_{\text{I}} + \underbrace{\Big|\sum_{M<k\leq n}(b_k - ca_k)\Big|\Big/A_n}_{\text{II}}.
\end{aligned}$$

For any fixed M, term I goes to 0 as $n \to \infty$ since $A_n \to \infty$. Given $\varepsilon > 0$ choose M such that, for $k \geq M$,

$$\left|\frac{b_k}{a_k} - c\right| \ < \ \varepsilon.$$

Then

$$\begin{aligned}\Big|\sum_{M<k\le n}(b_k - ca_k)\Big|\Big/A_n &\le \sum_{M<k\le n}|b_k - ca_k|\Big/A_n\\ &\le \sum_{M<k\le n}\varepsilon a_k\Big/A_n\\ &\le \varepsilon A_n/A_n = \varepsilon\end{aligned}$$

so $\left|\dfrac{B_n}{A_n} - c\right| \le$ (term I) $+\varepsilon$, and thus

$$\limsup_{n\to\infty}\left|\frac{B_n}{A_n} - c\right| \le \varepsilon.$$

This shows $B(n)/A(n) \to c$, and completes the proof. □

Lemma 3.7. *$\mathcal{A}$ and $\mathcal{A}^\star$ have the same global density function Δ, that is, any subset* B *of* A *has global density in $\mathcal{A}$ iff it has global density in $\mathcal{A}^\star$, and when these densities exist they are the same.*

PROOF. Use the fact that $B(n) = B^\star(n/d)$, where the right side counts in $\mathcal{A}^\star$. □

3.2. Dirichlet density

Let $\mathcal{A} = (\mathsf{A}, \mathsf{P}, +, 0, \|\ \|)$ be an additive number system with counting functions $a(n), p(n)$ and radius of convergence ρ.

Another approach to density was used successfully by Kevin Compton. It often agrees with asymptotic density, and seems easier to work with.

Definition 3.8. Suppose that the radius of convergence of $\mathcal{A}$ is positive, that is, $\rho > 0$. For B a subset of A, the *Dirichlet density*[1] $\partial(\mathsf{B})$ of B is defined as follows, whenever the indicated limit exists:

$$\partial(\mathsf{B}) = \lim_{x\to\rho}\frac{\mathbf{B}(x)}{\mathbf{A}(x)}.$$

Of course $\lim_{x\to\rho}$ means 'take the limit as x approaches ρ from the left', since $\mathbf{A}(x)$ is not convergent when $x > \rho$. In the future, an expression of the form '$\partial(\mathsf{B}) = \cdots$', without the qualifier 'if $\partial(\mathsf{B})$ exists', assumes '$\partial(\mathsf{B})$ exists'.

The collection of subsets of A that have Dirichlet density is closed under complement as well as union of disjoint members.

Lemma 3.9. *Suppose* B, C *are subsets of* A *such that* $\partial(\mathsf{B})$ *and* $\partial(\mathsf{C})$ *exist. Then*

(a) $\partial(\mathsf{A}\setminus\mathsf{B}) = 1 - \partial(\mathsf{B})$

(b) $\mathsf{B}\cap\mathsf{C} = \varnothing \quad\Rightarrow\quad \partial(\mathsf{B}\cup\mathsf{C}) = \partial(\mathsf{B}) + \partial(\mathsf{C})$.

[1]The choice of the name *Dirichlet density* is explained in §9.2 of Part 2 on multiplicative number systems.

PROOF. From $[\mathbf{A} \setminus \mathbf{B}](x) = \mathbf{A}(x) - \mathbf{B}(x)$, where $[\mathbf{A} \setminus \mathbf{B}](x)$ is the generating series for $\mathsf{A} \setminus \mathsf{B}$, one has

$$\frac{[\mathbf{A} \setminus \mathbf{B}](x)}{\mathbf{A}(x)} = 1 - \frac{\mathbf{B}(x)}{\mathbf{A}(x)},$$

so taking the limit as $x \longrightarrow \rho$ gives (a).

If B and C are disjoint then $[\mathbf{B} \cup \mathbf{C}](x) = \mathbf{B}(x) + \mathbf{C}(x)$, where $[\mathbf{B} \cup \mathbf{C}](x)$ is the generating series for $\mathsf{B} \cup \mathsf{C}$, so

$$\frac{[\mathbf{B} \cup \mathbf{C}](x)}{\mathbf{A}(x)} = \frac{\mathbf{B}(x)}{\mathbf{A}(x)} + \frac{\mathbf{C}(x)}{\mathbf{A}(x)}.$$

Now take the limit as $x \longrightarrow \rho$ to derive (b). □

As with asymptotic density, it turns out that one can concentrate on reduced additive number systems.

Lemma 3.10. *Suppose $\rho > 0$. Then $\mathcal{A}$ and $\mathcal{A}^\star$ have the same Dirichlet density function, that is, any subset B of A has Dirichlet density in $\mathcal{A}$ iff it has Dirichlet density in $\mathcal{A}^\star$, and when these densities exist they are the same.*

PROOF. Let $d = \gcd\big(\operatorname{supp} a(n)\big)$. For any $\mathsf{B} \subseteq \mathsf{A}$, let $\mathbf{B}^\star(x)$ be the generating series of B in $\mathcal{A}^\star$, and $b^\star(n)$ the counting function of B in $\mathcal{A}^\star$. Then, for $0 \le x < \rho^\star$,

$$\mathbf{B}^\star(x) = \sum b^\star(n) x^n = \sum b(nd) x^n,$$

so $\mathbf{B}^\star(x) = \mathbf{B}(x^{1/d})$. Since $\rho^\star = \rho^d$ by Lemma 2.45 (c), $\partial(\mathsf{B}) = \lim_{x \to \rho} \dfrac{\mathbf{B}(x)}{\mathbf{A}(x)}$ exists iff $\partial^\star(\mathsf{B}) = \lim_{x \to \rho^\star} \dfrac{\mathbf{B}^\star(x)}{\mathbf{A}^\star(x)}$ exists, and if so, they are equal. □

If $\mathbf{A}(\rho) < \infty$ then it is easy to see that *every* subset of A has Dirichlet density. Furthermore, the Dirichlet density of any nonempty subset is positive.

Lemma 3.11. *Assume $\mathbf{A}(\rho) < \infty$. Then, for $\mathsf{B} \subseteq \mathsf{A}$,*

$$\partial(\mathsf{B}) = \frac{\mathbf{B}(\rho)}{\mathbf{A}(\rho)}.$$

If $\mathsf{B} \neq \varnothing$ then $\partial(\mathsf{B}) > 0$.

PROOF. Since $\mathbf{A}(\rho) < \infty$ then, for any $\mathsf{B} \subseteq \mathsf{A}$, $\mathbf{B}(x)$ is continuous on $[0, \rho]$, so

$$\partial(\mathsf{B}) = \lim_{x \to \rho} \frac{\mathbf{B}(x)}{\mathbf{A}(x)} = \frac{\mathbf{B}(\rho)}{\mathbf{A}(\rho)}.$$

If $\mathsf{B} \neq \varnothing$ then $0 < \mathbf{B}(\rho) \le \mathbf{A}(\rho)$. □

If $\mathbf{A}(\rho) < \infty$ then the Dirichlet density and asymptotic density disagree on all nonempty finite subsets of A. This is because the Dirichlet density is positive, by Lemma 3.11, whereas the asymptotic density of any finite subset is 0.

If $\mathbf{A}(\rho) = \infty$ and the local asymptotic density of a set exists then the Dirichlet density also exists and agrees with the local asymptotic density. This is proved below. The condition $\mathbf{A}(\rho) = \infty$ always holds if $\rho = 1$, but one can easily construct examples of additive number systems with $0 < \rho < 1$ for which it fails, for example, take $p(n) = \lfloor 2^n/n^2 \rfloor$. Then $\rho = 1/2$ and $\mathbf{P}(1/2) \le \sum_n 1/n^2 < \infty$, so $\mathbf{A}(1/2) < \infty$. (Such examples are not central to this study since $\mathbf{A}(\rho) < \infty$ contradicts the desired property that all partition sets have asymptotic density.)

Definition 3.12. Given a power series $\mathbf{S}(x) = \sum_{n \ge 0} s(n)x^n$ let

$$\mathbf{S}^+(x) \;=\; \sum_{n \ge 0} S(n)x^n,$$

where $S(n) = s(0) + \cdots + s(n)$.

Proposition 3.13 (Compton, 1989). *Suppose* $\mathbf{A}(\rho) = \infty$. *Then one has*

$$\delta \;\subseteq\; \Delta \;\subseteq\; \partial\,,$$

that is, the function Δ *extends* δ*, and* ∂ *extends* Δ.

PROOF. Clearly $\mathbf{A}(\rho) = \infty$ implies $\rho > 0$. Proposition 3.6 says that Δ extends δ in all additive number systems. So assume $\Delta(\mathsf{B})$ exists and $\mathbf{A}(\rho) = \infty$. From

$$\mathbf{A}^+(x) \;=\; \sum_{n \ge 0} A(n)x^n \qquad \text{and} \qquad \mathbf{B}^+(x) \;=\; \sum_{n \ge 0} B(n)x^n$$

one has, for $0 \le x < \rho$,

$$\mathbf{A}^+(x) \;=\; (1-x)^{-1} \cdot \mathbf{A}(x) \qquad \text{and} \qquad \mathbf{B}^+(x) \;=\; (1-x)^{-1} \cdot \mathbf{B}(x),$$

so, for $0 \le x < \rho$,

$$\begin{aligned}
\left| \frac{\mathbf{B}(x)}{\mathbf{A}(x)} - \Delta(\mathsf{B}) \right| &= \left| \frac{\mathbf{B}^+(x)}{\mathbf{A}^+(x)} - \Delta(\mathsf{B}) \right| \\
&= \left| \mathbf{B}^+(x) - \Delta(\mathsf{B}) \cdot \mathbf{A}^+(x) \right| \Big/ \mathbf{A}^+(x) \\
&= \left| \Big(\sum_{n \ge 0} B(n)x^n \Big) - \Delta(\mathsf{B}) \cdot \Big(\sum_{n \ge 0} A(n)x^n \Big) \right| \Big/ \mathbf{A}^+(x) \\
&= \left| \sum_{n \ge 0} \Big(B(n) - \Delta(\mathsf{B}) \cdot A(n) \Big) x^n \right| \Big/ \mathbf{A}^+(x)
\end{aligned}$$

$$\leq \underbrace{\sum_{n\leq M}\left|\Big(B(n)-\Delta(\mathsf{B})\cdot A(n)\Big)x^n\right|\Big/\mathbf{A}^+(x)}_{\mathrm{I}}$$
$$+\underbrace{\sum_{n>M}\left|\Big(B(n)-\Delta(\mathsf{B})\cdot A(n)\Big)x^n\right|\Big/\mathbf{A}^+(x)}_{\mathrm{II}}.$$

For any choice of M, term I goes to 0 as $x \to \rho$ since $\mathbf{A}^+(\rho) = \infty$. So, it suffices to show that term II can be made less than any given ε by choosing M large enough.

Given $\varepsilon > 0$ choose M such that, for $n > M$,

$$\left|\frac{B(n)}{A(n)} - \Delta(\mathsf{B})\right| < \varepsilon.$$

Then, for $0 \leq x < \rho$,

$$\begin{aligned}\sum_{n>M}\left|\Big(B(n)-\Delta(\mathsf{B})\cdot A(n)\Big)x^n\right|\Big/\mathbf{A}^+(x) &< \sum_{n>M}\varepsilon A(n)x^n/\mathbf{A}^+(x)\\ &\leq \varepsilon\mathbf{A}^+(x)/\mathbf{A}^+(x) = \varepsilon.\end{aligned}$$

This suffices to prove the result. □

In view of later work on partition sets it will be possible to find examples of additive number systems $\mathcal{A}$ with $\mathbf{A}(\rho) = \infty$ but $\delta \neq \partial$. In particular, this happens when $\rho = 1$ but one cannot find the radius of convergence by the ratio test.

3.3. The standard assumption

Thanks to Lemmas 3.3, 3.7, and 3.10, for the remainder of Part 1 the assumption will be that $\mathcal{A}$ is reduced. The **standard assumption** will be:

> $\mathcal{A} = (\mathsf{A}, \mathsf{P}, +, 0, \|\ \|)$ is a *reduced* additive number system with local counting functions $a(n), p(n)$, global counting functions $A(x), P(x)$, generating series $\mathbf{A}(x), \mathbf{P}(x)$, and radius of convergence $\rho > 0$.

For the importance of the condition $\rho > 0$, see Remark 3.32.

3.4. The set of additives of an element

Definition 3.14. For $\mathsf{b} \in \mathsf{A}$ let

$$\mathsf{b} + \mathsf{A} = \{\mathsf{b} + \mathsf{a} : \mathsf{a} \in \mathsf{A}\},$$

the set of additives[2] *of* b.

[2]This nomenclature is chosen to parallel the standard use in number theory of *the set of multiples of* m to describe the set $m\mathbb{P}$.

Such sets always have Dirichlet density, and there is a simple formula for this density. If $a(n) \in \mathsf{RT}_\rho$ then such sets also have asymptotic density, and it agrees with the Dirichlet density.

Lemma 3.15. *For* $\mathsf{b} \in \mathsf{A}$,

$$|\mathsf{b}+\mathsf{A}|_n = \begin{cases} a(n-\|\mathsf{b}\|) & \text{if } n \geq \|\mathsf{b}\| \\ 0 & \text{otherwise.} \end{cases}$$

PROOF. The number of elements of $\mathsf{b}+\mathsf{A}$ of norm n is precisely the number of elements a in A such that $\|\mathsf{b}+\mathsf{a}\| = n$, and this is just the number of elements a with $\|\mathsf{a}\| = n - \|\mathsf{b}\|$. □

Proposition 3.16. *A set of additives* $\mathsf{b}+\mathsf{A}$ *has Dirichlet density given by*

$$\partial(\mathsf{b}+\mathsf{A}) = \rho^{\|\mathsf{b}\|}.$$

PROOF. For $0 \leq x < \rho$,

$$\begin{aligned} [\mathbf{b}+\mathbf{A}](x) &= \sum_{\mathsf{a}\in\mathsf{A}} x^{\|\mathsf{a}+\mathsf{b}\|} \\ &= x^{\|\mathsf{b}\|} \sum_{\mathsf{a}\in\mathsf{A}} x^{\|\mathsf{a}\|} \\ &= x^{\|\mathsf{b}\|}\mathbf{A}(x), \end{aligned}$$

so $\partial(\mathsf{b}+\mathsf{A}) = \rho^{\|\mathsf{b}\|}$. □

Lemma 3.17. *A set of additives* $\mathsf{b}+\mathsf{A}$ *has asymptotic density iff*

$$\lim_{n\to\infty} \frac{a(n-\|\mathsf{b}\|)}{a(n)} \quad \textit{exists},$$

in which case

$$\delta(\mathsf{b}+\mathsf{A}) = \lim_{n\to\infty} \frac{a(n-\|\mathsf{b}\|)}{a(n)}. \tag{3.17.1}$$

PROOF. This follows from Lemma 3.15. □

The expression on the right side of (3.17.1) reminds us of the ratio test, and indeed the ratio test is crucial to the study of asymptotic density.

Definition 3.18. $\mathcal{A} \in \mathsf{RT}_\rho$ means $a(n) \in \mathsf{RT}_\rho$, that is,

$$\lim_{n\to\infty} \frac{a(n-1)}{a(n)} = \rho.$$

(Note that the definition of $a(n) \in \mathsf{RT}_\rho$ says that $a(n)$ is eventually positive, and thus requires $\mathcal{A}$ to be reduced.)

Proposition 3.19. *Every set of additives* $\mathsf{b}+\mathsf{A}$ *has asymptotic density iff* $\mathcal{A} \in \mathsf{RT}_\rho$. *If* $\mathcal{A} \in \mathsf{RT}_\rho$ *then, for every* $\mathsf{b} \in \mathsf{A}$,

$$\delta(\mathsf{b}+\mathsf{A}) = \rho^{\|\mathsf{b}\|} = \partial(\mathsf{b}+\mathsf{A}).$$

PROOF. One always has $\partial(\mathsf{b}+\mathsf{A}) = \rho^{\|\mathsf{b}\|}$, by Proposition 3.16. So consider the asymptotic density, and assume $\mathcal{A} \in \mathsf{RT}_\rho$. From $\lim_{n\to\infty} \frac{a(n-1)}{a(n)} = \rho$ follows $\lim_{n\to\infty} \frac{a(n-k)}{a(n)} = \rho^k$, so, by Lemma 3.17,

$$\delta(\mathsf{b}+\mathsf{A}) \;=\; \lim_{n\to\infty} \frac{a\big(n-\|\mathsf{b}\|\big)}{a(n)} \;=\; \rho^{\|\mathsf{b}\|}.$$

Conversely, suppose that sets of additives of single elements have asymptotic density. For any $m \in \mathbb{N}$, and for $\mathsf{b} \in \mathsf{A}$, one has $\|m\mathsf{b}\| = m\|\mathsf{b}\|$, so, by Lemma 3.17, $\lim_{n\to\infty} \frac{a\big(n-m\|\mathsf{b}\|\big)}{a(n)}$ exists. Then this limit must exist for $m \in \mathbb{Z}$. Since $\mathcal{A}$ is reduced, by Lemma 2.42 the set of integers $\{\|\mathsf{a}\| : \mathsf{a} \in \mathsf{A}\}$ has gcd 1. So, it is possible to choose finitely many elements of A, say $\mathsf{a}_1, \dots, \mathsf{a}_k$, the gcd of whose norms equals 1. Then there are integers $m_1, \dots, m_k$ such that $1 = m_1\|\mathsf{a}_1\| + \cdots + m_k\|\mathsf{a}_k\|$. For n sufficiently large,

$$\frac{a(n-1)}{a(n)} \;=\; \frac{a\big(n \;-\; (m_1\|\mathsf{a}_1\| + \cdots + m_k\|\mathsf{a}_k\|)\big)}{a(n)}$$

$$(3.19.1) \qquad =\; \frac{a\big(n \;-\; (m_1\|\mathsf{a}_1\| + \cdots + m_k\|\mathsf{a}_k\|)\big)}{a\big(n \;-\; (m_2\|\mathsf{a}_2\| + \cdots + m_k\|\mathsf{a}_k\|)\big)} \cdots \frac{a\big(n - m_k\|\mathsf{a}_k\|\big)}{a(n)},$$

namely n needs to be large enough so that all $n \;-\; (m_i\|\mathsf{a}_i\| + \cdots + m_k\|\mathsf{a}_k\|)$, for $1 \le i \le k$, are nonnegative, and all $a\big(n \;-\; (m_i\|\mathsf{a}_i\| + \cdots + m_k\|\mathsf{a}_k\|)\big)$, for $1 < i \le k$, are positive.

Since each of the quotients of (3.19.1) approach a limit as $n \to \infty$, it follows that $\lim_{n\to\infty} \frac{a(n-1)}{a(n)}$ exists. By the ratio test this limit must be ρ, the radius of convergence of $\mathcal{A}$. Thus $\mathcal{A} \in \mathsf{RT}_\rho$. □

As with δ, the success of the ratio test is also connected with the global asymptotic density of the set of additives of an element b:

$$\Delta(\mathsf{b}+\mathsf{A}) \;=\; \lim_{n\to\infty} \frac{A\big(n-\|\mathsf{b}\|\big)}{A(n)},$$

whenever the limit exists. The result about the existence of the $\delta(\mathsf{b}+\mathsf{A})$, for $\mathsf{b} \in \mathsf{A}$, carries over to the global setting. (Note that one has $\rho_A = \rho$, by Lemma 2.31.)

Proposition 3.20. *Every set of additives* $\mathsf{b}+\mathsf{A}$ *has global asymptotic density iff* $A(n) \in \mathsf{RT}_\rho$, *that is,*

$$\lim_{n\to\infty} \frac{A(n-1)}{A(n)} \;=\; \rho.$$

If $A(n) \in \mathsf{RT}_\rho$ *then, for every* $\mathsf{b} \in \mathsf{A}$,

$$\Delta(\mathsf{b}+\mathsf{A}) \;=\; \rho^{\|\mathsf{b}\|}.$$

3.5. Partition sets

It is easy to construct sets $\mathsf{B} \subseteq \mathsf{A}$ such that B does not have asymptotic density, for example, by suitably choosing, for each n, B to have all or none of the elements of norm n in A. To develop a good theory of density one needs to focus on interesting candidates—for this purpose, the partition sets defined below are chosen. They are a much wider and more interesting collection of sets than the sets of the form $\mathsf{b}+\mathsf{A}$, and the existence of density of partition sets plays a major role in the establishment of logical limit laws for classes of structures.

Definition 3.21. For $\mathsf{B}_1, \ldots, \mathsf{B}_k \subseteq \mathsf{A}$, let

$$\mathsf{B}_1 + \cdots + \mathsf{B}_k \;=\; \{b_1 + \cdots + b_k : b_i \in \mathsf{B}_i\}.$$

The following simple estimate for the generating series of the sum of two subsets of A will play a key role in the development.

Lemma 3.22. *Suppose* $\mathsf{B}_1, \mathsf{B}_2$ *are two nonempty subsets of* A. *Then*

$$[\mathbf{B}_1 + \mathbf{B}_2](x) \;\le\; \mathbf{B}_1(x) \cdot \mathbf{B}_2(x), \qquad \textit{for } x \ge 0.$$

PROOF. This follows from observing, for all n, that

$$\big[x^n\big]\big([\mathbf{B}_1 + \mathbf{B}_2](x)\big) \;\le\; \big[x^n\big]\big(\mathbf{B}_1(x) \cdot \mathbf{B}_2(x)\big),$$

that is,

$$\Big|\big\{\mathsf{b}_1 + \mathsf{b}_2 : \mathsf{b}_i \in \mathsf{B}_i,\ \|\mathsf{b}_1 + \mathsf{b}_2\| = n\big\}\Big| \;\le\; \sum_{k=0}^{n} b_1(k) \cdot b_2(n-k).$$

□

Definition 3.23. For $\mathsf{B} \subseteq \mathsf{A}$,

$$\begin{aligned}
0\mathsf{B} &= \{0\}\\
m\mathsf{B} &= \underbrace{\mathsf{B} + \cdots + \mathsf{B}}_{m \text{ times}}, \qquad \text{for } m > 0\\
(\le m)\mathsf{B} &= 0\mathsf{B} \cup \cdots \cup m\mathsf{B}, \qquad \text{for } m \ge 0\\
(\ge m)\mathsf{B} &= \bigcup_{n \ge m} n\mathsf{B} \qquad \text{for } m \ge 0.
\end{aligned}$$

The coefficients m and $\le m$ are said to be *small*, and $\ge m$ is *large*.

Definition 3.24. $\mathsf{P}_1, \ldots, \mathsf{P}_k$ is a *partition* of P if each P_i is nonempty, $\mathsf{P}_i \cap \mathsf{P}_j = \varnothing$, for $i \ne j$, and $\mathsf{P} = \bigcup_{1 \le i \le k} \mathsf{P}_i$.

Definition 3.25. A subset B of A is a *partition set*[3] of $\mathcal{A}$ if B can be written in the form

$$\mathsf{B} \;=\; \gamma_1 \mathsf{P}_1 + \cdots + \gamma_k \mathsf{P}_k,$$

[3]The notion of a partition set is abstracted from the papers [**20**], [**21**], and [**22**] published by Compton in the late 1980s.

where $\mathsf{P}_1, \ldots, \mathsf{P}_k$ is a finite partition of P, and each coefficient γ_i is in one of the forms m, $(\leq m)$, $(\geq m)$.

So, for example,

$$\mathsf{B} \;=\; (\leq 5)\mathsf{P}_1 \;+\; (\geq 4)\mathsf{P}_2 \;+\; 2\mathsf{P}_3$$

says that B is the subset of elements of A which can be expressed as the sum of at most 5 elements from P_1 plus the sum of at least 4 elements from P_2 plus the sum of exactly 2 elements from P_3. (The elements used in such a sum do not have to be distinct—repeats are allowed.) There is one exception to this straightforward reading of the coefficients, namely if they are all 0—note that $0\mathsf{P}_1 + \cdots + 0\mathsf{P}_k = \{0\}$. The coefficients $(<m)$ and $(>m)$ will be used as shorthand for $(\leq m-1)$, respectively $(\geq m+1)$. The definition of partition set does not include the empty set. The above definition of $0\mathsf{B}$ is needed to make sure Corollary 3.34 works for all choices of coefficients.

Sometimes a given collection $\mathsf{P}_1, \ldots, \mathsf{P}_k$ is called a partition when in reality some of the P_i could be empty. For example, when Q is a nonempty subset of P, one speaks of the partition Q, $\mathsf{P} \smallsetminus \mathsf{Q}$, even though $\mathsf{P} \smallsetminus \mathsf{Q}$ could be empty. The convention, in such cases, is simply to ignore the empty P_i. Thus, when forming the sum $\gamma_1\mathsf{P}_1 + \cdots + \gamma_k\mathsf{P}_k$, what is really meant is the sum $\sum_{\mathsf{P}_i \neq \varnothing} \gamma_i \mathsf{P}_i$.

Definition 3.26. ω denotes the coefficient (≥ 0).

When working with partition sets the following notation will be useful.

Definition 3.27. For B a subset of A and n a nonnegative integer, let

$$\begin{aligned} \mathsf{B}_{(\leq n)} &= \{\mathsf{a} \in \mathsf{B} : \|\mathsf{a}\| \leq n\} \\ \mathsf{B}_{(n)} &= \{\mathsf{a} \in \mathsf{B} : \|\mathsf{a}\| = n\} \\ \mathsf{B}_{(\geq n)} &= \{\mathsf{a} \in \mathsf{B} : \|\mathsf{a}\| \geq n\}. \end{aligned}$$

It is easy to see that every nonempty subset Q of P is a partition set. P itself is a partition set and, when $\mathsf{Q} \neq \mathsf{P}$, one has a partition $\mathsf{Q}, \mathsf{P} \smallsetminus \mathsf{Q}$ of P, and $\mathsf{Q} = \mathsf{Q} + 0(\mathsf{P} \smallsetminus \mathsf{Q})$. From this the following proposition is easily proved.

Proposition 3.28 ([**17**], 1997). *If every partition set of $\mathcal{A}$ has asymptotic density then*

$$\delta(\mathsf{P}) \;=\; 0.$$

Proof. If P is finite then the proof is clear as $p(n)$ is eventually 0. So suppose P is infinite. Since P is a partition set, by assumption $\delta(\mathsf{P})$ exists. If it is not 0, say $\delta(\mathsf{P}) = c > 0$, let $\mathsf{P}_{(n_1)}, \mathsf{P}_{(n_2)}, \ldots$ be the nonempty $\mathsf{P}_{(j)}$, with $n_1 < n_2 < \cdots$. Then define $\mathsf{Q} \subset \mathsf{P}$ by $\mathsf{Q}_{(n_j)} = \mathsf{P}_{(n_j)}$, for j odd, and otherwise $\mathsf{Q}_{(n)} = \varnothing$. (The resulting set Q is pictured in Figure 2.) Then

$$\limsup_{n\to\infty} \frac{q(n)}{a(n)} \;=\; c \qquad \text{and} \qquad \liminf_{n\to\infty} \frac{q(n)}{a(n)} \;=\; 0.$$

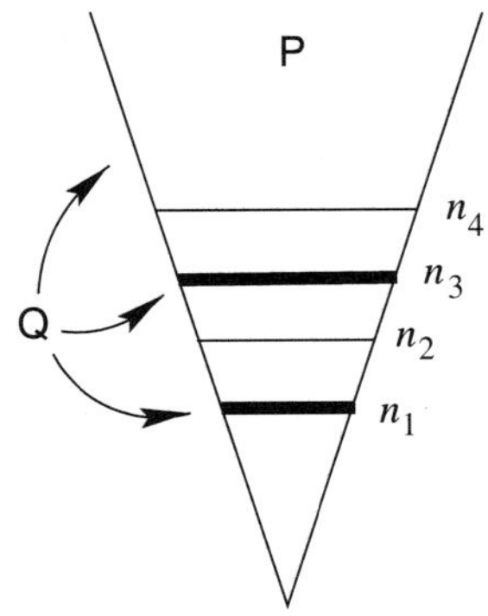

FIGURE 2

But this means Q does not have asymptotic density, which contradicts the assumption that all partition sets have asymptotic density. □

Lemma 3.29. *Partition sets include the sets of additives of single elements.*

PROOF. Given $\mathsf{b} \in \mathsf{A}$ let $\mathsf{b} = m_1\mathsf{p}_1 + \cdots + m_k\mathsf{p}_k$, where the p_i are indecomposable. Define a partition of P by $\mathsf{P}_1 = \{\mathsf{p}_1\}, \ldots, \mathsf{P}_k = \{\mathsf{p}_k\}$, and $\mathsf{P}_{k+1} = \mathsf{P} \setminus \{\mathsf{p}_1, \ldots, \mathsf{p}_k\}$. Then

$$\mathsf{b} + \mathsf{A} \;=\; (\geq m_1)\mathsf{P}_1 + \cdots + (\geq m_k)\mathsf{P}_k + \omega\mathsf{P}_{k+1}.$$

□

Corollary 3.30. *$\mathcal{A} \in \mathsf{RT}_\rho$ if all partition sets of $\mathcal{A}$ have asymptotic density.*

PROOF. This follows from Proposition 3.19 and Lemma 3.29. □

As will be seen in §3.7, the only assumption needed to show that all partition sets have Dirichlet density, in any additive number system, is that $\rho > 0$. The main goal of this study is to discover criteria to show that all partition sets have asymptotic density, and then to apply this to discover logical limit laws.

Remark 3.31. The condition $\mathcal{A} \in \mathsf{RT}_\rho$ is also sufficient in Corollary 3.30 if $\rho = 1$ (see Theorem 4.2). But, for $0 < \rho < 1$, the asymptotics to provide specific counterexamples to the sufficiency of $\mathcal{A} \in \mathsf{RT}_\rho$, such as the asymptotics of the additive number system derived from forests, or from N-free posets, have been known for a number of years (see [**18**], [**32**]). Knopfmacher and Knopfmacher [**32**] have recently proved that if $\mathcal{A}$ has

$$p(n) \;\sim\; K_1\beta^n/n^\mu$$

with $K_1 > 0$ and both $\beta, \mu > 1$, then

$$a(n) \;\sim\; K_2\beta^n/n^\mu,$$

for some $K_2 > 0$. Then $\rho = 1/\beta$, and $\mathcal{A} \in \mathsf{RT}_\rho$. As $\delta(\mathsf{P}) = K_1/K_2 > 0$, from Proposition 3.28 it follows that not all partition sets of $\mathcal{A}$ have asymptotic density. Thus $\mathcal{A} \in \mathsf{RT}_\rho$ is a sufficient condition for all partition sets to have asymptotic density *only* in the case $\rho = 1$.

Remark 3.32. Bell [**5**] proves that $\rho = 0$ implies that P does not have zero asymptotic density. Bell, Bender, Cameron, and Richmond [**7**] show that if $\mathbf{A}(\rho) < \infty$ then P does not have zero asymptotic density.[4] Thus the conditions $\rho > 0$ and $\mathbf{A}(\rho) = \infty$ are necessary conditions for all partition sets to have asymptotic density.

The condition $\mathbf{A}(\rho) < \infty$ can only occur when $\rho < 1$, and in such cases $\mathbf{A}(\rho) < \infty$ iff $\mathbf{P}(\rho) < \infty$. One can use this to easily create examples of additive number systems with specified ρ and with some partition set not having asymptotic density, for example, let $p(n) = \lfloor \rho^{-n}/n^2 \rfloor$.

3.6. Generating series of partition sets

One of the beautiful facts about generating series is that one has the following product theorem.

Proposition 3.33. *Let* $\mathsf{P}_1, \cdots, \mathsf{P}_k$ *be a partition of* P. *Then, for any nonempty* $\mathsf{B}_i \subseteq \omega\mathsf{P}_i$, *and for* $x \geq 0$, *the generating series* $[\mathbf{B}_1 + \cdots + \mathbf{B}_k](x)$ *converges iff each* $\mathbf{B}_i(x)$ *converges; and, for* $x \geq 0$,

$$[\mathbf{B}_1 + \cdots + \mathbf{B}_k](x) \quad = \quad \mathbf{B}_1(x) \ \cdots \ \mathbf{B}_k(x).$$

Proof. Let $\mathsf{B} = \mathsf{B}_1 + \cdots + \mathsf{B}_k$. Then, for $\mathsf{b} \in \mathsf{B}$, there are unique $\mathsf{b}_i \in \mathsf{B}_i$ such that $\mathsf{b} = \mathsf{b}_1 + \cdots + \mathsf{b}_k$ (by the unique decomposition property of $\mathcal{A}$), so

$$b(n) \quad = \quad \sum_{n_1 + \cdots + n_k = n} b_1(n_1) \ \cdots \ b_k(n_k).$$

Thus $b(n)$ is the coefficient of x^n in the Cauchy product

$$\Big(\sum_{m \geq 0} b_1(m) x^m\Big) * \cdots * \Big(\sum_{m \geq 0} b_k(m) x^m\Big).$$

For $x \geq 0$,

$$b(0) \ + \ \cdots \ + \ b(n)x^n \ \leq \ \Big(\sum_{m=0}^{n} b_1(m) x^m\Big) \ \cdots \ \Big(\sum_{m=0}^{n} b_k(m) x^m\Big) \ \leq \ \mathbf{B}(x),$$

so $\mathbf{B}(x)$ converges iff all $\mathbf{B}_i(x)$ converge; if so, $\mathbf{B}(x) = \mathbf{B}_1(x) \cdots \mathbf{B}_k(x)$. □

Corollary 3.34. *Let* $\mathsf{P}_1, \cdots, \mathsf{P}_k$ *be a partition of* P. *Then, for any partition set* $\mathsf{B} = \gamma_1\mathsf{P}_1 + \cdots + \gamma_k\mathsf{P}_k$, *and for* $x \geq 0$,

$$\begin{aligned} \mathbf{B}(x) \ &= \ [\gamma_1\mathbf{P}_1](x) \cdots [\gamma_k\mathbf{P}_k](x) \\ &= \ [\gamma_1\mathbf{P}_1](x) * \cdots * [\gamma_k\mathbf{P}_k](x). \end{aligned}$$

[4] See Appendix C for details of these results.

Two special cases of generating series are given next, using the simpler notation of $[m\mathbf{p}](x)$ for $[m\{\mathbf{p}\}](x)$, and $[\omega\mathbf{p}](x)$ for $[\omega\{\mathbf{p}\}](x)$.

Lemma 3.35. *For* $\mathsf{p} \in \mathsf{P}$ *and* $m \geq 0$,

$$\begin{aligned}[m\mathbf{p}](x) &= x^{m\|\mathsf{p}\|}\\ [\omega\mathbf{p}](x) &= \sum_{n\geq 0} x^{n\|\mathsf{p}\|}.\end{aligned}$$

PROOF. This follows immediately from

$$\begin{aligned}m\{\mathsf{p}\} &= \{m\mathsf{p}\}\\ \omega\{\mathsf{p}\} &= \{0, \mathsf{p}, 2\mathsf{p}, \cdots\},\end{aligned}$$

using $\|k\mathsf{p}\| = k\|\mathsf{p}\|$. □

3.7. Partition sets have Dirichlet density

In this section, based on Compton's paper [**21**], the fact that 'sets of additives of an element have Dirichlet density' is strengthened to the result that 'all partition sets have Dirichlet density'. Further hypotheses are necessary to ensure the existence of asymptotic density.

3.7.1. The Dirichlet density of $\gamma\mathsf{P}$.

Showing that the Dirichlet density of $\gamma\mathsf{P}$ exists will give the key idea used in §3.7.2 to show that every partition set has Dirichlet density.[5]

Corollary 3.36. *Given* $m \in \mathbb{N}$, *for* $x \geq 0$,

$$[m\mathbf{P}](x) \leq \mathbf{P}(x)^m.$$

PROOF. This follows immediately from Lemma 3.22. □

Lemma 3.37. *For* γ *a coefficient,*

$$\partial(\gamma\mathsf{P}) = \begin{cases} \dfrac{[\gamma\mathbf{P}](\rho)}{\mathbf{A}(\rho)} > 0 & \text{if } \mathbf{A}(\rho) < \infty\\ 0 & \text{if } \mathbf{A}(\rho) = \infty \text{ and } \gamma \text{ is small}\\ 1 & \text{if } \mathbf{A}(\rho) = \infty \text{ and } \gamma \text{ is large.}\end{cases}$$

PROOF. If $\mathbf{A}(\rho) < \infty$ then apply Lemma 3.11.

The interesting case is what happens when $\mathbf{A}(\rho) = \infty$. Assuming this, $\gamma = m$ will be considered first.

If P is finite, then $\rho = 1$, and $[m\mathbf{P}](x)$ is a polynomial since $m\mathsf{P}$ is finite. Then $\mathbf{A}(1) = \infty$, but $[m\mathbf{P}](1) < \infty$, so clearly the Dirichlet density is 0.

The situation when P is not finite remains. In this case, from $\mathbf{A}(\rho) = \infty$ follows $\mathbf{P}(\rho) = \infty$ when $\rho < 1$, and if $\rho = 1$ clearly $\mathbf{P}(\rho) = \infty$, so,

---

[5] A detailed treatment of the generating series of $m\mathsf{P}$ is given in Appendix B.

$$\begin{aligned}\partial(m\mathsf{P}) &= \lim_{x\to\rho}\frac{[m\mathbf{P}](x)}{\mathbf{A}(x)}\\ &\le \lim_{x\to\rho}\frac{\mathbf{P}(x)^m}{e^{\mathbf{P}(x)}} \qquad \text{by Corollaries 2.34 and 3.36}\\ &= 0.\end{aligned}$$

This finishes the case $\gamma = m$.

From the unique factorization property of $\mathcal{A}$, the sets $0\mathsf{P},\ldots,m\mathsf{P}$ are pairwise disjoint, and

$$\begin{aligned}(\le m)\mathsf{P} &= 0\mathsf{P}\cup\mathsf{P}\cup\cdots\cup m\mathsf{P}\\ (\ge m)\mathsf{P} &= \mathsf{A}\setminus(<m)\mathsf{P}.\end{aligned}$$

Thus, by Lemma 3.9,

$$\begin{aligned}\partial\big((\le m)\mathsf{P}\big) &= \partial(0)+\partial(\mathsf{P})+\cdots+\partial(m\mathsf{P}) = 0\\ \partial\big((\ge m)\mathsf{P}\big) &= 1-\partial\big((<m)\mathsf{P}\big) = 1.\end{aligned}$$

This proves the lemma. □

The asymptotic density of $\gamma\mathsf{P}$ is discussed in §3.10, under the assumptions $\mathcal{A}\in\mathsf{RT}_\rho$ and $\mathcal{A}(\rho)=\infty$.

3.7.2. General partition sets. The strategy to show the existence of $\lim_{x\to\rho}\frac{\mathbf{B}(x)}{\mathbf{A}(x)}$, where B is a partition set $\gamma_1\mathsf{P}_1+\cdots+\gamma_k\mathsf{P}_k$, is to note that

$$\frac{\mathbf{B}(x)}{\mathbf{A}(x)} = \prod_{1\le j\le m}\frac{[\gamma_j\mathbf{P}_j](x)}{[\omega\mathbf{P}_j](x)},$$

for then $\partial(\mathsf{B})$ exists if each $\lim_{x\to\rho}\frac{[\gamma_j\mathbf{P}_j](x)}{[\omega\mathbf{P}_j](x)}$ exists.

Definition 3.38. For Q a nonempty subset of P, let $\mathsf{A}_\mathsf{Q}=\omega\mathsf{Q}$ and, for any coefficient γ, let

$$\partial_\mathsf{Q}(\gamma\mathsf{Q}) = \lim_{x\to\rho}\frac{[\gamma\mathbf{Q}](x)}{\mathbf{A}_\mathsf{Q}(x)},$$

whenever the (left-) limit on the right side exists.

Thus A_Q is the universe of the subsystem $\mathcal{A}_\mathsf{Q}=\big(\mathsf{A}_\mathsf{Q},\mathsf{Q},+,0,\|\ \|\big)$ of $\mathcal{A}$ generated by the elements of Q. However, the definition of ∂_Q is not necessarily the Dirichlet density of $\mathcal{A}_\mathsf{Q}$ since the radius of convergence of $\mathcal{A}_\mathsf{Q}$ may be greater than that of $\mathcal{A}$.

Lemma 3.39. *For $\mathsf{Q}\subseteq\mathsf{P}$, $\partial_\mathsf{Q}(\gamma\mathsf{Q})$ exists, for any coefficient γ, and*

$$\partial_\mathsf{Q}(\gamma\mathsf{Q}) = \begin{cases}\dfrac{[\gamma\mathbf{Q}](\rho)}{\mathbf{A}_\mathsf{Q}(\rho)} > 0 & \textit{if } \mathbf{A}_\mathsf{Q}(\rho)<\infty\\ 0 & \textit{if } \mathbf{A}_\mathsf{Q}(\rho)=\infty \textit{ and } \gamma \textit{ is small}\\ 1 & \textit{if } \mathbf{A}_\mathsf{Q}(\rho)=\infty \textit{ and } \gamma \textit{ is large.}\end{cases}$$

PROOF. Just repeat the proof of Lemma 3.37, replacing P by Q, and $\mathbf{A}(x)$ by $\mathbf{A}_{\mathsf{Q}}(x)$. □

Theorem 3.40 (Compton, 1989). *All partition sets of $\mathcal{A}$ have Dirichlet density. If $\mathsf{B} = \gamma_1\mathsf{P}_1 + \cdots + \gamma_k\mathsf{P}_k$ then*

$$\partial(\mathsf{B}) \;=\; \partial_{\mathsf{P}_1}(\gamma_1\mathsf{P}_1)\;\cdots\;\partial_{\mathsf{P}_k}(\gamma_k\mathsf{P}_k).$$

PROOF. From Corollary 3.34, for $x \geq 0$,

$$\begin{aligned}\mathbf{B}(x) &= [\gamma_1\mathbf{P}_1](x)\;\cdots\;[\gamma_k\mathbf{P}_k](x)\\ \mathbf{A}(x) &= \mathbf{A}_{\mathsf{P}_1}(x)\;\cdots\;\mathbf{A}_{\mathsf{P}_k}(x),\end{aligned}$$

so,

$$\frac{\mathbf{B}(x)}{\mathbf{A}(x)} \;=\; \prod_{1\leq j\leq k}\frac{[\gamma_j\mathbf{P}_j](x)}{\mathbf{A}_{\mathsf{P}_j}(x)}, \qquad \text{for } 0 \leq x < \rho. \tag{3.40.1}$$

By Lemma 3.39, each $\partial_{\mathsf{P}_i}(\gamma_i\mathsf{P}_i)$ exists, so the limit as $x \to \rho$ of the right side of (3.40.1) exists, and thus equals the limit of the left side. □

3.7.3. Dirichlet density of subsystems. For each $\mathsf{Q} \subseteq \mathsf{P}$, there is a subsystem $\mathcal{A}_{\mathsf{Q}} = (\mathsf{A}_{\mathsf{Q}}, \mathsf{Q}, +, 0, \|\;\|)$ of $\mathcal{A}$, where $\mathsf{A}_{\mathsf{Q}} = \omega\mathsf{Q}$. Since $\omega\mathsf{Q}$ is a partition set, it follows that $\partial(\mathsf{A}_{\mathsf{Q}})$ must exist—the following gives a useful formula for this density. (Set $0^0 = 1$ if it appears in (3.41.1).)

Proposition 3.41. *For $\mathsf{Q} \subseteq \mathsf{P}$,*

$$\partial(\mathsf{A}_{\mathsf{Q}}) \;=\; \prod_{n\geq 1}(1-\rho^n)^{p(n)-q(n)}, \tag{3.41.1}$$

where $q(n)$ is the local counting function of Q.

PROOF. Since $\mathsf{A}_{\mathsf{Q}} = \omega\mathsf{Q} + 0(\mathsf{P} \setminus \mathsf{Q})$, apply Theorem 3.40 to obtain

$$\begin{aligned}\partial(\mathsf{A}_{\mathsf{Q}}) &= \partial_{\mathsf{Q}}(\mathsf{A}_{\mathsf{Q}}) \cdot \partial_{\mathsf{P}\setminus\mathsf{Q}}(\{0\})\\ &= 1 \cdot \lim_{x\to\rho} \mathbf{0}(x)/\mathbf{A}_{\mathsf{P}\setminus\mathsf{Q}}(x)\\ &= \lim_{x\to\rho}\Big(1\Big/\prod_{n\geq 1}(1-x^n)^{-\left(p(n)-q(n)\right)}\Big)\\ &= \lim_{x\to\rho}\prod_{n\geq 1}(1-x^n)^{p(n)-q(n)}\\ &= \prod_{n\geq 1}(1-\rho^n)^{p(n)-q(n)}.\end{aligned}$$

The last step follows from noting that, for $0 \leq x \leq 1$, the partial products

$$\prod_{n=1}^{N}\left(1-x^n\right)^{p(n)-q(n)}$$

are nonnegative, and nonincreasing in both N and x, so, for $0 \le x \le \rho$,

$$\begin{aligned}\prod_{n=1}^{N} \left(1 - x^n\right)^{p(n)-q(n)} &\ge \prod_{n\ge1} \left(1 - x^n\right)^{p(n)-q(n)} \\ &\ge \prod_{n\ge1} \left(1 - \rho^n\right)^{p(n)-q(n)}.\end{aligned}$$

This leads to

$$\begin{aligned}\prod_{n=1}^{N} \left(1 - \rho^n\right)^{p(n)-q(n)} &\ge \lim_{x\to\rho} \prod_{n\ge1} \left(1 - x^n\right)^{p(n)-q(n)} \\ &\ge \prod_{n\ge1} \left(1 - \rho^n\right)^{p(n)-q(n)}.\end{aligned}$$

Now take the limit as $N \to \infty$. □

3.7.4. Two-colored linear forests. Let us apply this to the example of two-colored linear forests by letting Q be the collection of chains without strictly alternating colors. Then, since $\rho = 1/2$, from (3.41.1)

$$\partial(\omega\mathsf{Q}) = \prod_{n\ge1} (1 - 2^{-n})^{p(n)-q(n)}.$$

Now $p(n) - q(n)$ is the number of n-element chains with strictly alternating colors, so, for $n \ge 1$,

$$\begin{aligned} p(n) - q(n) &= 2, && \text{for the poset case} \\ p(n) - q(n) &= \begin{cases} 2 & \text{if } n \text{ is odd} \\ 1 & \text{if } n \text{ is even,} \end{cases} && \text{for the graph case.} \end{aligned}$$

Thus, for $\mathsf{B} = \mathsf{A} \setminus \omega\mathsf{Q}$, the collection of linear forests with some component having strictly alternating colors, the calculations are

$$\begin{aligned}\partial(\mathsf{B}) &= 1 - \partial(\omega\mathsf{Q}) \\ &= 1 - \prod_{n\ge1} (1 - 2^{-n})^{p(n)-q(n)} \\ &= \begin{cases} 1 - \left(\prod_{n=1}^{\infty}(1 - 2^{-n})\right)^2 \approx 0.92 & \text{(poset case)} \\ 1 - \prod_{n=1}^{\infty}(1 - 2^{-2n+1})^2(1 - 2^{-2n}) \approx 0.88 & \text{(graph case).} \end{cases}\end{aligned}$$

These are the computations that were mentioned in §2.3.2. If the asymptotic density of B existed then Proposition 3.13 could be applied to conclude that $\delta(\mathsf{B}) = \partial(\mathsf{B})$, and this would give the results announced in §2.3.2. This is the most difficult step, and will be completed at the end of Chapter 5.

3.8. Schur's Tauberian Theorem

The theorems of this section and §5.1 will be used for additive number systems for which it is known that Dirichlet density extends asymptotic density (see Proposition 3.13). Since, for B a partition set, it is known that $\partial(\mathsf{B})$ exists (by §3.7), the goal is to prove that the asymptotic density $\delta(\mathsf{B})$ also exists, and therefore it is $\partial(\mathsf{B})$. Schur's Theorem[6] is a remarkably versatile tool in the study of density of additive number systems.

Theorem 3.42 (Schur, 1918). *Let* $\mathbf{S}(x)$ *and* $\mathbf{T}(x)$ *be two power series such that, for some* $\rho \geq 0$,

(a) $\mathbf{T}(x) \in \mathsf{RT}_\rho$, *and*
(b) $\mathbf{S}(x)$ *has radius of convergence* ρ_s *greater than* ρ.

Then

$$\lim_{n\to\infty} \frac{[x^n]\big(\mathbf{S}(x) * \mathbf{T}(x)\big)}{[x^n]\mathbf{T}(x)} = \mathbf{S}(\rho).$$

If $\mathbf{S}(\rho) > 0$ *this can be expressed as an asymptotic relationship:*

$$[x^n]\big(\mathbf{S}(x) * \mathbf{T}(x)\big) \sim \mathbf{S}(\rho) \cdot [x^n]\mathbf{T}(x). \tag{3.42.1}$$

Proof. Let $\mathbf{R}(x) = \mathbf{S}(x) * \mathbf{T}(x)$. Implicit in hypothesis (a) is the fact that the $t(n)$ are eventually positive. Choose λ such that $\rho < \lambda < \rho_s$, and choose M_0 such that $t(n) > 0$, for $n \geq M_0$, and

$$\frac{t(n-1)}{t(n)} < \lambda \qquad \text{if } n-1 \geq M_0.$$

Then

$$0 < \frac{t(n-k)}{t(n)} < \lambda^k \qquad \text{if } n-k \geq M_0. \tag{3.42.2}$$

Let

$$C = 1 + \max\Big(\Big|\frac{t(j)}{t(M_0)}\Big|\lambda^{j-M_0} : 0 \leq j < M_0\Big).$$

Claim: $$\Big|\frac{t(n-k)}{t(n)}\Big| \leq C\lambda^k, \qquad \text{for } 0 \leq k \leq n, \text{ provided } n \geq M_0.$$

[6]Schur's Theorem can be found in Pólya-Szegö [**44**] as Exercise 178 in Chapter 4 of Part I.

Clearly the claim holds if $n-k \geq M_0$ by (3.42.2) since $C \geq 1$. So only the case $n-k < M_0$ needs to be considered. For $n \geq M_0$,

$$\begin{aligned}
\left|\frac{t(n-k)}{t(n)}\right| &= \left|\frac{t(n-k)}{t(M_0)}\right| \cdot \left|\frac{t(M_0)}{t(n)}\right| \\
&\leq \left|\frac{t(n-k)}{t(M_0)}\right| \cdot \lambda^{n-M_0} && \text{by (3.42.2)} \\
&= \left|\frac{t(n-k)}{t(M_0)}\right| \cdot \lambda^{n-M_0-k} \cdot \lambda^k \\
&\leq C\lambda^k && \text{by the definition of } C.
\end{aligned}$$

From the triangle inequality, for $n \geq M_0$,

$$\left|\mathbf{S}(\rho) - \frac{r(n)}{t(n)}\right| \leq \underbrace{\left|\mathbf{S}(\rho) - \sum_{k\leq n} s(k)\rho^k\right|}_{\text{I}} + \underbrace{\left|\sum_{k\leq n} s(k)\rho^k - \frac{r(n)}{t(n)}\right|}_{\text{II}}.$$

Clearly term I goes to 0 as $n \to \infty$. Thus showing that term II vanishes as $n \to \infty$ suffices. For any $M \geq M_0$ and any $n \geq M$,

$$\begin{aligned}
\left|\sum_{k\leq n} s(k)\rho^k - \frac{r(n)}{t(n)}\right| &= \left|\sum_{k\leq n} s(k)\rho^k - \frac{1}{t(n)} \cdot \sum_{k\leq n} s(k)t(n-k)\right| \\
&= \left|\sum_{k\leq n} s(k)\left(\rho^k - \frac{t(n-k)}{t(n)}\right)\right| \\
&\leq \underbrace{\left|\sum_{k\leq M} s(k)\left(\rho^k - \frac{t(n-k)}{t(n)}\right)\right|}_{\text{III}} \\
&\quad + \underbrace{\left|\sum_{M<k\leq n} s(k)\left(\rho^k - \frac{t(n-k)}{t(n)}\right)\right|}_{\text{IV}}.
\end{aligned}$$

For any fixed M, term III goes to 0 as $n \to \infty$ since

$$\lim_{n\to\infty} \frac{t(n-k)}{t(n)} = \rho^k.$$

Thus showing that term IV goes to 0 as $M \to \infty$ suffices. For this term use the Claim to obtain

$$\left|\sum_{M<k\leq n} s(k)\left(\rho^k - \frac{t(n-k)}{t(n)}\right)\right| \leq \sum_{k>M} |s(k)|\rho^k + C \cdot \sum_{k>M} |s(k)|\lambda^k.$$

Each of the sums on the right side goes to 0 as $M \to \infty$ since they are the tail ends of convergent series. This finishes the proof. □

An immediate consequence of Schur's Theorem is that the Cauchy product of two power series is in RT_ρ if one of them is in RT_ρ, and the other, positive at ρ, has a larger radius of convergence.

Corollary 3.43. *Suppose* $\mathbf{T}(x) \in \mathsf{RT}_\rho$ *and the radius of convergence of* $\mathbf{S}(x)$ *is greater than* ρ. *If* $\mathbf{S}(\rho) > 0$ *then*

$$\mathbf{S}(x) * \mathbf{T}(x) \ \in \ \mathsf{RT}_\rho.$$

PROOF. Let $\mathbf{R}(x) = \mathbf{S}(x) * \mathbf{T}(x)$. From Schur's Theorem,

$$r(n) \ \sim \ \mathbf{S}(\rho) \cdot t(n)$$

so, by Proposition 1.26, $\mathbf{R}(x) \in \ \mathsf{RT}_\rho$. □

3.8.1. What is a Tauberian theorem? The subject of a Tauberian theorem, as originally introduced by G.H. Hardy, is that one wants to find $\lim_{n\to\infty} u_n$, for some class $\mathcal{U}$ of sequences (u_n), and one has discovered a 'generalized limit', let us call it $\mathsf{glim}(u_n)$, that has the property that it agrees with the ordinary $\lim_{n\to\infty} u_n$ when the latter exists, but $\mathsf{glim}(u_n)$ can exist for some sequences in $\mathcal{U}$ even when $\lim_{n\to\infty} u_n$ fails to exist. So glim is a proper extension of the concept of $\lim_{n\to\infty}$ with regard to the class $\mathcal{U}$. The essence of a Tauberian theorem is to find conditions that, when added to $\mathsf{glim}(u_n) = \lambda$, guarantee that $\lim_{n\to\infty} u_n = \lambda$.[7]

Conditions to allow the passage from $\partial(\mathsf{B}) = \lambda$ to $\delta(\mathsf{B}) = \lambda$ will provide the main Tauberian theorems of this study. In the applications of Schur's Theorem, the conditions $t(n) \geq 0$ and $\mathbf{T}(\rho) = \infty$ will be of primary interest. Given such a power series $\mathbf{T}(x)$, and any other power series $\mathbf{R}(x)$ with $\rho_r \geq \rho_t$, it is not difficult to show that

$$\lim_{n\to\infty} \frac{r(n)}{t(n)} = \lambda \quad \Rightarrow \quad \lim_{x\to\rho} \frac{\mathbf{R}(x)}{\mathbf{T}(x)} = \lambda.$$

Define $\mathsf{glim}\big(r(n)/t(n)\big)$ to be $\lim_{x\to\rho} \mathbf{R}(x)/\mathbf{T}(x)$. This gives a genuine extension of the concept of $\lim_{n\to\infty} r(n)/t(n)$. To have a Tauberian theorem, conditions are needed which, when added to $\mathsf{glim}\big(r(n)/t(n)\big) = \lambda$, guarantee that $\lim_{n\to\infty} r(n)/t(n) = \lambda$. This is what the hypotheses of Schur's Theorem provide. So, it is in the context of nonnegative sequences $t(n)$ with $\mathbf{T}(\rho) = \infty$ that Schur's Theorem is regarded as a Tauberian Theorem.

3.8.2. Adding and deleting indecomposables. Schur's Theorem will be applied to show that membership in RT_ρ is preserved when removing or adding sets of indecomposables that are not too large, provided $\rho < 1$. In particular, adding or removing finitely many indecomposables preserves membership in RT_ρ when $\rho < 1$.

[7] The original work of Tauber was aimed at finding conditions that, when added to

$$\lim_{x\to 1^-} \sum_n u_n x^n = \lambda$$

give $\sum_n u_n = \lambda$. Abel's Theorem says that the limit as $x \to 1^-$ of the above power series is an extension of ordinary convergence, that is, if $\sum_n u_n = \lambda$ then $\lim_{x\to 1^-} \sum_n u_n x^n = \lambda$. G.H. Hardy coined the phrase 'Tauberian theorem' to describe results like Tauber's (see Knopp [**34**], p. 486).

Proposition 3.44. *Suppose* $\mathsf{Q} \subseteq \mathsf{P}$ *is such that* $\mathbf{A}_{\mathsf{Q}}(x)$ *has radius of convergence* $\rho_q > \rho$. *Then*

$$\mathcal{A} \in \mathsf{RT}_\rho \quad \textit{iff} \quad \mathcal{A}_{\mathsf{P}\setminus\mathsf{Q}} \in \mathsf{RT}_\rho,$$

where $\mathcal{A}_{\mathsf{P}\setminus\mathsf{Q}}$ *is the subsystem of* $\mathcal{A}$ *generated by* $\mathsf{P} \setminus \mathsf{Q}$.

PROOF. By Corollary 2.39, the radius of convergence of $[\mathbf{1}/\mathbf{A}_{\mathsf{Q}}](x)$ is at least ρ_q, and $[\mathbf{1}/\mathbf{A}_{\mathsf{Q}}](\rho) > 0$. Then, using

$$\begin{aligned} \mathbf{A}(x) &= \mathbf{A}_{\mathsf{Q}}(x) * \mathbf{A}_{\mathsf{P}\setminus\mathsf{Q}}(x) \\ \mathbf{A}_{\mathsf{P}\setminus\mathsf{Q}}(x) &= [\mathbf{1}/\mathbf{A}_{\mathsf{Q}}](x) * \mathbf{A}(x), \end{aligned}$$

it follows from Corollary 3.43 that $\mathbf{A}(x) \in \mathsf{RT}_\rho$ iff $\mathbf{A}_{\mathsf{P}\setminus\mathsf{Q}}(x) \in \mathsf{RT}_\rho$. □

This gives a bit of insight into the structure of P when $\mathcal{A}$ is in RT_ρ and $\rho < 1$. Namely, for every prime number p, there must be infinitely many members of P whose norm is not divisible by p. For otherwise, after removing the finitely many members whose norm is not divisible by p, the subsystem generated by the remaining indecomposables would not be reduced, and therefore could not be in RT_ρ, giving a contradiction.

3.8.3. Comparing $\mathbf{A}(x)$ with $e^{\mathbf{P}(x)}$. The structure of the power series for $e^{\mathbf{P}(x)}$ can be more transparent than that of $\mathbf{A}(x)$, so it can be valuable to know how information about the former transfers to the latter.

Lemma 3.45. $\sum_{m\ge2} \mathbf{P}(x^m)/m$ *has a power series expansion with radius of convergence* $\sqrt{\rho}$.

PROOF. For $x \in [0, \sqrt{\rho})$ and $m \ge 2$, $\mathbf{P}(x^m)$ converges and

$$\begin{aligned} \mathbf{P}(x^m) &= \sum_{n\ge1} p(n)(x^m)^n \\ &= \sum_{n\ge1} p(n)(x^{m-2})^n(x^2)^n \\ &\le x^{m-2}\sum_{n\ge1} p(n)(x^2)^n \;=\; x^{m-2}\mathbf{P}(x^2). \end{aligned}$$

Thus, for $x \in [0, \sqrt{\rho})$,

$$\begin{aligned} \sum_{m\ge2} \mathbf{P}(x^m)/m &\le \sum_{m\ge2} x^{m-2}\mathbf{P}(x^2) \\ &= \Big(\sum_{m\ge0} x^m\Big)\mathbf{P}(x^2) \;=\; \frac{\mathbf{P}(x^2)}{1-x}, \end{aligned}$$

so $\sum_{m\ge2} \mathbf{P}(x^m)/m$ converges, for $x \in [0, \sqrt{\rho})$. Since $\mathbf{P}(x^2)$ diverges, whenever $x > \sqrt{\rho}$, $\sum_{m\ge2} \mathbf{P}(x^m)/m$ also diverges, for $x > \sqrt{\rho}$. □

Proposition 3.46. *Suppose* $\rho < 1$. *Then*

$$\mathcal{A} \in \mathsf{RT}_\rho \quad \textit{iff} \quad e^{\mathbf{P}(x)} \in \mathsf{RT}_\rho.$$

If $\mathcal{A} \in \mathsf{RT}_\rho$ *then*

$$a(n) \;\sim\; C \cdot \big[x^n\big] e^{\mathbf{P}(x)}, \quad \textit{where} \quad C = \exp\Big(\sum_{m\geq 2} \mathbf{P}(\rho^m)/m\Big).$$

Proof. For $0 \leq x < \rho$,

$$\mathbf{A}(x) \;=\; \exp\Big(\sum_{m\geq 1} \mathbf{P}(x^m)/m\Big)$$

$$(3.46.1) \qquad \;=\; e^{\mathbf{P}(x)} \cdot \exp\Big(\sum_{m\geq 2} \mathbf{P}(x^m)/m\Big)$$

and thus

$$(3.46.2) \qquad e^{\mathbf{P}(x)} \;=\; \mathbf{A}(x) \cdot \exp\Big(-\sum_{m\geq 2} \mathbf{P}(x^m)/m\Big).$$

Since $\rho < 1$, $\mathbf{P}(x)$, and hence $e^{\mathbf{P}(x)}$, have power series expansions with radius of convergence ρ, by Lemma 1.22 and Corollary 2.25 (a).

Now suppose $e^{\mathbf{P}(x)} \in \mathsf{RT}_\rho$. Apply Schur's Theorem as expressed in (3.42.1) to (3.46.1), in view of Lemma 3.45, to obtain

$$a(n) \;\sim\; C \cdot \big[x^n\big] e^{\mathbf{P}(x)}.$$

But then $\mathcal{A} \in \mathsf{RT}_\rho$.

A similar argument holds if it is assumed that $\mathcal{A} \in \mathsf{RT}_\rho$ and Schur's Theorem is applied to (3.46.2). □

3.8.4. An application to density. Schur's Theorem easily provides a basic density theorem for additive number systems, Theorem 3.49.

Definition 3.47. For B a subset of A, let $[\mathbf{B}/\mathbf{A}](x)$ be the power series given by

$$[\mathbf{B}/\mathbf{A}](x) \;=\; \mathbf{B}(x) * [\mathbf{1}/\mathbf{A}](x).$$

Lemma 3.48. *For* $x \geq 0$ *such that* $\mathbf{A}(x)$ *converges,*

$$[\mathbf{B}/\mathbf{A}](x) \;=\; \mathbf{B}(x)/\mathbf{A}(x)$$

and $[\mathbf{B}/\mathbf{A}](x)$ *converges absolutely.*

Proof. For $x \geq 0$ with $\mathbf{A}(x) < \infty$, one has $\mathbf{B}(x) < \infty$. Also, since $[\mathbf{1}/\mathbf{A}](x)$ converges absolutely, by Corollary 2.39,

$$\begin{aligned} [\mathbf{B}/\mathbf{A}](x) &= \mathbf{B}(x) * [\mathbf{1}/\mathbf{A}](x) && \text{definition} \\ &= \mathbf{B}(x) \cdot [\mathbf{1}/\mathbf{A}](x) && \text{by Proposition 1.4 (f)} \\ &= \frac{\mathbf{B}(x)}{\mathbf{A}(x)} && \text{by Corollary 2.39.} \end{aligned}$$

From the absolute convergence of $\mathbf{B}(x)$ and $[\mathbf{1}/\mathbf{A}](x)$, it follows that $[\mathbf{B}/\mathbf{A}](x)$ converges absolutely. □

Theorem 3.49 (Compton, 1989). *For $\mathcal{A} \in \mathsf{RT}_\rho$ and $\mathsf{B} \subseteq \mathsf{A}$, if the radius of convergence of $[\mathbf{B}/\mathbf{A}](x)$ is greater than ρ then*

$$\delta(\mathsf{B}) \;=\; \partial(\mathsf{B}) \;=\; [\mathbf{B}/\mathbf{A}](\rho).$$

PROOF. Let $\mathbf{S}(x) = [\mathbf{B}/\mathbf{A}](x)$ and $\mathbf{T}(x) = \mathbf{A}(x)$. Then the hypotheses of Schur's Theorem are satisfied, so

$$\begin{aligned}
\partial(\mathsf{B}) &= \lim_{x\to\rho} \frac{\mathbf{B}(x)}{\mathbf{A}(x)} && \text{definition}\\
&= \lim_{x\to\rho} [\mathbf{B}/\mathbf{A}](x) && \text{Lemma 3.48}\\
&= [\mathbf{B}/\mathbf{A}](\rho) && \text{as } \rho < \rho_s\\
&= \lim_{n\to\infty} \frac{b(n)}{a(n)} && \text{Theorem 3.42}\\
&= \delta(\mathsf{B}) && \text{definition.}
\end{aligned}$$

□

3.9. Simple partition sets

In this section it is proved that some of the simplest of the partition sets, namely those obtained by specifying the exact number of times that each member of a given finite set of indecomposables can occur in a decomposition, and otherwise leaving the number of occurrences of all other indecomposables unrestricted, have Dirichlet density that can be described by a simple formula. Also, the asymptotic density of such sets exists and agrees with the Dirichlet density if $\mathcal{A} \in \mathsf{RT}_\rho$.

Definition 3.50. Let $\mathsf{p}_1, \ldots, \mathsf{p}_k \in \mathsf{P}$ and let $m_1, \ldots, m_k$ be nonnegative integers. Then the partition set

$$\mathsf{B} \;=\; m_1\mathsf{p}_1 + \cdots + m_k\mathsf{p}_k + \omega\big(\mathsf{P} \smallsetminus \{\mathsf{p}_1, \ldots, \mathsf{p}_k\}\big)$$

is called a *simple* partition set.

Lemma 3.51 (Compton, 1989). *For a simple partition set*

$$\mathsf{B} \;=\; m_1\mathsf{p}_1 + \cdots + m_k\mathsf{p}_k + \omega\big(\mathsf{P} \smallsetminus \{\mathsf{p}_1, \ldots, \mathsf{p}_k\}\big),$$

the Dirichlet density is given by

$$\partial(\mathsf{B}) \;=\; [\mathbf{B}/\mathbf{A}](\rho) \;=\; \prod_{i=1}^{k} \rho^{m_i\|\mathsf{p}_i\|}\big(1 - \rho^{\|\mathsf{p}_i\|}\big).$$

PROOF. Since

$$\mathsf{A} \;=\; \omega\mathsf{p}_1 + \cdots + \omega\mathsf{p}_k + \omega\big(\mathsf{P} \smallsetminus \{\mathsf{p}_1, \ldots, \mathsf{p}_k\}\big),$$

from Corollary 3.34 and Lemma 3.35, for $0 \le x < 1$, one has

$$\begin{aligned}\mathbf{A}(x) &= [\omega\mathbf{p}_1](x)\cdots[\omega\mathbf{p}_k](x)\cdot\Big[\omega\big(\mathbf{P}\smallsetminus\{\mathbf{p}_1,\dots,\mathbf{p}_k\}\big)\Big](x)\\ &= \big(1-x^{\|\mathsf{p}_1\|}\big)^{-1}\cdots\big(1-x^{\|\mathsf{p}_k\|}\big)^{-1}\cdot\Big[\omega\big(\mathbf{P}\smallsetminus\{\mathbf{p}_1,\dots,\mathbf{p}_k\}\big)\Big](x),\\ \mathbf{B}(x) &= [m_1\mathbf{p}_1](x)\cdots[m_k\mathbf{p}_k](x)\cdot\Big[\omega\big(\mathbf{P}\smallsetminus\{\mathbf{p}_1,\dots,\mathbf{p}_k\}\big)\Big](x)\\ &= x^{m_1\|\mathsf{p}_1\|}\cdots x^{m_k\|\mathsf{p}_k\|}\cdot\Big[\omega\big(\mathbf{P}\smallsetminus\{\mathbf{p}_1,\dots,\mathbf{p}_k\}\big)\Big](x),\end{aligned}$$

and thus, for $0 \le x < \rho$,

$$[\mathbf{B}/\mathbf{A}](x) = \prod_{i=1}^{k} x^{m_i\|\mathsf{p}_i\|}\big(1-x^{\|\mathsf{p}_i\|}\big).$$

This is a continuous function on $\mathbb{R}$, so

$$\partial(\mathsf{B}) = [\mathbf{B}/\mathbf{A}](\rho) = \prod_{i=1}^{k} \rho^{m_i\|\mathsf{p}_i\|}\big(1-\rho^{\|\mathsf{p}_i\|}\big).$$

□

Proposition 3.52 (Compton, 1989). *Suppose $\mathcal{A} \in \mathsf{RT}_\rho$. Then every simple partition set*

$$\mathsf{B} = m_1\mathsf{p}_1 + \cdots + m_k\mathsf{p}_k + \omega\big(\mathsf{P}\smallsetminus\{\mathsf{p}_1,\dots,\mathsf{p}_k\}\big)$$

has asymptotic density equal to its Dirichlet density, that is,

$$\delta(\mathsf{B}) = \partial(\mathsf{B}) = [\mathbf{B}/\mathbf{A}](\rho) = \prod_{i=1}^{k} \rho^{m_i\|\mathsf{p}_i\|}\big(1-\rho^{\|\mathsf{p}_i\|}\big). \tag{3.52.1}$$

Proof. From Lemma 3.51, the last two equalities in (3.52.1) hold. Now Theorem 3.49 applies since $[\mathbf{B}/\mathbf{A}](x)$ is a polynomial. □

3.10. The asymptotic density of $\gamma\mathsf{P}$

In §3.7.1 the Dirichlet density of $\gamma\mathsf{P}$ was determined. Now the asymptotic density is determined, given that $\mathcal{A} \in \mathsf{RT}_\rho$ and $\mathbf{A}(\rho) = \infty$.

Lemma 3.53. *Suppose $\mathsf{B} \subseteq \mathsf{C} \subseteq \mathsf{A}$ and $\delta(\mathsf{C}) = 0$. Then $\delta(\mathsf{B}) = 0$.*

Proof. This follows from $b(n) \le c(n)$ and $\displaystyle\lim_{n\to\infty}\frac{c(n)}{a(n)} = 0.$ □

Lemma 3.54. *If $\mathsf{B} \subseteq \mathsf{B}_k \subseteq \mathsf{A}$ and $\displaystyle\lim_{k\to\infty}\delta(\mathsf{B}_k) = 0$ then $\delta(\mathsf{B}) = 0$.*

Proof. This follows from $b(n) \le b_k(n)$ and $\displaystyle\lim_{k\to\infty}\lim_{n\to\infty}\frac{b_k(n)}{a(n)} = 0.$ □

Lemma 3.55. *If $\mathcal{A} \in \mathsf{RT}_\rho$ and $\mathbf{A}(\rho) = \infty$ then, for γ a coefficient,*

$$\delta(\gamma\mathsf{P}) \;=\; \begin{cases} 0 & \text{if } \gamma \text{ is small} \\ 1 & \text{if } \gamma \text{ is large.} \end{cases}$$

PROOF. First the case $\rho = 1$ is considered. Given $m \in \mathbb{N}$, choose $\mathsf{p} \in \mathsf{P}$. Then

$$\delta\big(m\mathsf{p} + \mathsf{A}\big) \;=\; 1$$

by the formula in Proposition 3.19. As

$$m\mathsf{p} + \mathsf{A} \;\subseteq\; (\geq m)\mathsf{P},$$

it follows that the density of $(\geq m)\mathsf{P}$ is also 1. The other cases of $\delta(\gamma\mathsf{P})$, when $\rho = 1$, follow from this.

Now suppose $\rho < 1$. Define P_n to be $\mathsf{P}_{(\leq n)}$, the (finite) set of indecomposables in P of norm at most n. For $m, n \in \mathbb{N}$, let

$$\mathsf{B}_{m,n} \;=\; m\mathsf{P}_n + \omega\big(\mathsf{P} \setminus \mathsf{P}_n\big).$$

From

$$\begin{aligned} \mathbf{B}_{m,n}(x) \;&=\; [m\mathbf{P}_n](x) \cdot [\omega\big(\mathbf{P} \setminus \mathbf{P}_n\big)](x) \\ \mathbf{A}(x) \;&=\; [\omega\mathbf{P}_n](x) \cdot [\omega\big(\mathbf{P} \setminus \mathbf{P}_n\big)](x) \end{aligned}$$

follows, for $0 \leq x < \rho$,

$$\begin{aligned} \frac{\mathbf{B}_{m,n}(x)}{\mathbf{A}(x)} \;&=\; \frac{[m\mathbf{P}_n](x)}{[\omega\mathbf{P}_n](x)} \\ &=\; \Big(\sum_{\mathsf{a} \in m\mathsf{P}_n} x^{\|\mathsf{a}\|}\Big) \cdot \prod_{\mathsf{p} \in \mathsf{P}_n} \big(1 - x^{\|\mathsf{p}\|}\big). \end{aligned}$$

Since the last line expands into a polynomial it follows, by Lemma 3.48 and Theorem 3.49, that the asymptotic density of $\mathsf{B}_{m,n}$ exists and

$$\delta(\mathsf{B}_{m,n}) \;=\; \partial(\mathsf{B}_{m,n}) \;=\; [\mathbf{B}_{m,n}/\mathbf{A}](\rho).$$

Thus,

$$\begin{aligned} \delta(\mathsf{B}_{m,n}) \;&=\; \frac{[m\mathbf{P}_n](\rho)}{[\omega\mathbf{P}_n](\rho)} && \text{from above} \\ &\leq\; \frac{\mathbf{P}_n(\rho)^m}{e^{\mathbf{P}_n(\rho)}} && \text{by Corollaries 2.34 and 3.36} \\ &\to\; 0 \qquad \text{as } n \to \infty, \end{aligned}$$

the last line following from the fact that $\mathbf{P}(\rho) = \infty$, and

$$\mathbf{P}_n(\rho) \;=\; p(1)\rho + \cdots + p(n)\rho^n \;\to\; \mathbf{P}(\rho) \quad \text{as } n \to \infty.$$

Then $\delta(m\mathsf{P}) = 0$ follows from $m\mathsf{P} \subseteq \mathsf{B}_{0,n} \cup \cdots \cup \mathsf{B}_{m,n}$.

From this it is easy to show that $\delta(\gamma\mathsf{P})$ is as described. □

Corollary 3.56. *Suppose* $\mathcal{A} \in \mathsf{RT}_\rho$ *and* $\mathbf{A}(\rho) = \infty$. *Then, for any partition set* $\mathsf{B} = \gamma_1\mathsf{P}_1 + \cdots + \gamma_k\mathsf{P}_k$ *with all coefficients* γ_i *small,* $\delta(\mathsf{B}) = 0$.

PROOF. Let each γ_i be either m_i or $(\leq m_i)$, and let $m = \sum m_i$. Then,

$$\mathsf{B} \subseteq (\leq m)\mathsf{P},$$

and since $\delta\big((\leq m)\mathsf{P}\big) = 0$, the result follows. □

Remark 3.57. Suppose $\mathcal{A} \in \mathsf{RT}_\rho$. Lemma 3.55 shows that $\mathbf{A}(\rho) = \infty$ implies $\delta(\mathsf{P}) = 0$. From [**7**] one finds that $\delta(\mathsf{P}) = 0$ implies $\mathbf{A}(\rho) = \infty$. Consequently, given $\mathcal{A} \in \mathsf{RT}_\rho$,

$$\delta(\mathsf{P}) = 0 \quad \text{iff} \quad \mathbf{A}(\rho) = \infty.$$

Thus, for example, in Corollary 3.56 or in Proposition 5.6, the hypothesis $\mathbf{A}(\rho) = \infty$ can be replaced by $\delta(\mathsf{P}) = 0$.

3.11. Adding an indecomposable

Given an additive number system $\mathcal{A}$ and a new indecomposable q one can form the additive number system $\mathcal{A}_\mathsf{q}$ by simply adding q to P to give $\mathsf{P}_\mathsf{q} = \mathsf{P} \cup \{\mathsf{q}\}$, by letting A_q be the set of formal sums $m\mathsf{q} + \mathsf{a}$, for m a nonnegative integer and $\mathsf{a} \in \mathsf{A}$, and extending the norm by the definition

$$\|m\mathsf{q} + \mathsf{a}\| = m\|\mathsf{q}\| + \|\mathsf{a}\|.$$

Let $a_\mathsf{q}, p_\mathsf{q}$ be the counting functions of $\mathcal{A}_\mathsf{q}$. Then

$$p_\mathsf{q}(n) = \begin{cases} p(n) + 1 & \text{if} \quad n = \|\mathsf{q}\| \\ p(n) & \text{if} \quad n \neq \|\mathsf{q}\|. \end{cases}$$

Lemma 3.58. *Let* ρ_q *be the radius of convergence of* $\mathcal{A}_\mathsf{q}$. *Then* $\rho_\mathsf{q} = \rho$.

PROOF. Clearly $\rho_p = \rho_{p_\mathsf{q}}$, so use Corollary 2.25 (b). □

From the fundamental identity of $\mathcal{A}_\mathsf{q}$, for $0 \leq x < \rho$,

$$\begin{aligned}\sum_{n\geq 0} a_\mathsf{q}(n)x^n &= \prod_{n\geq 1}(1 - x^n)^{-p_\mathsf{q}(n)} \\ &= \big(1 - x^{\|\mathsf{q}\|}\big)^{-1} \cdot \prod_{n\geq 1}(1 - x^n)^{-p(n)}\end{aligned}$$

and thus

$$\mathbf{A}_\mathsf{q}(x) = \big(1 - x^{\|\mathsf{q}\|}\big)^{-1} \cdot \mathbf{A}(x). \tag{3.58.1}$$

Theorem 3.59. *Suppose all partition sets of* $\mathcal{A}$ *have asymptotic density. Then, for any extension* $\mathcal{A}_\mathsf{q}$ *by a new indecomposable* q, *all partition sets have asymptotic density.*

PROOF. $\mathcal{A}_\mathsf{q}$ is reduced since $\mathcal{A}$ is reduced. Without loss of generality, an arbitrary partition of $\mathsf{P}_\mathsf{q} = \mathsf{P} \cup \{\mathsf{q}\}$ is written in the form

$$\mathsf{P}_1 \cup \{\mathsf{q}\}, \mathsf{P}_2, \dots, \mathsf{P}_k,$$

where $\mathsf{P}_1, \dots, \mathsf{P}_k$ is a partition of P, that is, the new element q is in the first partition class. The goal is to show that any partition set

$$\mathsf{B} \;=\; \gamma_1(\mathsf{P}_1 \cup \{\mathsf{q}\}) \;+\; \cdots \;+\; \gamma_k \mathsf{P}_k$$

has asymptotic density in $\mathcal{A}_\mathsf{q}$.

From (3.58.1),

$$a_\mathsf{q}(n) \;=\; a(n) \;+\; a\big(n - \|\mathsf{q}\|\big) \;+\; a\big(n - 2\|\mathsf{q}\|\big) \;+\; \cdots.$$

Claim 1:
$\mathcal{A}_\mathsf{q} \in \mathsf{RT}_\rho$.

Proposition 3.19 shows that $\mathcal{A} \in \mathsf{RT}_\rho$. For $\varepsilon > 0$, choose M such that, for $n \geq M$,

$$|a(n-1) \;-\; \rho a(n)| \;<\; \varepsilon a(n).$$

Then

$$\begin{aligned}
|a_\mathsf{q}(n-1) - \rho a_\mathsf{q}(n)| \;&=\; \Big|\sum_{k\geq 0} a\big(n-1-k\|\mathsf{q}\|\big) \;-\; \rho a\big(n-k\|\mathsf{q}\|\big)\Big| \\
&\leq\; \sum_{n-k\|\mathsf{q}\|\geq M} \Big|a\big(n-1-k\|\mathsf{q}\|\big) \;-\; \rho a\big(n-k\|\mathsf{q}\|\big)\Big| \\
&\qquad + \sum_{j<M} |a(j-1) \;-\; \rho a(j)| \\
&\leq\; \sum_{n-k\|\mathsf{q}\|\geq M} \varepsilon a\big(n-k\|\mathsf{q}\|\big) \;+\; \sum_{j<M} |a(j-1) - \rho a(j)| \\
&\leq\; \varepsilon a_\mathsf{q}(n) \;+\; \sum_{j<M} |a(j-1) - \rho a(j)|.
\end{aligned}$$

Since $\lim_{n\to\infty} a_\mathsf{q}(n) = \infty$ it follows that $\lim_{n\to\infty} \dfrac{a_\mathsf{q}(n-1)}{a_\mathsf{q}(n)} \;=\; \rho.$

Claim 2:
Let δ_q be the asymptotic density function of $\mathcal{A}_\mathsf{q}$. Then

$$\delta_\mathsf{q}(\mathsf{A}) \;=\; 1 - \rho^{\|\mathsf{q}\|}.$$

This follows from Proposition 3.52 since A is a simple partition set of $\mathcal{A}_\mathsf{q}$ (note that $\mathsf{A} = 0\mathsf{q} + \omega\mathsf{P}$) and since $\mathcal{A}_\mathsf{q} \in \mathsf{RT}_\rho$ by Claim 1.

Claim 3:
For B a partition set of $\mathcal{A}$, and for any nonnegative integer m,

$$\delta_\mathsf{q}(m\mathsf{q} + \mathsf{B}) \;=\; \delta(\mathsf{B}) \cdot \rho^{m\|\mathsf{q}\|} \cdot \big(1 - \rho^{\|\mathsf{q}\|}\big).$$

This follows from

$$\begin{aligned}\frac{|m\mathsf{q}+\mathsf{B}|_n}{a_{\mathsf{q}}(n)} &= \frac{b(n-m\|\mathsf{q}\|)}{a_{\mathsf{q}}(n)}\\ &= \frac{b(n-m\|\mathsf{q}\|)}{a(n-m\|\mathsf{q}\|)}\cdot\frac{a(n-m\|\mathsf{q}\|)}{a(n)}\cdot\frac{a(n)}{a_{\mathsf{q}}(n)}\end{aligned}$$

since

$$\begin{aligned}\lim_{n\to\infty}\frac{b(n-m\|\mathsf{q}\|)}{a(n-m\|\mathsf{q}\|} &= \delta(\mathsf{B}) && \text{as } \delta(\mathsf{B}) \text{ exists}\\ \lim_{n\to\infty}\frac{a(n-m\|\mathsf{q}\|)}{a(n)} &= \rho^{m\|\mathsf{q}\|} && \text{as } \mathcal{A}\in\mathsf{RT}_\rho\\ \lim_{n\to\infty}\frac{a(n)}{a_{\mathsf{q}}(n)} &= 1-\rho^{\|\mathsf{q}\|} && \text{by Claim 2.}\end{aligned}$$

Claim 4:
With the same assumptions as in the previous claim,

$$\delta_{\mathsf{q}}\big((\leq m)\mathsf{q}\ +\ \mathsf{B}\big)\ =\ \delta(\mathsf{B})\cdot\Big(1-\rho^{(m+1)\|\mathsf{q}\|}\Big).$$

This follows from noting that $(\leq m)\mathsf{q}+\mathsf{B}$ is the disjoint union of the $j\mathsf{q}+\mathsf{B}$, for $0\leq j\leq m$, and thus

$$\begin{aligned}\delta_{\mathsf{q}}\big((\leq m)\mathsf{q}\ +\ \mathsf{B}\big) &= \sum_{0\leq j\leq m}\delta_{\mathsf{q}}(j\mathsf{q}+\mathsf{B})\\ &= \sum_{0\leq j\leq m}\delta(\mathsf{B})\cdot\rho^{j\|\mathsf{q}\|}\cdot\Big(1-\rho^{\|\mathsf{q}\|}\Big) \qquad \text{by Claim 3}\\ &= \delta(\mathsf{B})\cdot\Big(1-\rho^{(m+1)\|\mathsf{q}\|}\Big).\end{aligned}$$

Claim 5:
For any partition set B of $\mathcal{A}$,

$$\delta_{\mathsf{q}}(\omega\mathsf{q}+\mathsf{B})\ =\ \delta(\mathsf{B}).$$

Since $\delta(\mathsf{B})$ exists it is possible to choose, for $\varepsilon>0$, M such that, for $n\geq M$,

$$\Big|b(n)-\delta(\mathsf{B})\cdot a(n)\Big|\ <\ \varepsilon a(n).$$

Then

$$\begin{aligned}\Big||\omega\mathsf{q}+\mathsf{B}|_n-\delta(\mathsf{B})\cdot a_{\mathsf{q}}(n)\Big| &= \Big|\sum_{k\geq 0}\Big(b(n-k\|\mathsf{q}\|)-\delta(\mathsf{B})\cdot a(n-k\|\mathsf{q}\|\Big)\Big|\\ &\leq \sum_{n-k\|\mathsf{q}\|\geq M}\Big|b(n-k\|\mathsf{q}\|)-\delta(\mathsf{B})\cdot a(n-k\|\mathsf{q}\|)\Big|\\ &\qquad + \sum_{j<M}\Big|b(j)-\delta(\mathsf{B})\cdot a(j)\Big|\end{aligned}$$

$$
\begin{aligned}
&\le \sum_{n-k\|\mathsf{q}\|\ge M} \varepsilon a(n-k\|\mathsf{q}\|) \\
&\qquad\qquad + \sum_{j<M} \big|b(j) - \delta(\mathsf{B})\cdot a(j)\big| \\
&\le \varepsilon a_{\mathsf{q}}(n) + \sum_{j<M} \big|b(j) - \delta(\mathsf{B})\cdot a(j)\big|,
\end{aligned}
$$

which establishes the claim since $a_{\mathsf{q}}(n) \to \infty$.

Claim 6:
Partition sets of $\mathcal{A}_{\mathsf{q}}$ have asymptotic density.

Let B be a partition set of $\mathcal{A}_{\mathsf{q}}$, say

$$\mathsf{B} = \gamma_1\big(\mathsf{P}_1 \cup \{\mathsf{q}\}\big) + \gamma_2\mathsf{P}_2 + \cdots + \gamma_k\mathsf{P}_k.$$

Then B is in the closure under disjoint union and set difference (between comparable sets) of partition sets of the form

$$\gamma\mathsf{q} + \gamma_1\mathsf{P}_1 + \gamma_2\mathsf{P}_2 + \cdots + \gamma_k\mathsf{P}_k$$

since

$$
\begin{aligned}
m\big(\mathsf{P}_1 \cup \{\mathsf{q}\}\big) &= \bigcup_{0\le j\le m} j\mathsf{q} + (m-j)\mathsf{P}_1 \\
(\le m)\big(\mathsf{P}_1 \cup \{\mathsf{q}\}\big) &= \bigcup_{0\le j\le m} j\big(\mathsf{P}_1 \cup \{\mathsf{q}\}\big) \\
\omega\big(\mathsf{P}_1 \cup \{\mathsf{q}\}\big) &= \omega\mathsf{q} + \omega\mathsf{P}_1 \\
(>m)\big(\mathsf{P}_1 \cup \{\mathsf{q}\}\big) &= \omega\big(\mathsf{P}_1 \cup \{\mathsf{q}\}\big) \setminus (\le m)\big(\mathsf{P}_1 \cup \{\mathsf{q}\}\big).
\end{aligned}
$$

From Claims 3, 4 and 5 it can be shown that each partition set of the form $\gamma\mathsf{q} + \gamma_1\mathsf{P}_1 + \gamma_2\mathsf{P}_2 + \cdots + \gamma_k\mathsf{P}_k$ has an asymptotic density, so $\delta_{\mathsf{q}}(\mathsf{B})$ exists.

□

CHAPTER 4

The Case $\rho = 1$

When $\rho = 1$ the Dirichlet density of each partition set is either 0 or 1, and there is a simple criterion to determine which is the case. Furthermore, the Dirichlet density extends the asymptotic density, and the fortunate situation exists (discovered by Compton [**20**]) that the condition

$$\mathcal{A} \in \mathsf{RT}_1$$

is necessary and sufficient for partition sets to have asymptotic density. The interesting work on density, when $\rho = 1$, focuses on criteria to determine when $\mathcal{A} \in \mathsf{RT}_1$.

The 1918 analysis of the classic fundamental identity

$$\sum_{n \geq 0} a(n) x^n \ = \ \prod_{n \geq 1} (1 - x^n)^{-1}$$

by Hardy and Ramanujan, using their circle method in complex analysis, was a milestone in the beautiful results achieved with fundamental identities. They found an asymptotic expression for $a(n)$, namely

$$a(n) \ \sim \ e^{\pi\sqrt{\frac{2}{3}n}} \Big/ \big(4n\sqrt{3}\big), \tag{4.0.1}$$

as well as efficient methods to actually compute $a(n)$.[1] (4.0.1) can be applied to any additive number system $\mathcal{A}$ with $p(n) = 1$, for $n \geq 1$, for example, to equivalence relations, etc., from §2.2. Using (4.0.1) it is easily seen that $a(n) \in \mathsf{RT}_1$. It has been difficult to obtain asymptotic results like (4.0.1) for other additive number systems. Nonetheless, there have been some major successes with proving $a(n) \in \mathsf{RT}_1$.

4.1. The fundamental results when $\rho = 1$

When $\mathcal{A} \in \mathsf{RT}_1$, the Dirichlet density and asymptotic density of partition sets are the same, and they can be determined by a particularly simple rule. First, Dirichlet density is considered.

Theorem 4.1 (Compton, 1987). *If $\rho = 1$ then, for any partition set $\sum \gamma_i \mathsf{P}_i$, the Dirichlet density is given by*

$$\partial\Big(\sum \gamma_i \mathsf{P}_i\Big) \ = \ \begin{cases} 0 & \textit{if some } \gamma_i \textit{ is small} \\ 1 & \textit{otherwise.} \end{cases}$$

[1] These methods were improved by Rademacher in 1937.

PROOF. First apply Proposition 3.16 to show that, for any $\mathsf{b} \in \mathsf{A}$,

$$\partial(\mathsf{b} + \mathsf{A}) = 1. \tag{4.1.1}$$

Now suppose $\mathsf{B} = \sum \gamma_i \mathsf{P}_i$ is a partition set with all γ_i large. Choose any $\mathsf{b} \in \mathsf{B}$. Then

$$\mathsf{b} + \mathsf{A} \subseteq \mathsf{B}$$

ensures, by (4.1.1), that $\partial(\mathsf{B}) = 1$.

If some γ_i is small, say γ_j, then γ_j is either m or $(\leq m)$, for some m. Thus

$$\mathsf{B} \subseteq (\leq m)\mathsf{P}_j + \omega(\mathsf{P} \setminus \mathsf{P}_j). \tag{4.1.2}$$

The complement of the right side of (4.1.2) is

$$(>m)\mathsf{P}_j + \omega(\mathsf{P} \setminus \mathsf{P}_j),$$

which, by the above, has Dirichlet density 1 since it is a partition set with all coefficients large. Thus $(\leq m)\mathsf{P}_j + \omega(\mathsf{P} \setminus \mathsf{P}_j)$ has Dirichlet density 0, and then, by (4.1.2), it follows that B has Dirichlet density 0. □

Theorem 4.2 (Compton, 1987). *If $\rho = 1$ then $\mathcal{A} \in \mathsf{RT}_1$ is necessary and sufficient for all partition sets to have (local) asymptotic density. If $\mathcal{A} \in \mathsf{RT}_1$ then, for any partition set $\sum \gamma_i \mathsf{P}_i$, the asymptotic density is given by*

$$\delta\Big(\sum \gamma_i \mathsf{P}_i\Big) = \begin{cases} 0 & \text{if some } \gamma_i \text{ is small} \\ 1 & \text{otherwise,} \end{cases}$$

so asymptotic density agrees with Dirichlet density on partition sets.

PROOF. From Proposition 3.19 the condition $\mathcal{A} \in \mathsf{RT}_1$ is necessary for all partition sets to have asymptotic density. To show the sufficiency, suppose that $\mathcal{A} \in \mathsf{RT}_1$. By Lemma 3.17, for any $\mathsf{b} \in \mathsf{A}$,

$$\delta(\mathsf{b} + \mathsf{A}) = 1.$$

Now proceed as in the proof of Theorem 4.1. □

The corresponding global results concerning asymptotic density are as follows.

Theorem 4.3. $A(n) \in \mathsf{RT}_1$, *that is,*

$$\lim_{n\to\infty} \frac{A(n-1)}{A(n)} = 1,$$

is necessary and sufficient for all partition sets to have global asymptotic density when $\rho = 1$. If $A(n) \in \mathsf{RT}_1$ then, for any partition set $\sum \gamma_i \mathsf{P}_i$,

$$\Delta\Big(\sum \gamma_i \mathsf{P}_i\Big) = \begin{cases} 0 & \text{if some } \gamma_i \text{ is small} \\ 1 & \text{otherwise.} \end{cases}$$

There is a particularly simple test for $A(n) \in \mathsf{RT}_1$.

Lemma 4.4. *If* $\Delta(\mathsf{b}+\mathsf{A}) = 1$, *for* some $\mathsf{b} \in \mathsf{A}$, $\mathsf{b} \neq \mathsf{0}$, *then* $A(n) \in \mathsf{RT}_1$.

PROOF. This follows from

$$\Delta(\mathsf{b}+\mathsf{A}) \;=\; \lim_{n\to\infty} \frac{A(n-\|\mathsf{b}\|)}{A(n)}$$

and from

$$\frac{A(n-\|\mathsf{b}\|)}{A(n)} \;\le\; \frac{A(n-1)}{A(n)} \;\le\; 1,$$

which follows from the monotonicity of $A(n)$. □

4.2. The Bateman and Erdös Result

In the 1956 paper [4] Bateman and Erdös studied the growth behavior of $a(n)$ under the assumption that $p(n) \le 1$, for $n \le 1$. In terms of additive number systems this means looking at systems with at most one indecomposable with norm n, for each positive integer n. Their main goal was to present a simple necessary and sufficient condition for $a(n)$ to be eventually strictly increasing—this result is presented in Appendix D. A side result of their investigations was that $p(n) \le 1$, for $n \le 1$, leads to $a(n) \in \mathsf{RT}_1$. This is the result presented in this section.

Lemma 4.5. *If* $\mathcal{A}$ *has infinite rank then, for all* c,

$$\lim_{n\to\infty} \frac{n^c}{A(n)} \;=\; 0.$$

PROOF. This follows easily from Theorem 2.52 (d). □

Definition 4.6. Define $a_q(n)$ to be the number of elements of A of norm n that can be written as a sum $\sum_{i=1}^{q} m_i\mathsf{p}_i$, with each $m_i > 0$, and with the p_i being distinct indecomposables.

Lemma 4.7. *Suppose* $p(n) \le 1$, *for* $n \ge 1$. *Then, for* $n \ge 1$,

$$\sum_{q=1}^{n} q \cdot a_q(n) \;\le\; A(n-1). \tag{4.7.1}$$

PROOF. For each element b of A of norm n that is a linear combination of exactly q indecomposables, say

$$\mathsf{b} \;=\; m_1\mathsf{p}_1 + \cdots + m_q\mathsf{p}_q, \qquad \text{where all } m_i > 0,$$

the set $\mathsf{S}_\mathsf{b} = \{\mathsf{b}-\mathsf{p}_1, \ldots, \mathsf{b}-\mathsf{p}_q\}$ has q elements,[2] and each element of S_b has norm less than n.

[2]Additive number systems do not have subtraction as a fundamental operation. The notation $\mathsf{b}-\mathsf{p}_i$ refers to the element obtained by expressing b as a sum of indecomposables, and then decreasing the number of copies of p_i by one.

Note that if $\mathsf{a} \in \mathsf{A}$ and $\|\mathsf{a}\| < n$ then a belongs to at most one $\mathsf{S_b}$ with $\|\mathsf{b}\| = n$. To see this, suppose that $\|\mathsf{b}\| = \|\mathsf{c}\|$ and that $\mathsf{a} \in \mathsf{S_b} \cap \mathsf{S_c}$. Then, for some $\mathsf{p}_j, \mathsf{p}_k$,

$$\mathsf{a} = \mathsf{b} - \mathsf{p}_j$$
$$\mathsf{a} = \mathsf{c} - \mathsf{p}_k.$$

Thus $\|\mathsf{p}_j\| = \|\mathsf{p}_k\|$ since $\|\mathsf{b}\| = \|\mathsf{c}\|$, and then $\mathsf{p}_j = \mathsf{p}_k$ as $p(\|\mathsf{p}_j\|) = 1$, so $\mathsf{b} = \mathsf{c}$.

Now each element b of norm n, which can be written as a linear sum of q indecomposables, is associated with the set $\mathsf{S_b}$ of exactly q elements of norm less than n, and, by the disjointness of the $\mathsf{S_b}$ as b ranges over the elements of norm n, one has $\sum_{q=1}^{n} q \cdot a_q(n) \leq A(n-1)$. □

Remark 4.8. Warlimont has pointed out that one can improve this lemma to read: for any count function $p(n)$ one has

$$\sum_{q=1}^{n} q \cdot a_q(n) = \sum_{k=1}^{n} p(k) a(n-k).$$

Lemma 4.9. *If $p(n) \leq 1$, for $n \geq 1$, then*

$$\lim_{n\to\infty} \frac{a(n)}{A(n)} = 0.$$

PROOF. If the rank of $\mathcal{A}$ is finite then use Corollary 2.49 (c).

So now consider the more challenging case, when the rank is infinite. Let $S = \operatorname{supp} p(n)$. Observe that, for q a positive integer, $a_q(n)$ is the coefficient of x^n in the expansion of

$$\sum_{\substack{s_1 < \cdots < s_q \leq n \\ s_i \in S}} \left(x^{s_1} + x^{2s_1} + \cdots\right) \cdots \left(x^{s_q} + x^{2s_q} + \cdots\right).$$

The coefficient of x^n in the expansion of each of the products

$$\left(x^{s_1} + x^{2s_1} + \cdots\right) \cdots \left(x^{s_q} + x^{2s_q} + \cdots\right)$$

does not exceed the coefficient of x^n in

$$(x + x^2 + \cdots)^q = \left(\frac{x}{1-x}\right)^q = \sum_{m=q}^{\infty} \binom{m-1}{q-1} x^m,$$

and the number of choices of $s_1, \ldots, s_q$ from S, with $1 \leq s_1 < \cdots < s_q \leq n$, is at most $\binom{n}{q}$. Thus we have

$$a_q(n) \leq \binom{n}{q}\binom{n-1}{q-1} \leq n^{2q-1}. \tag{4.9.1}$$

Let t be a fixed positive integer. Then, for $n \geq t$,

$$\begin{aligned}
A(n) &= a(n) + A(n-1) \\
&\geq a(n) + \sum_{q=1}^{n} q \cdot a_q(n) && \text{by (4.7.1)} \\
&= (t+1) \cdot a(n) + \sum_{q=1}^{n} q \cdot a_q(n) - t \cdot a(n) \\
&= (t+1) \cdot a(n) + \sum_{q=1}^{n} q \cdot a_q(n) - t \cdot \sum_{q=1}^{n} a_q(n) \\
&= (t+1) \cdot a(n) + \sum_{q=1}^{n} (q-t) \cdot a_q(n) \\
&\geq (t+1) \cdot a(n) + \sum_{q=1}^{t-1} (q-t) \cdot a_q(n) && \text{as } n \geq t \\
&\geq (t+1) \cdot a(n) - (t-1) \cdot \sum_{q=1}^{t-1} a_q(n) \\
&\geq (t+1) \cdot a(n) - (t-1) \cdot \sum_{q=1}^{t-1} n^{2q-1} && \text{by (4.9.1)} \\
&\geq (t+1) \cdot a(n) - (t-1)^2 n^{2t-3}.
\end{aligned}$$

Therefore

$$1 \geq (t+1) \cdot \frac{a(n)}{A(n)} - (t-1)^2 \cdot \frac{n^{2t-3}}{A(n)},$$

so, by Lemma 4.5,

$$\limsup_{n\to\infty} \frac{a(n)}{A(n)} \leq \frac{1}{t+1}.$$

Since t can be arbitrarily large this gives $\lim_{n\to\infty} \dfrac{a(n)}{A(n)} = 0.$ □

From Lemma 4.9 it follows that $A(n) \in \mathsf{RT}_1$ since

$$\frac{A(n-1)}{A(n)} = \frac{A(n) - a(n)}{A(n)} = 1 - \frac{a(n)}{A(n)}.$$

A further analysis will show that $a(n) \in \mathsf{RT}_1$.

Definition 4.10. For $f : \mathbb{N} \to \mathbb{N}$ let

$$\begin{aligned}
f^{-}(n) &= f(n) - f(n-1) \qquad \text{for } n > 0 \\
f^{-}(0) &= f(0).
\end{aligned}$$

Before proving the Bateman and Erdös result, two simple lemmas are needed.

Lemma 4.11. *For $x \in [0, \rho)$,*

$$\begin{aligned}\sum_{n\geq 0} a^-(n)x^n &= (1-x)\cdot\sum_{n\geq 0} a(n)x^n \\ &= (1-x)\cdot\prod_{n\geq 1}(1-x^n)^{-p(n)}.\end{aligned}$$

PROOF. The first equality follows from simply multiplying out the right side, and then the second equality is from the fundamental identity of $\mathcal{A}$. □

Lemma 4.12. *If $\mathsf{P}_1, \mathsf{P}_2$ is a partition of P, let $\mathcal{A}_i$ be the subsystem of $\mathcal{A}$ generated by P_i, and let $a_i(n), p_i(n)$ be the counting functions of $\mathcal{A}_i$. Then, for $x \in [0, \rho)$,*

$$\sum_{n\geq 0} a(n)x^n = \sum_{n\geq 0} a_1(n)x^n \cdot \sum_{n\geq 0} a_2(n)x^n \tag{4.12.1}$$

$$\sum_{n\geq 0} a^-(n)x^n = \sum_{n\geq 0} a_1^-(n)x^n \cdot \sum_{n\geq 0} a_2(n)x^n. \tag{4.12.2}$$

PROOF. Using the fundamental identities one has

$$\begin{aligned}\sum_{n\geq 0} a(n)x^n &= \prod_{\mathsf{p}\in\mathsf{P}}(1-x^{\|\mathsf{p}\|})^{-1} \\ &= \prod_{\mathsf{p}\in\mathsf{P}_1}(1-x^{\|\mathsf{p}\|})^{-1}\cdot\prod_{\mathsf{p}\in\mathsf{P}_2}(1-x^{\|\mathsf{p}\|})^{-1} \\ &= \sum_{n\geq 0} a_1(n)x^n\cdot\sum_{n\geq 0} a_2(n)x^n.\end{aligned}$$

This establishes the first equation.

For the second equation, one only needs to multiply the first equation by $1 - x$ in view of Lemma 4.11:

$$\begin{aligned}\sum_{n\geq 0} a^-(n)x^n &= (1-x)\cdot\sum_{n\geq 0} a(n)x^n \\ &= (1-x)\cdot\sum_{n\geq 0} a_1(n)x^n\cdot\sum_{n\geq 0} a_2(n)x^n \\ &= \sum_{n\geq 0} a_1^-(n)x^n\cdot\sum_{n\geq 0} a_2(n)x^n.\end{aligned}$$

□

Theorem 4.13 (Bateman and Erdös, 1956). *If $p(n) \leq 1$, for $n \geq 1$, then $\mathcal{A} \in \mathsf{RT}_1$.*

PROOF. If the rank of $\mathcal{A}$ is finite then the result follows from the fact that $a(n) \sim Cn^{r-1}$ (Theorem 2.48).

Now suppose $S = \operatorname{supp} p(n)$ is infinite. Choose S_1 to be a finite subset of S with $|S_1| \geq 2$ and $\gcd(S_1) = 1$, and let $S_2 = S \setminus S_1$.

Define $a_1(n)$ and $a_2(n)$ by the fundamental identities

$$\sum_{n\geq 0} a_i(n)x^n = \prod_{n\in S_i}(1-x^n)^{-1} \qquad \text{for } i=1,2.$$

Let t be a positive integer. Since $a_1(n) \sim Cn^{r-1}$, where $r=|S_1|$, it follows that

$$\lim_{n\to\infty}\frac{a_1(n)}{1+\left|a_1^-(n)\right|} = \lim_{n\to\infty}\frac{a_1(n)}{1+\left|a_1(n)-a_1(n-1)\right|} = \infty.$$

Consequently there exists a positive integer M such that, for the t specified above,

$$\frac{a_1(n)}{1+\left|a_1^-(n)\right|} \geq t+1 \qquad \text{for } n\geq M. \tag{4.13.1}$$

Let

$$K = \max_{0\leq n\leq M-1}\left(1+\left|a_1^-(n)\right|\right). \tag{4.13.2}$$

From Lemma 4.9 one has $\dfrac{a_2(n)}{A_2(n)} \to 0$, so there is an N such that

$$\sum_{m=n-M+1}^{n} a_2(m)\Big/A_2(n) \leq \frac{1}{(t+1)K} \qquad \text{for } n\geq N. \tag{4.13.3}$$

For $n\geq \max(M,N)$,

$$\begin{aligned}
a(n) &= \sum_{m=0}^{n} a_1(n-m)\cdot a_2(m) \qquad \text{by (4.12.1)}\\
&\geq \sum_{m=0}^{n-M} a_1(n-m)\cdot a_2(m)\\
&\geq (t+1)\cdot\sum_{m=0}^{n-M}\left(1+\left|a_1^-(n-m)\right|\right)\cdot a_2(m) \qquad \text{by (4.13.1)}\\
&= (t+1)\cdot\sum_{m=0}^{n}\left(1+\left|a_1^-(n-m)\right|\right)\cdot a_2(m)\\
&\qquad -(t+1)\cdot\sum_{m=n-M+1}^{n}\left(1+\left|a_1^-(n-m)\right|\right)\cdot a_2(m)\\
&\geq (t+1)\cdot\sum_{m=0}^{n}\left(1+\left|a_1^-(n-m)\right|\right)\cdot a_2(m)\\
&\qquad -(t+1)\cdot\sum_{m=n-M+1}^{n} Ka_2(m) \qquad \text{by (4.13.2)}
\end{aligned}$$

$$
\begin{aligned}
&= (t+1)\cdot\sum_{m=0}^{n}\Big(1+\big|a_1^-(n-m)\big|\Big)\cdot a_2(m) \\
&\qquad -(t+1)\cdot K\cdot\sum_{m=n-M+1}^{n} a_2(m) \\
&\geq \left[(t+1)\cdot\sum_{m=0}^{n}\Big(1+\big|a_1^-(n-m)\big|\Big)\cdot a_2(m)\right] - A_2(n) \quad \text{by (4.13.3)} \\
&\geq t\cdot\sum_{m=0}^{n}\Big(1+\big|a_1^-(n-m)\big|\Big)\cdot a_2(m) \\
&\geq t\cdot\left|\sum_{m=0}^{n} a_1^-(n-m)\cdot a_2(m)\right| \\
&= t\cdot|a^-(n)| \qquad \text{by (4.12.2).}
\end{aligned}
$$

Thus, for $n \geq \max(M,N)$, $\dfrac{|a^-(n)|}{a(n)} \leq \dfrac{1}{t}$. Since t can be arbitrarily large,

$$\lim_{n\to\infty}\frac{a^-(n)}{a(n)} = 0.$$

Consequently $\displaystyle\lim_{n\to\infty}\frac{a(n-1)}{a(n)} = 1.$ □

4.2.1. Application to equivalence relations, etc. By the Bateman and Erdös Theorem, all of the examples from §2.2 with $p(n) = 1$ (equivalence relations, etc.) are in RT_1.

Consider the following question, given $k \geq 0$ and $m \geq 1$: *What is the probability that a finite equivalence relation chosen at random has at least k equivalence classes of size m?* To answer this let p be the indecomposable equivalence relation on m elements, that is, with all elements equivalent. Then the asymptotic density of the set

$$\mathsf{B} = (\geq k)\mathsf{p} + \omega(\mathsf{P} \setminus \{\mathsf{p}\})$$

gives the answer. By Theorem 4.2 the answer is $\delta(\mathsf{B}) = 1$ since all coefficients are large. So, a randomly chosen finite equivalence relation will almost surely have at least k equivalence classes of size m.

This same argument shows that a randomly chosen finite permutation will almost certainly have at least k m-cycles, and a randomly chosen finite linear forest will almost certainly have at least k chains of size m.

The properties that have been considered in this section (with respect to their probability of occurring) can be expressed in first-order logic. In Chapter 6 it will be seen, when the additive number system of a class of relational structures is in RT_1, that any monadic second-order sentence has a probability of either 0 or 1 of holding in a randomly selected finite member of the class.

4.3. Stewart's Sum Theorem

Definition 4.14. Given two additive number systems $\mathcal{A}_1$ and $\mathcal{A}_2$, let $\mathcal{A}_1 + \mathcal{A}_2$, the *sum* of $\mathcal{A}_1$ and $\mathcal{A}_2$, be the additive number system whose underlying monoid is the direct product of the two monoids, so the addition is coordinatewise, and the norm is the sum of the coordinate norms, that is,

$$\|(\mathsf{a}_1, \mathsf{a}_2)\| \;=\; \|\mathsf{a}_1\|_1 \;+\; \|\mathsf{a}_2\|_2.$$

Thus the set of indecomposables of $\mathcal{A}_1 + \mathcal{A}_2$ is given by

$$\big\{(\mathsf{p}_1, 0) \;:\; \mathsf{p}_1 \in \mathsf{P}_1\big\} \;\cup\; \big\{(0, \mathsf{p}_2) \;:\; \mathsf{p}_2 \in \mathsf{P}_2\big\}.$$

Let $a_i(n), p_i(n)$ be the counting functions of $\mathcal{A}_i$, and let $a(n), p(n)$ be the counting functions of $\mathcal{A}_1 + \mathcal{A}_2$. Then

$$\begin{aligned} p(n) &= p_1(n) \;+\; p_2(n) \\ a(n) &= \sum_{k \le n} a_1(k) \cdot a_2(n-k). \end{aligned}$$

In case $\mathcal{A}_1$ and $\mathcal{A}_2$ are subsystems of a given additive number system $\mathcal{A}$ with $\mathsf{P}_1 \cap \mathsf{P}_2 = \varnothing$ then $\mathcal{A}_1 + \mathcal{A}_2$ is isomorphic to the subsystem of $\mathcal{A}$ generated by $\mathsf{P}_1 \cup \mathsf{P}_2$.

Theorem 4.15 (Stewart, 1997). *Let $\mathcal{A}_1, \ldots, \mathcal{A}_n$ be additive number systems in RT_1. Then $\mathcal{A}_1 + \cdots + \mathcal{A}_n$ is also in RT_1.*

PROOF. It suffices to prove the theorem for $n = 2$. So assume one has $\mathcal{A}_1, \mathcal{A}_2 \in \mathsf{RT}_1$ with counting functions $a_1(n), a_2(n)$, and let $a(n)$ be the counting function for $\mathcal{A} = \mathcal{A}_1 + \mathcal{A}_2$.

First, for any integer k,

$$a_2(n-k) \;=\; \mathrm{o}\big(a(n)\big) \tag{4.15.1}$$

will be proved. For $n \ge N > 0$, clearly

$$a(n) \;\ge\; \sum_{j=0}^{N} a_1(j) \cdot a_2(n-j). \tag{4.15.2}$$

Since $\mathcal{A}_2 \in \mathsf{RT}_1$, for any fixed integer j,

$$\lim_n \frac{a_2(n-j)}{a_2(n-k)} \;=\; 1,$$

so, from the inequality (4.15.2), it follows that

$$\liminf_n \frac{a(n)}{a_2(n-k)} \;\ge\; \sum_{j=0}^{N} a_1(j).$$

Thus $\displaystyle\lim_n \frac{a(n)}{a_2(n-k)} \;=\; \infty$, as $\mathbf{A}_1(1) = \infty$. This proves (4.15.1).

Given $\varepsilon > 0$ choose M such that

$$n \geq M \Rightarrow \left| \frac{a_1(n-1)}{a_1(n)} - 1 \right| < \varepsilon.$$

Then

$$\begin{aligned}
\left| a(n-1) - a(n) \right| &= \left| \sum_{k=0}^{n} a_1(k)a_2(n-k) - \sum_{k=1}^{n} a_1(k-1)a_2(n-k) \right| \\
&\leq a_2(n) + \sum_{k=1}^{n} \left| a_1(k) - a_1(k-1) \right| a_2(n-k) \\
&= \mathrm{o}\big(a(n)\big) + \sum_{k=M}^{n} \left| a_1(k) - a_1(k-1) \right| a_2(n-k) \\
&\leq \mathrm{o}\big(a(n)\big) + \sum_{k=M}^{n} \varepsilon a_1(k)a_2(n-k) \\
&\leq \mathrm{o}\big(a(n)\big) + \varepsilon \sum_{k=0}^{n} a_1(k)a_2(n-k) \\
&= \mathrm{o}\big(a(n)\big) + \varepsilon a(n).
\end{aligned}$$

Thus $\mathcal{A} \in \mathsf{RT}_1$. □

4.4. $\rho = 1$ does not imply RT_1

In this section a method is given to construct additive number systems $\mathcal{A}$ with $\rho = 1$, but $\mathcal{A} \notin \mathsf{RT}_1$. The idea of the construction is to occasionally add enough indecomposables to disrupt the ratio test, but not enough to influence the radius of convergence. The construction depends on knowing the asymptotics for finitely generated systems.

An increasing sequence N_j of positive integers is selected as described below. The indecomposables for the system under construction will have sizes N_j only. $a_j(n)$ will be the count function for a system $\mathcal{A}_j$ with $p_j(n) = p(n)$, for $n \leq N_j$, and $p_j(n) = 0$ otherwise.

Let $N_1 = 1$, $p(N_1) = 1$. Then $a_1(n) = 1$, the count function for the system $\mathcal{A}_1$ with exactly one indecomposable, it being of size one.

Having defined N_j, $p(N_j)$, and $a_j(n)$, for $j \leq k$, the inductive step in this construction is first to choose N_{k+1} larger than N_k and satisfying

$$\text{(4.15.3)} \qquad \Big(a_k(N_{k+1})\Big)^{1/N_{k+1}} < 1 + \frac{1}{k+1}.$$

This is possible because $a_k(n)$ has polynomial growth, being the count function for a finitely generated system $\mathcal{A}_k$. Next, augment the indecomposables by letting $p(N_{k+1}) = a_k(N_{k+1})$, and then define $a_{k+1}(n)$ to be the count function for all indecomposables obtained so far, that is,

$$\sum_{n\geq 0} a_{k+1}(n)x^n = \prod_{j=1}^{k+1}(1-x^{N_j})^{-p(N_j)}. \tag{4.15.4}$$

See Figure 1.

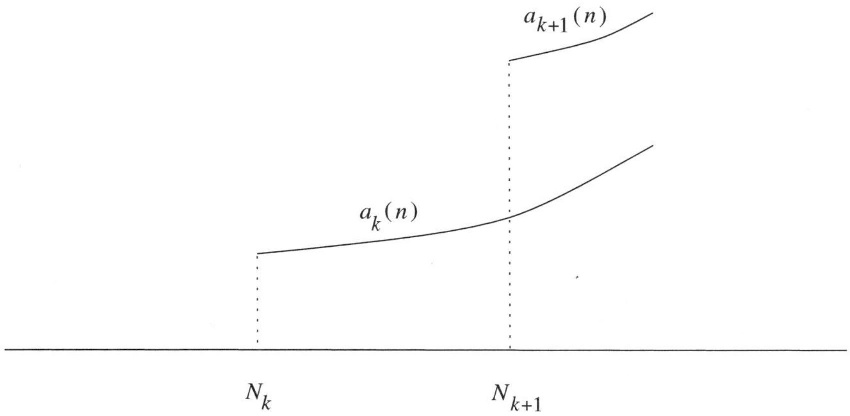

FIGURE 1. Constructing an additive number system

At the conclusion of this process one has a function $p(n)$ whose support is $\{N_1, N_2, \dots\}$. Let $\mathcal{A}$ be an additive number system with $p(n)$ being the count function for the indecomposables of $\mathcal{A}$. Then, by (4.15.3),

$$p(n)^{1/n} = 1 + \mathrm{o}(n).$$

Consequently $\mathbf{P}(x)$ has radius of convergence 1, and thus the radius of convergence of $\mathcal{A}$ is 1. However, $\mathcal{A} \notin \mathsf{RT}_1$ since $a_k(n)$ is monotonic leads to

$$\frac{a(N_k-1)}{a(N_k)} = \frac{a_{k-1}(N_k-1)}{2a_{k-1}(N_k)} \leq \frac{1}{2}.$$

Constructions such as the above are currently the only way known to produce an additive number system $\mathcal{A} \notin \mathsf{RT}_1$ with $\rho = 1$.

Problem 4.16. Find a natural example of an $\mathcal{A} \notin \mathsf{RT}_1$ with $\rho = 1$.

4.5. Further results on RT_1

Compton's Theorem 4.2 led to a search for simple conditions that guarantee $\mathcal{A} \in \mathsf{RT}_1$. The first results he found were in Bateman and Erdös [**4**], featured in this chapter, and in Meinardus [**36**]. The former can be generalized to $p(n) \leq c$, for $n \geq 1$. Meinardus was concerned with finding asymptotic estimates for $a(n)$, using the complex variable saddle-point method. With his asymptotics one could show a few examples, like $p(n) = n$, lead to $\mathcal{A} \in \mathsf{RT}_1$. In 1987 Compton asked ([**20**], Question 8.3): does $p(n) = \mathrm{O}(n^c)$ lead to $a(n-1)/a(n) \to 1$?

The next method was to apply the work of Bruce Richmond ([**45**]–[**50**]) from the 1970s (based on Roth and Szekeres [**51**]) to obtain the following conditions, any one of which guarantees $\mathcal{A} \in \mathsf{RT}_1$ (recall that $P(x)$ is the global counting function of the indecomposables):

1. $\left(\limsup_{x\to\infty} \frac{\log P(x)}{\log x}\right) \Big/ \left(\liminf_{x\to\infty} \frac{\log P(x)}{\log x}\right) \;<\; \frac{3}{2}$, or
2. $\log P(x) \;\sim\; C \log\log x$, or
3. $P(2x) \;=\; \mathrm{O}\big(P(x)\big)$.

Recently Bell [**6**] announced that the answer to Compton's question is 'yes': $p(n)$ is polynomially bounded implies $\mathcal{A} \in \mathsf{RT}_1$. This covers the results mentioned above, and many others. For example, consider the class of linear forests (as posets) in which each chain has a distinguished node, as in Figure 2. Such structures are of the form $(S, \leq, G)$, where G is a unary predicate that indicates a color (green). In this case $p(n) = n$, so a randomly selected forest of this kind almost certainly has at least k copies of any given component, for any choice of $k \geq 0$.

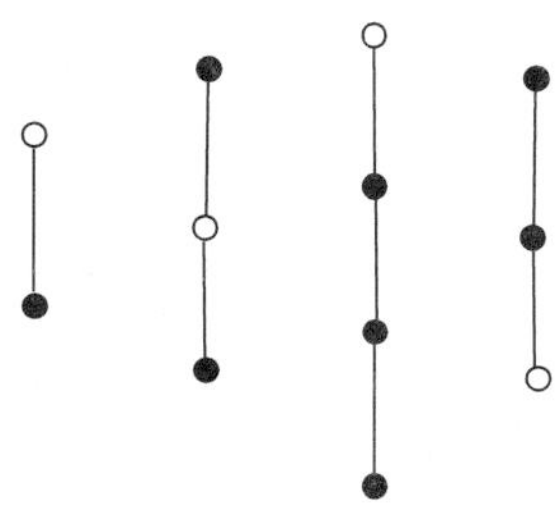

FIGURE 2. A Linear Forest with a Distinguished Node on each Chain

This example can be generalized further, to the class of linear forests with r distinguished nodes. In this case, for $n \geq r$,

$$p(n) \;=\; \binom{n}{r},$$

a polynomially bounded function. So a randomly selected forest of this kind almost certainly has at least k copies of any given component, for any choice of $k \geq 0$.

It is likely very difficult to find a simple characterization of all possible count functions $p(n)$ for which $a(n) \in \mathsf{RT}_1$. Recently Bell and Burris [**8**] announced that if $p(n) \in \mathsf{RT}_1$ then $a(n) \in \mathsf{RT}_1$, giving a large number of new examples.

Problem 4.17. Extend the class of additive number systems $\mathcal{A}$ for which one can show that $\mathcal{A} \in \mathsf{RT}_1$.

CHAPTER 5

The Case $0 < \rho < 1$

In this chapter the question of when all partition sets of $\mathcal{A}$ have asymptotic density is examined, under the assumption that $0 < \rho < 1$.[1] From Corollary 3.30 it follows that $\mathcal{A} \in \mathsf{RT}_\rho$ is a necessary condition, but, as stated in Remark 3.31, this is sufficient only when $\rho = 1$. The goal is to find solutions to

$$\mathcal{A} \in \mathsf{RT}_\rho + \Big(\quad\Big) \qquad \text{implies} \qquad \delta(\mathsf{B}) \text{ exists, for all partition sets } \mathsf{B} \text{ of } \mathcal{A}.$$

5.1. Compton's Tauberian Theorem

Modifications of Schur's Tauberian Theorem are used to obtain asymptotic density results. The following is the basis of Compton's Density Theorem.

Theorem 5.1 (Compton, 1989). *Let $\mathbf{S}(x), \mathbf{T}(x)$ be two power series such that, for some $\rho > 0$,*

(a) $\mathbf{T}(x) \in \mathsf{RT}_\rho$,
(b) $\mathbf{S}(x)$ *converges absolutely at $x = \rho$, and,*
(c) *for some K and C,*

$$\left|\frac{t(n-k)}{t(n)}\right| \;\le\; C\rho^k, \qquad \text{for } K \le k \le n.$$

Then

$$\lim_{n\to\infty} \frac{[x^n]\big(\mathbf{S}(x) * \mathbf{T}(x)\big)}{[x^n]\mathbf{T}(x)} \;=\; \mathbf{S}(\rho).$$

If $\mathbf{S}(\rho) > 0$ this can be expressed as an asymptotic relationship:

$$[x^n]\big(\mathbf{S}(x) * \mathbf{T}(x)\big) \;\sim\; \mathbf{S}(\rho)\cdot[x^n]\mathbf{T}(x). \tag{5.1.1}$$

PROOF. The proof is almost identical to the first part of the proof of Schur's Tauberian Theorem. Let $\mathbf{R}(x) = \mathbf{S}(x) * \mathbf{T}(x)$. Implicit in condition (a) is the assumption that $t(n)$ is eventually nonzero.

[1] Most of the results apply to the case $\rho = 1$, but they are not as strong as the results of the previous chapter. This chapter is essentially an exposition of the main results of Compton from [**21**], augmented by the asymptotics of Knopfmacher, Knopfmacher and Warlimont [**31**] that show a wide range of count functions are covered by Compton's work.

First,

$$\left|\mathbf{S}(\rho)-\frac{r(n)}{t(n)}\right| \le \underbrace{\left|\mathbf{S}(\rho)-\sum_{k\le n}s(k)\rho^k\right|}_{\text{I}} + \underbrace{\left|\sum_{k\le n}s(k)\rho^k-\frac{r(n)}{t(n)}\right|}_{\text{II}}.$$

Clearly term I goes to 0 as $n\to\infty$. Thus it is sufficient to show that term II vanishes as $n\to\infty$. For any $M\in\mathbb{N}$, and any $n\ge M$,

$$\begin{aligned}\left|\sum_{k\le n}s(k)\rho^k-\frac{r(n)}{t(n)}\right| &= \left|\sum_{k\le n}s(k)\rho^k-\sum_{k\le n}\frac{s(k)t(n-k)}{t(n)}\right|\\ &= \left|\sum_{k\le n}s(k)\left(\rho^k-\frac{t(n-k)}{t(n)}\right)\right|\\ &\le \underbrace{\left|\sum_{k\le M}s(k)\left(\rho^k-\frac{t(n-k)}{t(n)}\right)\right|}_{\text{III}}\\ &\qquad+\underbrace{\left|\sum_{M<k\le n}s(k)\left(\rho^k-\frac{t(n-k)}{t(n)}\right)\right|}_{\text{IV}}.\end{aligned}$$

For any fixed M, term III goes to 0 as $n\to\infty$ since (a) guarantees, for any fixed k, that

$$\lim_{n\to\infty}\frac{t(n-k)}{t(n)} = \rho^k.$$

Thus it is sufficient to show that term IV goes to 0 as $M\to\infty$. Applying the triangle inequality to term IV gives

$$\left|\sum_{M<k\le n}s(k)\left(\rho^k-\frac{t(n-k)}{t(n)}\right)\right| \le \underbrace{\left|\sum_{k>M}s(k)\rho^k\right|}_{\text{V}} + \underbrace{\sum_{M<k\le n}|s(k)|\left|\frac{t(n-k)}{t(n)}\right|}_{\text{VI}}.$$

Term V goes to 0 as $M\to\infty$ since it is the tail end of the convergent series $\sum_k s(k)\rho^k$. Thus it suffices to show that term VI goes to 0 as $M\to\infty$.

Assume $M\ge K$. Then, from (c),

$$\sum_{M<k\le n}|s(k)|\left|\frac{t(n-k)}{t(n)}\right| \le C\cdot\sum_{M<k\le n}|s(k)|\rho^k.$$

Since $\sum s(n)\rho^n$ is absolutely convergent, the expression on the right side goes to 0 as $M\to\infty$, so term VI goes to 0 as $M\to\infty$. □

The following gives a practical tool to verify that a given series $\sum t(n)x^n$ satisfies condition (c) in Theorem 5.1.

Lemma 5.2. *Suppose* $t(n) \sim \rho^{-n} t_0(n)$, *where*

(a) $\rho > 0$,
(b) $t_0(n) \in \mathsf{RT}_1$, *and*
(c) $t_0(n)$ *is eventually nondecreasing.*

Then $t(n) \in \mathsf{RT}_\rho$, *and there are positive constants* C *and* K *such that*

$$\left|\frac{t(n-k)}{t(n)}\right| \leq C\rho^k, \qquad \text{for } K \leq k \leq n.$$

PROOF. By changing only finitely many initial values of $t_0(n)$, it is possible to assume:

(c′) $t_0(n)$ is always positive and nondecreasing.

This change does not affect the validity of the hypotheses. One easily has $t(n) \in \mathsf{RT}_\rho$ from Proposition 1.26. So it suffices to show that the desired constants C and K exist.

Choose K such that $n \geq K$ implies

$$\left|t(n) - \rho^{-n} t_0(n)\right| \leq \tfrac{1}{2}\rho^{-n} t_0(n) \qquad \text{and} \quad t(n) > 0.$$

Then, for $n \geq K$,

$$\tfrac{1}{2}\rho^{-n} t_0(n) \leq t(n) \leq \tfrac{3}{2}\rho^{-n} t_0(n). \tag{5.2.1}$$

Choose D such that, for any $j < K$,

$$\left|t(j)\right| \leq D\rho^{-j} t_0(j). \tag{5.2.2}$$

Let $C = \max(2D, 3)$, and let $K \leq k \leq n$.
Case 1: $n - k \geq K$. Then

$$\begin{aligned}
\left|\frac{t(n-k)}{t(n)}\right| &\leq \Big(\tfrac{3}{2}\rho^{-(n-k)} t_0(n-k)\Big)\Big/\Big(\tfrac{1}{2}\rho^{-n} t_0(n)\Big) && \text{by (5.2.1)}\\
&\leq 3\rho^k && \text{by (c}'\text{)}\\
&\leq C\rho^k && \text{definition of } C.
\end{aligned}$$

Case 2: $n - k < K$. Then

$$\begin{aligned}
\left|\frac{t(n-k)}{t(n)}\right| &\leq \Big(D\rho^{-(n-k)} t_0(n-k)\Big)\Big/\Big(\tfrac{1}{2}\rho^{-n} t_0(n)\Big) && \text{by (5.2.1), (5.2.2)}\\
&\leq 2D\rho^k && \text{by (c}'\text{)}\\
&\leq C\rho^k && \text{definition of } C.
\end{aligned}$$

This conclues the proof. □

The simplest case of Lemma 5.2 occurs when $t_0(n) = C_0 > 0$.

5.2. When is $\delta(\mathsf{B}) = 0$?

In this section it will be shown that if $\mathcal{A} \in \mathsf{RT}_\rho$ then any partition set with Dirichlet density 0 also has asymptotic density 0. Recall that $\mathsf{A}_\mathsf{Q} = \omega\mathsf{Q}$.

Lemma 5.3. *Suppose $\mathcal{A} \in \mathsf{RT}_\rho$ and $\mathsf{Q} \subseteq \mathsf{P}$. If $\mathbf{A}_\mathsf{Q}(\rho) = \infty$ then every partition set of the form*

$$\mathsf{B} \;=\; \gamma\mathsf{Q} \;+\; \omega(\mathsf{P} \setminus \mathsf{Q})$$

has asymptotic density, and

$$\delta(\mathsf{B}) \;=\; \partial(\mathsf{B}) \;=\; \begin{cases} 0 & \text{if } \gamma \text{ is small} \\ 1 & \text{if } \gamma \text{ is large.} \end{cases}$$

PROOF. Just repeat the proof of Lemma 3.55, replacing P_n by Q_n. □

Remark 5.4. In this chapter the condition $\mathbf{A}_\mathsf{Q}(\rho) = \infty$ is equivalent to $\mathbf{Q}(\rho) = \infty$ since $\rho < 1$ (see Corollary 2.25).

The cases when $\partial_\mathsf{Q}(\gamma\mathsf{Q})$ is zero are determined next.

Lemma 5.5. *Let Q be a nonempty subset of P. Then*

$$\partial_\mathsf{Q}(\gamma\mathsf{Q}) \;=\; 0 \quad \textit{iff} \quad \gamma \textit{ is small and } \mathbf{A}_\mathsf{Q}(\rho) \;=\; \infty.$$

PROOF. If $\mathbf{A}_\mathsf{Q}(\rho) < \infty$ then, by Lemma 3.39, $\partial_\mathsf{Q}(\gamma\mathsf{Q}) > 0$. Thus $\partial_\mathsf{Q}(\gamma\mathsf{Q}) = 0$ implies $\mathbf{A}_\mathsf{Q}(\rho) = \infty$. Consequently, proving the lemma reduces to assuming $\mathbf{A}_\mathsf{Q}(\rho) = \infty$ and proving $\partial_\mathsf{Q}(\gamma\mathsf{Q}) = 0$ iff γ is small, which is clear from Lemma 3.39. □

Proposition 5.6. *Suppose $\mathcal{A} \in \mathsf{RT}_\rho$ and $\mathbf{A}(\rho) = \infty$. For B a partition set, say $\mathsf{B} = \gamma_1\mathsf{P}_1 + \cdots + \gamma_k\mathsf{P}_k$, the following are equivalent:*

(a) $\delta(\mathsf{B}) \;=\; 0.$
(b) $\partial(\mathsf{B}) \;=\; 0.$
(c) *For some i, γ_i is small and $\mathbf{A}_{\mathsf{P}_i}(\rho) = \infty$.*

PROOF. (a) $\Rightarrow$ (b): By Lemma 3.13.
(b) $\Rightarrow$ (c): From $\partial(\mathsf{B}) = 0$ it follows that, for some i, $\partial_{\mathsf{P}_i}(\gamma_i\mathsf{P}_i) = 0$, by Theorem 3.40. For such an i, γ_i is small and $\mathbf{A}_{\mathsf{P}_i}(\rho) = \infty$, by Lemma 5.5.
(c) $\Rightarrow$ (a): Assume (c) holds. By Lemma 5.3,

$$\delta\big(\gamma_i\mathsf{P}_i \;+\; \omega(\mathsf{P} \setminus \mathsf{P}_i)\big) \;=\; 0.$$

Since $\mathsf{B} \subseteq \gamma_i\mathsf{P}_i + \omega(\mathsf{P} \setminus \mathsf{P}_i)$, by Lemma 3.53 it follows that $\delta(\mathsf{B}) = 0$. □

Corollary 5.7. *Suppose $\mathcal{A} \in \mathsf{RT}_\rho$. For any partition set B,*

$$\partial(\mathsf{B}) \;=\; 0 \quad \textit{implies} \quad \delta(\mathsf{B}) \;=\; 0.$$

PROOF. $\partial(\mathsf{B}) = 0$ implies $\mathbf{A}(\rho) = \infty$, so use Proposition 5.6. □

5.3. The density of $\overline{\mathsf{B}}$

Assuming $\mathcal{A} \in \mathsf{RT}_\rho$, conditions that guarantee that the asymptotic density of any partition set B exists, and equals the Dirichlet density, are sought. For partition sets with zero Dirichlet density no extra conditions are needed since $\partial(\mathsf{B}) = 0$ implies $\delta(\mathsf{B}) = 0$.

The more difficult case occurs when $\partial(\mathsf{B}) > 0$. The Tauberian theorem used to show that $\delta(\mathsf{B})$ exists requires that $[\mathbf{B}/\mathbf{A}](x) \in \mathcal{PS}_\rho$. One method to fulfill this requirement is to consider a possibly larger partition set $\overline{\mathsf{B}}$ that has the same Dirichlet density as B, and $[\overline{\mathbf{B}}/\mathbf{A}](x)$ is in $\mathcal{PS}_\rho$. The Tauberian theorem gives, under suitable hypotheses, $\delta(\overline{\mathsf{B}}) = \partial(\overline{\mathsf{B}}) = [\overline{\mathbf{B}}/\mathbf{A}](\rho)$. Then, using the zero density of $\overline{\mathsf{B}} \smallsetminus \mathsf{B}$, one has $\delta(\mathsf{B}) = \partial(\mathsf{B}) = [\overline{\mathbf{B}}/\mathbf{A}](\rho)$.

Lemma 5.8. *Let* Q *be a nonempty subset of* P. *If* $\mathbf{A}_\mathsf{Q}(\rho) < \infty$ *then, for any coefficient* γ,

$$[\gamma\mathbf{Q}/\mathbf{A}_\mathsf{Q}](x) \in \mathcal{PS}_\rho \quad \textit{and} \quad \partial_\mathsf{Q}(\gamma\mathsf{Q}) \;=\; [\gamma\mathbf{Q}/\mathbf{A}_\mathsf{Q}](\rho).$$

PROOF. Lemma 3.39 shows that $\partial_\mathsf{Q}(\gamma\mathsf{Q})$ is $[\gamma\mathbf{Q}](\rho)/\mathbf{A}_\mathsf{Q}(\rho)$. Then, applying Lemma 3.48 to $\mathcal{A}_\mathsf{Q}$ gives the desired conclusions. □

Lemma 5.9. *Suppose* $\mathsf{B} = \gamma_1\mathsf{P}_1 + \cdots + \gamma_k\mathsf{P}_k$ *is a partition set such that* $\partial(\mathsf{B}) > 0$. *If* $\gamma_i = \omega$ *whenever* $\mathsf{A}_{\mathsf{P}_i}(\rho) = \infty$ *then*

$$[\mathbf{B}/\mathbf{A}](x) \in \mathcal{PS}_\rho \quad \textit{and} \quad \partial(\mathsf{B}) \;=\; [\mathbf{B}/\mathbf{A}](\rho).$$

PROOF. From Compton's Dirichlet Density Theorem 3.40,

$$\partial(\mathsf{B}) \;=\; \partial_{\mathsf{P}_1}(\gamma_1\mathsf{P}_1) \;\cdots\; \partial_{\mathsf{P}_k}(\gamma_k\mathsf{P}_k),$$

and, since $\partial(\mathsf{B}) > 0$, it follows that each $\partial_{\mathsf{P}_i}(\gamma_i\mathsf{P}_i) > 0$. Thus, for $x \in [0, \rho)$,

$$\begin{aligned}
[\mathbf{B}/\mathbf{A}](x) &= \frac{\mathbf{B}(x)}{\mathbf{A}(x)} && \text{by Lemma 3.48}\\
&= \prod_{j=1}^{k} \frac{[\gamma_j\mathbf{P}_j](x)}{\mathbf{A}_{\mathsf{P}_j}(x)} && \text{by Proposition 3.33}\\
&= \prod_{j=1}^{k} [\gamma_j\mathbf{P}_j/\mathbf{A}_{\mathsf{P}_j}](x) && \text{by Lemma 3.48.}
\end{aligned}$$

Each $[\gamma_j\mathbf{P}_j/\mathbf{A}_{\mathsf{P}_j}](x)$ is in $\mathcal{PS}_\rho$ since first, the large coefficients γ_j, where one has $\mathsf{A}_{\mathsf{P}_j}(\rho) = \infty$, are ω, and clearly $[\omega\mathbf{P}_j/\mathbf{A}_{\mathsf{P}_j}](x) = 1$ is in $\mathcal{PS}_\rho$, and secondly, Lemma 5.8 applies to the other coefficients. Then Lemma 1.18 gives $[\mathbf{B}/\mathbf{A}](x) \in \mathcal{PS}_\rho$, so $\partial(\mathsf{B}) = [\mathbf{B}/\mathbf{A}](\rho)$. □

Definition 5.10. For Q a nonempty subset of P and γ a coefficient, let

$$\overline{\gamma\mathsf{Q}} \;=\; \begin{cases} \omega\mathsf{Q} & \text{if } \gamma \text{ is large and } \mathbf{A}_\mathsf{Q}(\rho) = \infty \\ \gamma\mathsf{Q} & \text{otherwise.} \end{cases}$$

For any partition set $\mathsf{B} = \gamma_1 \mathsf{P}_1 + \cdots + \gamma_k \mathsf{P}_k$, let

$$\overline{\mathsf{B}} = \overline{\gamma_1 \mathsf{P}_1} + \cdots + \overline{\gamma_k \mathsf{P}_k}.$$

$\overline{\mathsf{B}}$ is also a partition set and $\mathsf{B} \subseteq \overline{\mathsf{B}}$. Any large coefficient γ_i of B is promoted to ω in $\overline{\mathsf{B}}$, provided $\mathsf{A}_{\mathsf{P}_i}(\rho) = \infty$.

Lemma 5.11. *Suppose* $\mathcal{A} \in \mathsf{RT}_\rho$. *Then, for any partition set* B,

(a) $\delta(\overline{\mathsf{B}} \setminus \mathsf{B}) = \partial(\overline{\mathsf{B}} \setminus \mathsf{B}) = 0$, *and*
(b) $\partial(\mathsf{B}) > 0$ *implies*

$$[\overline{\mathbf{B}}/\mathbf{A}](x) \in \mathcal{PS}_\rho \quad \text{and} \quad \partial(\overline{\mathsf{B}}) = [\overline{\mathbf{B}}/\mathbf{A}](\rho).$$

PROOF. Let $\mathsf{B} = \gamma_1 \mathsf{P}_1 + \cdots + \gamma_k \mathsf{P}_k$.

(a) For γ a large coefficient that is not ω, say $\gamma = (\geq m)$ with $m > 0$, let $\widehat{\gamma}$ be $(<m)$, a small coefficient. Then

$$\overline{\mathsf{B}} \setminus \mathsf{B} \subseteq \bigcup_{\substack{\gamma_i \text{ large}, \neq \omega \\ \mathbf{P}_i(\rho) = \infty}} \Big(\widehat{\gamma_i} \mathsf{P}_i + \omega(\mathsf{P} \setminus \mathsf{P}_i)\Big).$$

Now, for γ_i large, $\neq \omega$, with $\mathbf{P}_i(\rho) = \infty$, by Lemma 5.3 one has

$$\delta\big(\widehat{\gamma_i} \mathsf{P}_i + \omega(\mathsf{P} \setminus \mathsf{P}_i)\big) = \partial\big(\widehat{\gamma_i} \mathsf{P}_i + \omega(\mathsf{P} \setminus \mathsf{P}_i)\big) = 0,$$

as $\widehat{\gamma_i}$ is small. Thus $\delta(\overline{\mathsf{B}} \setminus \mathsf{B}) = \partial(\overline{\mathsf{B}} \setminus \mathsf{B}) = 0$.

(b) This follows from Lemma 5.9 since $\partial(\overline{\mathsf{B}}) \geq \partial(\mathsf{B}) > 0$. □

Lemma 5.12. *Suppose* $\mathcal{A} \in \mathsf{RT}_\rho$. *Then, for any partition set* B,

(a) $\partial(\overline{\mathsf{B}}) = \partial(\mathsf{B})$, *and*
(b) $\delta(\overline{\mathsf{B}}) = \delta(\mathsf{B})$ (*if either side exists*).

PROOF. $\overline{\mathsf{B}}$ is the disjoint union of B and $\overline{\mathsf{B}} \setminus \mathsf{B}$, and, by Lemma 5.11 (a), $\delta(\overline{\mathsf{B}} \setminus \mathsf{B}) = \partial(\overline{\mathsf{B}} \setminus \mathsf{B}) = 0$. □

5.4. Compton's Density Theorem

The next result is based on Theorem 6.6.(ii) of [**21**].

Theorem 5.13 (Compton, 1989). *Suppose* $\mathcal{A} \in \mathsf{RT}_\rho$. *If there exist constants* K *and* C *such that*

$$\frac{a(n-k)}{a(n)} \leq C\rho^k, \qquad \text{for } K \leq k \leq n,$$

then all partition sets B *of* $\mathcal{A}$ *have asymptotic density, and* $\delta(\mathsf{B}) = \partial(\mathsf{B})$.

PROOF. Let B be a partition set. By Theorem 3.40, $\partial(\mathsf{B})$ exists.

Case 1: $\partial(\mathsf{B}) = 0$.
From Corollary 5.7, $\delta(\mathsf{B}) = 0$.

Case 2: $\partial(\mathsf{B}) > 0$.
Then $[\overline{\mathbf{B}}/\mathbf{A}](\rho)$ is absolutely convergent by Lemma 5.11 (b), so $\mathbf{A}(x)$ and $[\overline{\mathbf{B}}/\mathbf{A}](x)$ satisfy the hypotheses, for the power series $\mathbf{T}(x)$ and $\mathbf{S}(x)$ respectively, of Theorem 5.1. Thus $\lim_{n\to\infty} \dfrac{\overline{b}(n)}{a(n)}$ exists and equals $[\overline{\mathbf{B}}/\mathbf{A}](\rho)$, so

$$\delta(\overline{\mathsf{B}}) \;=\; \partial(\overline{\mathsf{B}}) \;=\; [\overline{\mathbf{B}}/\mathbf{A}](\rho).$$

Now apply Lemma 5.12. □

Remark 5.14. The hypothesis $\dfrac{a(n-k)}{a(n)} \le C\rho^k$, for $K \le k \le n$, implies (by setting k equal to n) that $a(n) \ge \rho^{-n}/C$ holds, for $n \ge K$, and thus clearly $\mathbf{A}(\rho) = \infty$.

The following corollary of Theorem 5.13 applies to numerous situations.

Corollary 5.15. *Suppose* $a(n) \sim \beta^n a_0(n)$, *where*

(a) $\beta \ge 1$,
(b) $a_0(n) \in \mathsf{RT}_1$, *and*
(c) $a_0(n)$ *is eventually nondecreasing.*

Then all partition sets of $\mathcal{A}$ *have asymptotic density which equals the Dirichlet density.*

PROOF. This follows from Lemma 5.2 and Theorem 5.13. □

Remark 5.16. Cameron [**18**] notes that $a(n) \sim C\beta^n/n^\mu$ is fairly common among examples of additive number systems from combinatorics. If $C > 0$, $\beta > 1$, and $\mu \le 0$ then Corollary 5.15 shows that all partition sets have asymptotic density which equals the Dirichlet density. For the cases $C > 0$ and $\beta, \mu > 1$, see Remark 3.31.

5.4.1. Corresponding global results. The theorem of Compton in the preceding section, and the corollaries to Compton's Density Theorem, have global asymptotic density versions—one simply replaces $a(n)$ by $A(n)$ and δ by Δ.

To prove these global theorems: for the case $\partial(\mathsf{B}) = 0$ one still has $\delta(\mathsf{B}) = 0$, and thus $\Delta(\mathsf{B}) = 0$; and for the case $\partial(\mathsf{B}) > 0$, replace $\mathbf{A}(x)$ by $\mathbf{A}^+(x) = (1-x)^{-1} \cdot \mathbf{A}(x)$ and $\mathbf{B}(x)$ by $\mathbf{B}^+(x) = (1-x)^{-1} \cdot \mathbf{B}(x)$ so that the coefficients of these power series give the global counts of A and B.

5.5. The Knopfmacher, Knopfmacher, Warlimont asymptotics

At the time Compton published his Density Theorem, in 1989, no examples of additive number systems that satisfied the hypotheses of his theorem were known. This situation changed in 1992 when John Knopfmacher, Arnold Knopfmacher, and Richard Warlimont [**31**] (later referred to as KKW) gave results on additive number systems that parallel the classical results on counting the number of ways to factorize numbers (see §11.5), a problem that is often referred to by its Latin name *Factorisatio Numerorum.*

The KKW result was for discrete multiplicative, hence additive, number systems (see §9.8); and, because of the application in mind, they required that the counting function $p(n)$ of the set of indecomposables also be the counting function $\widetilde{a}(n)$ of some additive number system $\widetilde{\mathcal{A}}$. However, this condition on $p(n)$ is not used in their proof, so it is discarded to obtain the following elegant asymptotics that were used to provide the first examples for Compton's Density Theorem.

Theorem 5.17 (KKW, 1992). *Let $\mathcal{A}$ be an additive number system with*

$$p(n) \;=\; a\beta^n \;+\; \mathrm{O}(\eta^n),$$

where $a > 0$, $\beta > 1$, and $0 < \eta < \beta$. Then

$$a(n) \;\sim\; C\beta^n \,\frac{e^{2\sqrt{an}}}{n^{3/4}},$$

for a suitable positive constant C. Consequently, all partition sets of $\mathcal{A}$ have asymptotic density which equals the Dirichlet density.

The proof presented will be somewhat simpler than the one given in [**31**], as only the dominant term of the asymptotics is of interest, and not the error term. Also the magnitude of the leading constant coefficient does not play any role, so the notation

$$f_1(n) \;\overset{*}{=}\; f_2(n)$$

is used to mean that, for some nonzero constant c, $f_1(n) = c \cdot f_2(n)$; and likewise

$$f_1(n) \;\overset{*}{\sim}\; f_2(n)$$

is used to mean that $f_1(n) \sim c \cdot f_2(n)$, for some nonzero constant c.

Note that the hypotheses give $\rho = 1/\beta$. John Knopfmacher [**30**] refers to the same hypotheses, when applied to $a(n)$ instead of $p(n)$, as **Axiom $\mathbf{A}^{\#}$** in his work on abstract analytic additive number theory.

Before proving this theorem, a technical lemma on asymptotics is given that is used both in the proof of this theorem and in a proof of a generalization of Oppenheim's asymptotics in §11.5.

In this section it will be most useful to introduce an additional role for the 'big O' notation, namely in an integral

$$\int_a^b \cdots \mathrm{O}(g(u)) \cdots du,$$

the expression $\mathrm{O}\big(g(u)\big)$ refers to a function $f(u)$ such that $\big|f(u)\big| \le Cg(u)$, for some constant C, and for $u \in [a, b]$.

Note that, since $e^{-u^2} < \dfrac{1}{1+u^2}$ on $[0,\infty)$,

$$\int_0^\infty e^{-u^2}du \ < \ \infty.$$

Lemma 5.18.

$$\int_{-\pi/2}^{\pi/2} e^{\lambda\cos\theta}\cos\theta d\theta \ \stackrel{*}{=} \ \frac{e^\lambda}{\sqrt{\lambda}}\cdot\Big(1 \ + \ \mathrm{O}\big(1/\sqrt{\lambda}\big)\Big) \tag{5.18.1}$$

$$\int_{-\pi/2}^{\pi/2} e^{\lambda\cos\theta}\cdot|\theta|\cdot\cos\theta d\theta \ = \ \mathrm{O}\big(e^\lambda/\lambda\big). \tag{5.18.2}$$

PROOF. Given $c>0$,

$$\int_0^c e^{-\lambda u^2}du \ \stackrel{*}{=} \ \frac{1}{\sqrt{\lambda}}\cdot\Big(1 \ + \ \mathrm{O}\big(1/\sqrt{\lambda}\big)\Big)$$

since, from

$$\int_c^\infty e^{-\lambda u^2}du \ < \ \int_c^\infty \frac{du}{\lambda u^2} \ = \ O\big(1/\lambda\big),$$

it follows that

$$\begin{aligned}\int_0^c e^{-\lambda u^2}du \ &= \ \int_0^\infty e^{-\lambda u^2}du \ - \ \int_c^\infty e^{-\lambda u^2}du \\ &= \ \frac{1}{\sqrt{\lambda}}\int_0^\infty e^{-v^2}dv \ - \ \mathrm{O}\big(1/\lambda\big) \\ &\stackrel{*}{=} \ \frac{1}{\sqrt{\lambda}} \ + \ \mathrm{O}\big(1/\lambda\big) \\ &= \ \frac{1}{\sqrt{\lambda}}\cdot\Big(1+\mathrm{O}\big(1/\sqrt{\lambda}\big)\Big).\end{aligned}$$

Also

$$\int_0^c e^{-\lambda u^2}udu \ = \ \mathrm{O}\big(1/\lambda\big)$$

holds since

$$\begin{aligned}\int_0^c e^{-\lambda u^2}udu \ &= \ \frac{1}{\lambda}\int_0^{c\sqrt{\lambda}} e^{-v^2}vdv \\ &= \ -\frac{1}{2\lambda}\Big[e^{-v^2}\Big]_0^{c\sqrt{\lambda}} \\ &= \ \mathrm{O}\big(1/\lambda\big).\end{aligned}$$

Next, it is proved that

$$\int_0^{\pi/2} e^{-\lambda\sin^2(\theta/2)}\cos\theta d\theta \ \stackrel{*}{=} \ \frac{1}{\sqrt{\lambda}}\cdot\Big(1 \ + \ \mathrm{O}\big(1/\sqrt{\lambda}\big)\Big). \tag{5.18.3}$$

Let $w = \sin\frac{\theta}{2}$. Then

$$\begin{aligned} dw &= \frac{1}{2}\cos\frac{\theta}{2}d\theta \\ \cos\theta &= 1 - 2\sin^2\frac{\theta}{2} = 1 - 2w^2. \end{aligned}$$

Using this change of variable gives

$$\begin{aligned} \int_0^{\pi/2} e^{-\lambda\sin^2(\theta/2)}\cos\theta d\theta &= 2\int_0^{\frac{1}{\sqrt{2}}} e^{-\lambda w^2}\cdot\frac{1-2w^2}{\sqrt{1-w^2}}dw \\ &= 2\int_0^{\frac{1}{\sqrt{2}}} e^{-\lambda w^2}\cdot\big(1+\mathrm{O}(w)\big)dw \\ &\stackrel{*}{=} \frac{1}{\sqrt{\lambda}}\cdot\Big(1+\mathrm{O}\big(1/\sqrt{\lambda}\big)\Big) + \mathrm{O}(1/\lambda) \\ &= \frac{1}{\sqrt{\lambda}}\cdot\Big(1+\mathrm{O}\big(1/\sqrt{\lambda}\big)\Big). \end{aligned}$$

(5.18.1) then follows from (5.18.3):

$$\begin{aligned} \int_0^{\pi/2} e^{\lambda\cos\theta}\cos\theta d\theta &= \int_0^{\pi/2} e^{\lambda(1-2\sin^2\frac{\theta}{2})}\cos\theta d\theta \\ &= e^{\lambda}\int_0^{\pi/2} e^{-2\lambda\sin^2(\theta/2)}\cos\theta d\theta \\ &\stackrel{*}{=} e^{\lambda}\frac{1}{\sqrt{\lambda}}\cdot\Big(1+\mathrm{O}\big(1/\sqrt{\lambda}\big)\Big). \end{aligned}$$

For (5.18.2),

$$\begin{aligned} \int_0^{\pi/2} e^{\lambda\cos\theta}\cdot\theta\cdot\cos\theta d\theta &\le \pi\int_0^{\pi/2} e^{\lambda\cos\theta}\sin\theta\cos\theta d\theta \\ &= -\frac{\pi}{\lambda}\cdot\Big[e^{\lambda\cos\theta}\cos\theta\Big]_0^{\pi/2} \\ &\qquad -\frac{\pi}{\lambda}\cdot\int_0^{\pi/2} e^{\lambda\cos\theta}\sin\theta d\theta \\ &= \frac{\pi}{\lambda}\cdot e^{\lambda} + \frac{\pi}{\lambda^2}\cdot\Big[e^{\lambda\cos\theta}\Big]_0^{\pi/2} \\ &= \frac{\pi}{\lambda}\cdot e^{\lambda} + \frac{\pi}{\lambda^2}\cdot\big(1-e^{\lambda}\big) \\ &= \mathrm{O}(e^{\lambda}/\lambda). \end{aligned}$$

□

With this preparation the theorem can now be proved.

PROOF OF THE THEOREM. From the hypotheses on $p(n)$,

$$p(n) = a\beta^n + f(n),$$

where, for some constant c, $|f(n)| \leq c\eta^n$. With $\varepsilon(z)$ defined by

$$\varepsilon(z) \;=\; \sum_{n\geq 1} f(n)z^n,$$

$\varepsilon(z)$ has radius of convergence at least $1/\eta$, which is greater than $\rho = 1/\beta$. Now

$$\begin{aligned}\mathbf{P}(z) \;&=\; \sum_{n\geq 0} a\beta^n z^n \;+\; \varepsilon(z)\\ &=\; \frac{a}{1-\beta z} \;+\; \varepsilon(z),\end{aligned}$$

so,

$$e^{\mathbf{P}(z)} \;=\; e^{a/(1-\beta z)} \cdot e^{\varepsilon(z)}. \tag{5.18.4}$$

The strategy to prove the theorem is first to use complex analysis to find the asymptotics of $[z^n]e^{a/(1-z)}$, and then to use this to show that $[z^n]e^{a/(1-\beta z)} \in \mathsf{RT}_\rho$. This allows Schur's Tauberian Theorem 3.42, as expressed in (3.42.1), to be used to find the asymptotics of $[z^n]e^{\mathbf{P}(z)}$, and to show that this is also in RT_ρ. Then Proposition 3.46 can be applied to find the desired asymptotics of $a(n)$. The asymptotics of $[z^n]e^{a/(1-z)}$ were determined by Perron [**42**] in 1914. The approach used here is essentially that of Knopfmacher, et al, with some simplifications since only the dominant term is needed.

To determine the asymptotics of

$$J_n \;=\; [z^n]e^{a/(1-z)},$$

the Cauchy integral formula

$$\begin{aligned}J_n \;&=\; \frac{1}{2\pi i}\int_\Gamma e^{a/(1-z)} \cdot \frac{dz}{z^{n+1}}\\ &\stackrel{*}{=}\; \int_\Gamma e^{a/(1-z)} \cdot \frac{dz}{z^{n+1}}\end{aligned}$$

is used, where Γ is a simple closed curve with 0, but not 1, in its interior. Let Γ be the following curve composed of a rectilinear portion Γ_1 and a semicircular portion Γ_0 of radius r, as in Figure 1.

Since the integrand

$$e^{a/(1-z)} \cdot \frac{1}{z^{n+1}}$$

of J_n is continuous on the rectilinear portion Γ_1 of the boundary, and since it is bounded by e^a on Γ_1 (independent of the choice of r, for $0 < r < 1$), it follows that

$$J_n \;\stackrel{*}{=}\; \int_{\Gamma_0} e^{a/(1-z)} \cdot \frac{dz}{z^{n+1}} \;+\; \mathrm{O}(1).$$

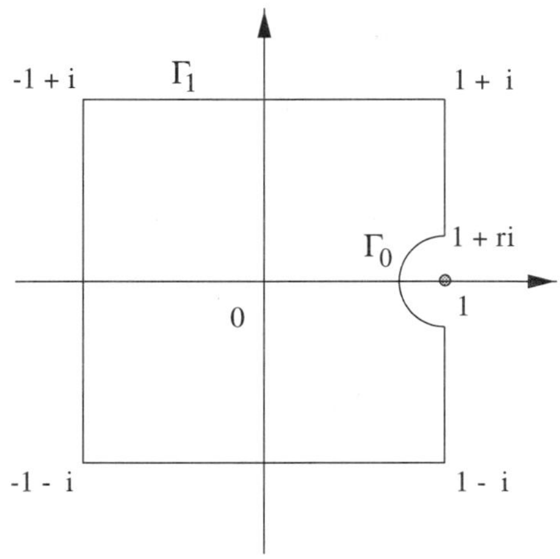

FIGURE 1

For the parametrization of Γ_0 given by

$$z = 1 - re^{-i\theta} \qquad -\pi/2 \le \theta \le \pi/2,$$

one has

$$\begin{aligned} J_n &\overset{*}{=} r\int_{-\pi/2}^{\pi/2} e^{(a/r)e^{i\theta}} \cdot (1 - re^{-i\theta})^{-(n+1)} \cdot e^{-i\theta} d\theta + \mathrm{O}(1) \\ &= r\int_{-\pi/2}^{\pi/2} \exp\Big(\frac{a}{r}e^{i\theta} + (n+1)\cdot \sum_{k\ge 1}(re^{-i\theta})^k/k\Big) \cdot e^{-i\theta} d\theta + \mathrm{O}(1) \\ &= r\int_{-\pi/2}^{\pi/2} \exp\Big(\frac{a}{r}e^{i\theta} + (n+1)\cdot re^{-i\theta} + (n+1)r^2e^{-2i\theta}/2 \\ &\qquad + (n+1)\cdot \sum_{k\ge 3}(re^{-i\theta})^k/k\Big) \cdot e^{-i\theta} d\theta + \mathrm{O}(1). \end{aligned}$$

Choose

$$r_n = \sqrt{\frac{a}{n+1}}$$

so that $a/r_n = (n+1)r_n = \sqrt{a(n+1)}$ and $(n+1)r_n^2 = a$. Then

$$\begin{aligned} J_n &\overset{*}{=} r_n \int_{-\pi/2}^{\pi/2} \exp\Big(\frac{a}{r_n}(e^{i\theta} + e^{-i\theta}) + ae^{-2i\theta}/2 \\ &\qquad + ar_n \sum_{k\ge 3} r_n^{k-3} e^{-ik\theta}/k\Big) \cdot e^{-i\theta} d\theta + \mathrm{O}(1) \\ &= r_n \int_{-\pi/2}^{\pi/2} \exp\Big(2\frac{a}{r_n}\cos\theta + a/2 + \mathrm{O}(|\theta|) + \mathrm{O}(r_n)\Big) \cdot e^{-i\theta} d\theta + \mathrm{O}(1) \end{aligned}$$

$$
\begin{aligned}
&\overset{*}{=} r_n \int_{-\pi/2}^{\pi/2} e^{(2a/r_n)\cos\theta} \cdot e^{\mathrm{O}(|\theta|)} \cdot e^{\mathrm{O}(r_n)} \cdot e^{-i\theta} d\theta \;+\; \mathrm{O}(1) \\
&= r_n \int_{-\pi/2}^{\pi/2} e^{(2a/r_n)\cos\theta} \cdot \big(1+\mathrm{O}(|\theta|)\big) \cdot \big(1+\mathrm{O}(r_n)\big) \cdot e^{-i\theta} d\theta \;+\; \mathrm{O}(1) \\
&= r_n \int_{-\pi/2}^{\pi/2} e^{(2a/r_n)\cos\theta} \cdot \big(1+\mathrm{O}(|\theta|)+\mathrm{O}(r_n)\big) \cdot e^{-i\theta} d\theta \;+\; \mathrm{O}(1) \\
&= \underbrace{r_n \int_{-\pi/2}^{\pi/2} e^{(2a/r_n)\cos\theta} \cdot e^{-i\theta} d\theta}_{K_1} \\
&\qquad + \underbrace{r_n \int_{-\pi/2}^{\pi/2} e^{(2a/r_n)\cos\theta} \cdot \mathrm{O}(|\theta|) \cdot e^{-i\theta} d\theta}_{K_2} \\
&\qquad\qquad + \underbrace{r_n \int_{-\pi/2}^{\pi/2} e^{(2a/r_n)\cos\theta} \cdot \mathrm{O}(r_n) \cdot e^{-i\theta} d\theta}_{K_3} \;+\; \mathrm{O}(1).
\end{aligned}
$$

Using (5.18.1),

$$
\begin{aligned}
K_1 &= \int_{-\pi/2}^{\pi/2} e^{(2a/r_n)\cos\theta} \cdot e^{-i\theta} d\theta \\
&= \int_{-\pi/2}^{\pi/2} e^{(2a/r_n)\cos\theta} \cdot \cos\theta d\theta \\
&\overset{*}{=} e^{2a/r_n}/\sqrt{2a/r_n} \cdot \Big(1+\mathrm{O}\big(1/\sqrt{2a/r_n}\big)\Big) \\
&\overset{*}{=} e^{2a/r_n}\sqrt{r_n} \cdot \Big(1+\mathrm{O}\big(\sqrt{r_n}\big)\Big).
\end{aligned}
$$

Now, using (5.18.2),

$$
K_2 \;=\; \int_{-\pi/2}^{\pi/2} e^{(2a/r_n)\cos\theta} \cdot \mathrm{O}(|\theta|) \cdot e^{-i\theta} d\theta \;=\; \mathrm{O}\big(e^{2a/r_n}/(2a/r_n)\big) \;=\; \mathrm{o}(K_1)
$$

and also

$$
K_3 \;=\; \int_{-\pi/2}^{\pi/2} e^{(2a/r_n)\cos\theta} \cdot \mathrm{O}(r_n) \cdot e^{-i\theta} d\theta \;=\; \mathrm{O}(r_n)K_1 = \;\mathrm{o}(K_1).
$$

Thus

$$
\begin{aligned}
J_n &\overset{*}{\sim} r_n K_1 \\
&\overset{*}{\sim} r_n e^{2a/r_n} \cdot \sqrt{r_n} \\
&= e^{2a/r_n} \cdot r_n^{3/2}
\end{aligned}
$$

$$\begin{aligned}
&= e^{2\sqrt{a(n+1)}} \cdot \left(\sqrt{\frac{a}{n+1}}\right)^{3/2} \\
&\stackrel{*}{=} e^{2\sqrt{a(n+1)}}/(n+1)^{3/4} \\
&\sim e^{2\sqrt{an}}/n^{3/4}.
\end{aligned}$$

Consequently

$$[z^n]e^{a/(1-\beta z)} = \beta^n [z^n]e^{a/(1-z)} \stackrel{*}{\sim} \beta^n e^{2\sqrt{an}}/n^{3/4},$$

so

$$[z^n]e^{a/(1-\beta z)} \in \mathsf{RT}_\rho.$$

This means Schur's Tauberian Theorem can be applied, as expressed in (3.42.1), to (5.18.4) to show that

$$\begin{aligned}
[z^n]e^{\mathbf{P}(z)} &\stackrel{*}{\sim} [z^n]e^{a/(1-\beta z)} \\
&\stackrel{*}{\sim} \beta^n e^{2\sqrt{an}}/n^{3/4}
\end{aligned}$$

as $\rho = 1/\beta$ and

$$\begin{aligned}
e^{a/(1-\beta z)} &\in \mathsf{RT}_\rho \\
\rho_\varepsilon &\geq 1/\eta > \rho.
\end{aligned}$$

Then it follows that $e^{\mathbf{P}(z)} \in \mathsf{RT}_\rho$, so Proposition 3.46 can be applied to obtain

$$\begin{aligned}
a(n) &\stackrel{*}{\sim} [z^n]e^{\mathbf{P}(z)} \\
&\stackrel{*}{\sim} \beta^n e^{2\sqrt{an}}/n^{3/4}.
\end{aligned}$$

It follows that $a(n) = \beta^n a_0(n)$, with $a_0(n) \sim Cn^{-3/4}e^{2\sqrt{an}}$, where C is a suitable positive constant. Now apply Corollary 5.15. □

5.5.1. Two-colored linear forests. In both examples (posets, graphs) of these forests one has, for $a = 1/2$,

$$p(n) = a2^n + \mathrm{O}\big(\sqrt{2}^{\,n}\big).$$

So, Theorem 5.17 guarantees that any partition set B has asymptotic density that equals the Dirichlet density. Finally, the computations first presented in §2.3.2, for the probability that a randomly chosen finite two-colored linear forest will have at least one component with strictly alternating colors, have been justified.

Consider the probability that a given component of size m appears exactly k times in a randomly chosen two-colored linear forest.[2] Letting p be a component of norm m, the subset B of A whose members have exactly k occurrences of p in their decomposition is described as a partition set by

$$\mathsf{B} \;=\; k\mathsf{p} + \omega(\mathsf{P} \setminus \{\mathsf{p}\}).$$

Thus the generating series of B is given by

$$\begin{aligned}\mathbf{B}(x) &= [k\mathbf{p}](x) \cdot \omega(\mathbf{P} \setminus \{\mathbf{p}\})(x) \\ &= x^{km} \cdot (1 - x^m) \prod_{n \geq 1} (1 - x^n)^{-p(n)} \\ &= x^{km} \cdot (1 - x^m) \mathbf{A}(x).\end{aligned}$$

Since the asymptotic density of B is given by its Dirichlet density,

$$\begin{aligned}\delta(\mathsf{B}) &= \lim_{x \to 1/2} \frac{\mathbf{B}(x)}{\mathbf{A}(x)} \\ &= \lim_{x \to 1/2} x^{km}(1 - x^m) \\ &= 2^{-km}(1 - 2^{-m}),\end{aligned}$$

and this is the desired probability. Similarly, the probability that a given component of size m appears at least k times is 2^{-km}.

5.5.2. Further results. Having the radius of convergence ρ of $\mathbf{A}(z)$ strictly between 0 and 1 seems to offer simpler possibilities to analyze $a(n)$ than does the case that $\rho = 1$, the reason being that if $\rho = 1$ then all the terms $\mathbf{P}(z^m)$ also have radius of convergence 1. Then, not only can $\mathbf{P}(z)$ contribute a pole at 1, these other terms contribute to this pole as well and also distribute the singularity around the unit circle at all the roots of unity. This was the challenge faced by Hardy and Ramanujan in the analysis of the classical partition problem where $p(n) = 1$, for all n.

The analysis of this section succeeded because $e^{\mathbf{P}(z)}$ contributed the dominant term to the asymptotics, and the asymptotics of the coefficients $[z^n]e^{\mathbf{P}(z)}$ could be found because $\mathbf{P}(z)$ has a simple pole at ρ.[3] It would

[2] In the case of equivalence relations, etc., considered in Chapter 4, the probability is 0 since the probability that a component appears at least k times was proved to be 1.

[3] The method used is closely related to the famous 'saddle point' method—for modern surveys of asymptotics see Bender [**9**] and Odlyzko [**39**].

seem quite reasonable to extend this result to higher-order poles[4] by finding asymptotics for the coefficients of the power series expansion of $e^{a/(1-z)^\nu}$.

Indeed, these asymptotics have been found by Knopfmacher and Warlimont [**33**], when $a, \nu > 0$, using Hayman's work on admissible functions. They show, for $a, \nu > 0$ and $\beta > 1$, with $\rho = 1/\beta$, that

$$e^{a/(1-\beta z)^\nu} \in \mathsf{RT}_\rho.$$

More recently, Bell and Burris [**8**] have announced that $p(n) \in \mathsf{RT}_\rho$ and $\liminf_n np(n)\rho^n > 1$ suffice to prove that all partition sets of $\mathcal{A}$ have asymptotic density. This goes well beyond the additive number systems covered by Theorem 5.17.

For an important variation on Compton's results of this chapter, see Appendix E for results of Woods.

Problem 5.19. Extend the class of additive number systems $\mathcal{A}$ with $\rho < 1$ for which one can show that all partition sets of $\mathcal{A}$ have asymptotic density.

Problem 5.20. Find necessary and sufficient conditions on two additive number systems $\mathcal{A}_1$ and $\mathcal{A}_2$ for all partition sets of $\mathcal{A}_1 + \mathcal{A}_2$ to have asymptotic density.

---

[4]According to Knopp, "Theory and Application of Infinite Series", Dover, 1990, p. 547, Perron found asymptotics for $[s^n]e^{\alpha/(1-s)^\rho}$, for $\alpha > 0$ and $\rho > 0$. This seems to be incorrect. Perron (1914, [**42**]) actually analyzed

$$[s^n]\Phi(s) \cdot \frac{e^{\alpha/(1-s)}}{(1-s)^\rho} \cdot \left(\log \frac{1}{1-s}\right)^k,$$

where $\Phi(s)$ is regular, for $|s| \le 1$, α and ρ are arbitrary complex numbers, and k is a nonnegative integer. He was generalizing a 1909 result of Fejér that treated the case $\alpha < 0$ and ρ real, and without the log factor above. In 1921, in [**43**], Perron was able to improve the error estimates in the cases

$$[s^n]\frac{e^{\alpha/(1-s)}}{(1-s)^\rho}$$

by applying his asymptotic results on hypergeometric series.

CHAPTER 6

Monadic Second-Order Limit Laws

This chapter is devoted to applying the previous results to obtain monadic second-order limit laws for certain classes $\mathcal{K}$ of graphs, posets, etc.

6.1. First-order logic

To construct a *first-order*[1] *logic*, start with a *language* L which consists of a set $\mathcal{R}$ of *relation symbols*, a set $\mathcal{F}$ of *function symbols*, a set $\mathcal{C}$ of *constant symbols*, and the *equality symbol* $\approx$. Each relation symbol has a positive integer assigned to it, called the *arity* of the relation. The arity of the relation symbol specifies how many arguments it takes. Likewise, each function symbol has an *arity*.

Definition 6.1. An L-*structure* $\mathbf{S}$ consists of a set S, the *universe* of $\mathbf{S}$, with an assignment of the constant symbols to elements of S, an assignment of the n-ary relation symbols to n-ary relations on S, that is, to subsets of S^n, and an assignment of the n-ary function symbols to n-ary functions on S, that is, to functions from S^n to S.

Thus an L-structure $\mathbf{S}$ consists of an appropriate *interpretation* of the symbols of L in the universe S. For example, when working with ordered fields a popular choice of language would be $\mathsf{L} = \{+, \cdot, -, \leq, 0, 1\}$, and an ordered field would be an L-structure $\mathbf{S} = (S, +, \cdot, -, \leq, 0, 1)$ that satisfies the commutative, associative, etc., laws.[2]

In this part, on additive number systems, the focus is on first-order *relational* languages L, meaning that there are no function symbols. For the results on asymptotic density, the languages are narrowed even further, to *purely* relational languages, meaning that constant symbols are not permitted either. However, to carry out induction proofs it is convenient to work with relational languages that include constant symbols.

To describe the first-order logic for a language L, in addition to the symbols of L the following are needed:

[1]The phrase *first-order* refers to the fact that the variables are thought of as ranging over the *elements* of a structure. If there are variables that range over the subsets, or relations, or functions on the elements of a structure, then one speaks of a *second-order* logic. One can also have variables ranging over sets of sets, etc. All these possibilities belong to the *higher-order* logics.

[2]These laws can be expressed as first-order sentences in this language. For example, $\forall x\,((x \not\approx 0) \rightarrow \exists y\,(x \cdot y \approx 1))$ expresses the fact that any nonzero element has a multiplicative inverse.

- symbols for *propositional connectives*, say $\neg$ (not), $\wedge$ (and), $\vee$ (or), $\rightarrow$ (implies), $\leftrightarrow$ (iff) ;
- the *quantifier* symbols $\forall$ (for all) and $\exists$ (there exists) ;
- a set $\mathcal{X}$ of *first-order variables.*

Definition 6.2. Let L be a relational language. Then the *first-order* L*-formulas* are defined as follows, by induction:

- the *atomic* L*-formulas* are expressions of the form
  $R(\mu_1, \ldots, \mu_n)$, for R an n-ary relation symbol; or
  $(\mu_1 \approx \mu_2)$,
  where each μ_i is either a variable or a constant symbol;
- if φ and ψ are first-order L-formulas then so are

$$(\neg\varphi) \qquad (\varphi \vee \psi) \qquad (\varphi \wedge \psi) \qquad (\varphi \rightarrow \psi) \qquad (\varphi \leftrightarrow \psi);$$

- if φ is a first-order L-formula then so are $(\forall x\, \varphi)$ and $(\exists x\, \varphi)$.

It is assumed that the reader is familiar with the notions of *free* and *bound* variables in a formula. For example, the occurrence of x in $(\exists y(x < y))$ is free, those of y are bound.

Definition 6.3. A *sentence* is a formula with no free variables. For φ a sentence, the notation $\mathbf{S} \models \varphi$, read $\mathbf{S}$ *satisfies* φ, means that φ *holds*, or *is true*, in $\mathbf{S}$. Also, for a formula $\varphi(\vec{x})$ and elements $\vec{s}$ of S, the notation $\mathbf{S} \models \varphi(\vec{s})$, read $\mathbf{S}$ *satisfies* $\varphi(\vec{s})$, or $\vec{s}$ *satisfies* φ in $\mathbf{S}$, means that $\varphi(\vec{x})$ is true in a structure $\mathbf{S}$ when the free variables are interpreted as the elements $\vec{s}$ of S.

6.2. Monadic second-order logic

One can extend a first-order logic with language L to a *monadic*[3] *second-order logic* with language L by adding

- a set $\mathcal{U}$ of *monadic second-order variables.*

In the case of a relational language L, the *monadic second-order* L*-formulas* are obtained by augmenting the inductive definition of first-order L-formulas by:

- $U(\mu)$ is an atomic L-formula, for any $U \in \mathcal{U}$ and $\mu \in \mathcal{X} \cup \mathcal{C}$;
- if φ is a monadic second-order L-formula then so are $(\forall U\, \varphi)$ and $(\exists U\, \varphi)$.

This chapter is concerned with monadic second-order logic. The reader can obtain the corresponding results for first-order logic by simply deleting all references to second-order variables.

[3] The adjective *monadic* means that one is dealing with variables for *unary* predicates, that is, predicates with only one argument. Such a predicate is often thought of in terms of its extension, that is, as the *set of elements* that satisfy the predicate. In monadic second-order logic one does not have, for example, variables for binary predicates, that is, one cannot quantify over the binary relations on S.

The notation $\varphi(x_1, \ldots, x_m, U_1, \ldots, U_n)$, abbreviated as $\varphi(\vec{x}, \vec{U})$, means that the free variables of the formula φ are among the first-order variables $x_1, \ldots, x_m$ and the second-order variables $U_1, \ldots, U_n$.

The notation $\mathbf{S} \models \varphi$ is used as before, to express the fact that a sentence φ holds in $\mathbf{S}$, and $\mathbf{S} \models \varphi(\vec{s}, \vec{Q})$ means that the elements $\vec{s}$ and the unary predicates $\vec{Q}$ satisfy the formula $\varphi(\vec{x}, \vec{U})$ in $\mathbf{S}$.

A slight-of-hand technique will be used in the induction proofs that allow one to decrease the number of free variables by increasing the number of symbols in the language. When writing $(\mathbf{S}, s)$, where $s \in S$, what is meant is a structure for the language L (of $\mathbf{S}$) augmented by a new constant symbol c, and that the interpretation of c in $\mathbf{S}$ is the element s. By doing this one has, for an L-formula $\varphi(x)$,

$$\mathbf{S} \models \varphi(s) \quad \text{iff} \quad (\mathbf{S}, s) \models \varphi(s),$$

where the $\varphi(s)$ on the right is treated as a *sentence* in the language $\mathsf{L} \cup \{c\}$ of $(\mathbf{S}, s)$. Likewise, when writing $(\mathbf{S}, Q)$, where Q is a unary relation on S, what is meant is a structure for the language L of $\mathbf{S}$ augmented by a new unary relation symbol R, and that the interpretation of R on S is the relation Q. By doing this one likewise has, for an L-formula $\varphi(U)$,

$$\mathbf{S} \models \varphi(Q) \quad \text{iff} \quad (\mathbf{S}, Q) \models \varphi(Q),$$

where the $\varphi(Q)$ on the right is treated as a *sentence* in the language $\mathsf{L} \cup \{R\}$ of $(\mathbf{S}, Q)$.

6.2.1. Two-colored linear forests. In the previous chapters the probability that a randomly chosen finite two-colored linear forest has at least one component with strictly alternating colors was discussed. Now, it will be shown that this property can be expressed in monadic second-order logic—so this is an example of the kind of property that can be proved to have a probability of holding by the main result of Compton from this chapter, Theorem 6.29. The case of two-colored linear forests as graphs is given below. The case of posets is handled similarly.

Let us start by introducing a number of abbreviations:

Abbreviation	Definition
$x \in U$	Ux
$U \not\approx \varnothing$	$\exists x\, Ux$
$U \subseteq V$	$\forall x\, (Ux \rightarrow Vx)$
$U \approx V \cup W$	$\forall x\, \big(Ux \leftrightarrow (Vx \vee Wx)\big)$
$\mathrm{Sep}(U, V)$	$\forall x \forall y\, \big((Ux \wedge Vy) \rightarrow \neg(xRy) \wedge \neg(yRx)\big)$
$\mathrm{Split}(U)$	$\exists V \exists W \Big[(U \approx V \cup W) \wedge (V \not\approx \varnothing) \wedge (W \not\approx \varnothing) \wedge (V \cap W \approx \varnothing) \wedge \mathrm{Sep}(V, W)\Big]$
$\mathrm{SameCol}(x, y)$	$Gx \leftrightarrow Gy$
$\mathrm{AltCol}(U)$	$\forall x \forall y \Big[\big((x \in U) \wedge (y \in U) \wedge (xRy)\big) \rightarrow \neg\,\mathrm{SameCol}(x, y)\Big]$

Looking at the abbreviations, the first four express what is indicated on the left. Sep(U,V) says that the relation R does not hold between members of U and members of V. Split(U) says that U is not connected. SameCol(x, y) says that x and y have the same color (blue or gold). AltCol(U) says that related elements of U have different colors.

From this a monadic second-order sentence can be put together, namely

$$\exists U \Big[U \not\approx \varnothing \wedge \neg\,\mathrm{Split}(U) \wedge \mathrm{AltCol}(U) \wedge \forall V \big(U \cap V \approx \varnothing \rightarrow \mathrm{Sep}(U,V) \big) \Big],$$

which says that there is a subset U of the forest that is a strictly alternately colored component.

6.3. Asymptotic density of subsets of $\mathcal{K}$

Definition 6.4. For $\mathcal{H}$ a class of finite structures, and $n \geq 0$, let

$$\mathcal{H}_{(n)} \;=\; \{\mathbf{S} \in \mathcal{H} : |S| = n\},$$

the subclass of structures of size n in $\mathcal{H}$; and let

$$\mathcal{H}_{(\leq n)} \;=\; \{\mathbf{S} \in \mathcal{H} : |S| \leq n\},$$

the subclass of structures of size at most n in $\mathcal{H}$.

Definition 6.5. For $\mathcal{H}$ a class of structures, let $\widetilde{\mathcal{H}}$ be the set of *isomorphism types* $\widetilde{\mathbf{S}}$ of members $\mathbf{S}$ of $\mathcal{H}$.

Definition 6.6. For $\mathcal{H} \subseteq \mathcal{K}$, the *local asymptotic density* and the *global asymptotic density* of $\mathcal{H}$ in $\mathcal{K}$ (if they exist) are given by

$$\delta(\mathcal{H}) \;=\; \lim_{\substack{n\to\infty \\ \mathcal{K}_{(n)} \neq \varnothing}} \frac{|\widetilde{\mathcal{H}_{(n)}}|}{|\widetilde{\mathcal{K}_{(n)}}|}$$

$$\Delta(\mathcal{H}) \;=\; \lim_{n\to\infty} \frac{|\widetilde{\mathcal{H}_{(\leq n)}}|}{|\widetilde{\mathcal{K}_{(\leq n)}}|}.$$

The existence of such asymptotic densities has been a favorite topic of finite model theory, where the goal is to find interesting logics Λ (such as first-order and monadic second-order logic) and corresponding interesting classes of structures $\mathcal{K}$ such that, for any Λ-sentence φ, the subclass of $\mathcal{K}$ defined by φ, namely

$$\mathcal{K}_{\varphi} \;=\; \{\mathbf{S} \in \mathcal{K} : \mathbf{S} \models \varphi\},$$

has an asymptotic density. In this situation, $\mathcal{K}$ is said to have a Λ *limit law.* If the asymptotic densities of the $\mathcal{K}_{\varphi}$ exist and are always 0 or 1 then one speaks of a Λ 0–1 *law.* The favorite choice of Λ is *first-order logic.*

This subject started when Glebskij, Logan, Liogonkij and Talanov (1969), and independently Fagin (1976) (see [**23**], p. 93), showed that *graphs have a local first-order 0–1 law.* The basic idea of the proof is to guess the φ such that $\mathcal{K}_{\varphi}$ has density 1, the correct guess being the sentences true of the countable random graph. Then one shows that $\mathcal{K}_\varphi$ does indeed have density 1, for φ in a natural basis for the sentences true of the random graph.

This method is rather ad hoc, and can be applied to only a few classes. In the late 1980s Kevin Compton showed how to use the results developed in previous chapters to obtain monadic second-order limit laws for classes $\mathcal{K}$ of finite purely relational structures that are closed under components as well as unions of components, and are such that $\mathcal{K}$ has a suitable growth. The purpose of this chapter is to give a detailed treatment of the logical aspects of Compton's study.

6.4. Quantifier rank and equivalent formulas

Each monadic second-order formula has a visual description by a tree, for example, consider the formula

$$\forall x \left[Q(x,y) \rightarrow \exists U \left(U(y) \wedge \neg U(x)\right)\right].$$

The tree of this formula is shown in Figure 1.

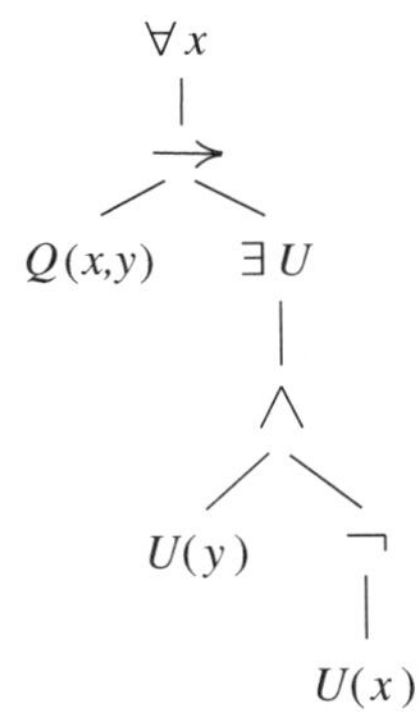

FIGURE 1

The quantifier rank of φ, $\mathrm{qr}(\varphi)$, is the maximum number of quantifiers that occur in a path in the tree of φ. In the above example the quantifier rank is 2.

Definition 6.7. Given a language L the *quantifier rank* of a formula is defined inductively by:

- $\mathrm{qr}(\varphi) = 0$, for φ an atomic formula,
- $\mathrm{qr}(\neg\varphi) = \mathrm{qr}(\varphi)$,
- $\mathrm{qr}(\varphi \,\square\, \psi) = \max\left(\mathrm{qr}(\varphi), \mathrm{qr}(\psi)\right)$, for $\square$ any binary propositional connective, and
- $\mathrm{qr}(\mathsf{Q}\,\mu\,\varphi) = 1 + \mathrm{qr}(\varphi)$, where Q is a quantifier and μ a variable.

Definition 6.8. For L a relational language,

- $\Phi(x_1, \ldots, x_m, U_1, \ldots, U_n)$ is the set of L-formulas of the form

$$\varphi(x_1, \ldots, x_m, U_1, \ldots, U_n),$$

- and Φ_r is the set of L-formulas with quantifier rank at most r.

Two L-formulas in $\Phi(x_1, \ldots, x_m, U_1, \ldots, U_n)$, say

$$\varphi(x_1, \ldots, x_m, U_1, \ldots, U_n) \quad \text{and} \quad \psi(x_1, \ldots, x_m, U_1, \ldots, U_n),$$

are *equivalent*, written $\varphi \equiv \psi$, if, for every L-structure $\mathbf{S}$, they are satisfied by the same tuples $(s_1, \ldots, s_m, Q_1, \ldots, Q_n)$. Of course this is an equivalence relation on $\Phi(x_1, \ldots, x_m, U_1, \ldots, U_n)$.

Lemma 6.9. *Given a finite relational language* L*, for each choice of* m, n, r*, there are only finitely many equivalence classes in the set of formulas* $\Phi_r \cap \Phi(x_1, \ldots, x_m, U_1, \ldots, U_n)$.

PROOF. It will be proved that the lemma holds, for any choice of m, n, by induction on r.

For the Ground Case, $r = 0$, note that every formula in $\Phi_0 \cap \Phi(\vec{x}, \vec{U})$ is equivalent to a Boolean combination of atomic formulas in $\Phi_0 \cap \Phi(\vec{x}, \vec{U})$. But there are only finitely many atomic formulas in $\Phi_0 \cap \Phi(\vec{x}, \vec{U})$ since the language is finite and relational. Thus there are only finitely many equivalence classes in $\Phi_0 \cap \Phi(\vec{x}, \vec{U})$.

The Induction Hypothesis is that, for any choice of m, n, there are only finitely many equivalence classes in $\Phi_r \cap \Phi(\vec{x}, \vec{U})$. Now, the goal is to show that the same holds replacing r by $r + 1$.

Observe that any formula in $\Phi_{r+1} \cap \Phi(\vec{x}, \vec{U})$ is equivalent to a Boolean combination of formulas that are in one of the forms

$$\begin{array}{l} \psi(\vec{x}, \vec{U}) \\ \exists y\, \psi(y, \vec{x}, \vec{U}) \\ \exists V\, \psi(V, \vec{x}, \vec{U}), \end{array}$$

where $\mathrm{qr}(\psi) \leq r$. By the induction hypothesis there are only finitely many equivalence classes in $\Phi_r \cap \Phi(y, \vec{x}, V, \vec{U})$, and thus there are only finitely many inequivalent formulas in the above three forms. Thus there are only finitely many equivalence classes in $\Phi_{r+1} \cap \Phi(\vec{x}, \vec{U})$. □

Remark 6.10. The above lemma fails if there are function symbols in L, for example, if f is a unary function symbol in L then the following atomic formulas in $\Phi_0 \cap \Phi(x)$ are pairwise inequivalent:

$$\begin{array}{rcl} f(x) & \approx & x \\ f(f(x)) & \approx & x \\ f(f(f(x))) & \approx & x \\ & \textit{etc.} & \end{array}$$

Definition 6.11. Given two L-structures $\mathbf{S}_1$ and $\mathbf{S}_2$,

$$\mathbf{S}_1 \equiv_n \mathbf{S}_2$$

means that $\mathbf{S}_1$ and $\mathbf{S}_2$ satisfy precisely the same sentences with quantifier rank at most n.

Lemma 6.12. *Given a finite relational language* L *there are, for each* n, *only finitely many* $\equiv_n$ *equivalence classes of* L*-structures.*

PROOF. By Lemma 6.9 there are only finitely many equivalence classes of monadic second-order L-sentences of quantifier rank at most n. Let $\varphi_1, \ldots, \varphi_k$ be a complete set of representatives of these equivalence classes. Then the $\equiv_n$ classes are each defined by a subset of $\varphi_1, \ldots, \varphi_k$, and thus there are at most 2^k $\equiv_n$ equivalence classes of L-structures. □

6.5. Ehrenfeucht-Fraissé games

Next the notion of an Ehrenfeucht-Fraissé game of length n is described. This gives a valuable aid to understanding $\equiv_n$. Given two L-structures $\mathbf{S}_1$ and $\mathbf{S}_2$ the game $\Gamma_n(\mathbf{S}_1, \mathbf{S}_2)$ is played by two players, Player I and Player II. First take the collection G_0 of pairs $(c^{\mathbf{S}_1}, c^{\mathbf{S}_2})$ obtained as the interpretation of the constant symbols c in $\mathbf{S}_1$, respectively $\mathbf{S}_2$. Then Player I is allowed to choose

- an element $s_i \in S_i$, where i is 1 or 2; or
- a unary relation $Q_i \subseteq S_i$, where i is 1 or 2.

In the first case Player II must choose an element s_j from the other S_j, and in the second case a subset Q_j from the other S_j. In the first case, respectively the second case, the pair (s_1, s_2), respectively (Q_1, Q_2), constitutes a *round of play*. This continues for n rounds. The n rounds constitute a *play of the game* $\Gamma_n(\mathbf{S}_1, \mathbf{S}_2)$. (See Figure 2.)

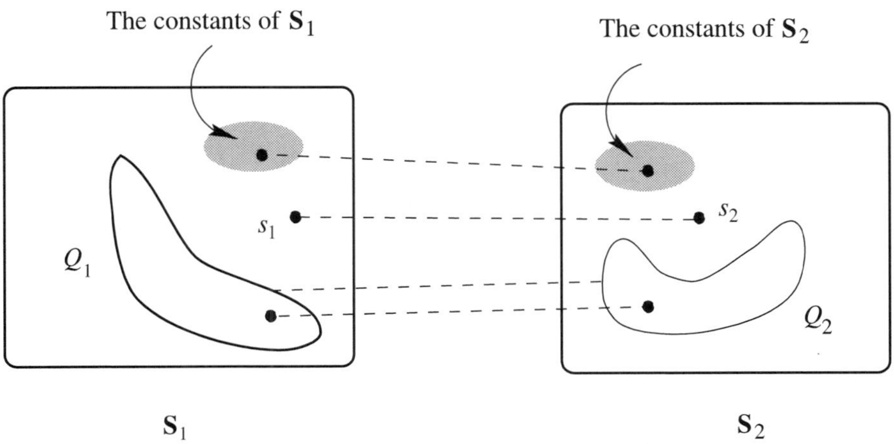

FIGURE 2. Some Rounds of a Game

At the end all the pairs from $S_1 \times S_2$ that are either in G_0 or resulted from a round of play constitute the set G; and all the pairs from $\mathrm{Su}(S_1) \times \mathrm{Su}(S_2)$, where $\mathrm{Su}(S)$ is the set of subsets of S that resulted from a round of play constitute the set H.

After the kth round in a play of the game consider the subsets G_k and H_k of pairs from G, respectively H, obtained so far. Let $\mathsf{L}^{[k]}$ be the language

L augmented by as many new unary relation symbols R_π as there have been rounds $\pi = (Q_1, Q_2)$ played that involve unary predicates. Let $\mathbf{S}_i^{[k]}$ be the expansions of the $\mathbf{S}_i$ to this augmented language, where R_π, for $\pi = (Q_1, Q_2)$, is interpreted in $\mathbf{S}_i^{[k]}$ as Q_i. Let $\mathbf{T}_1^{[k]}$ be the substructure of $\mathbf{S}_1^{[k]}$ whose universe is the domain of the relation G_k; and let $\mathbf{T}_2^{[k]}$ be the substructure of $\mathbf{S}_2^{[k]}$ whose universe is the range of the relation G_k.

In a given play of the game, Player II is *successful for k rounds* in a given play of the game $\Gamma_n(\mathbf{S}_1, \mathbf{S}_2)$ if, after playing k rounds, G_k is an isomorphism between the $\mathsf{L}^{[k]}$-structures $\mathbf{T}_1^{[k]}$ and $\mathbf{T}_2^{[k]}$. If so, then $G_k \cup H_k$ is called a *partial isomorphism* between $\mathbf{S}_1$ and $\mathbf{S}_2$.

If Player II is successful after n rounds in a given play of $\Gamma_n(\mathbf{S}_1, \mathbf{S}_2)$ then Player II *wins* the given play of the game. Player II has a *winning strategy* for the game $\Gamma_n(\mathbf{S}_1, \mathbf{S}_2)$ if, for any choice of moves by Player I, it is always possible for Player II to win.

$\mathbf{S}_1 \sim_n \mathbf{S}_2$ means Player II has a winning strategy for the game $\Gamma_n(\mathbf{S}_1, \mathbf{S}_2)$. Clearly, $\sim_n$ is reflexive and symmetric. Of course, if Player II has a winning strategy for the game $\Gamma_{n+1}(\mathbf{S}_1, \mathbf{S}_2)$ then Player II also has a winning strategy for the game $\Gamma_n(\mathbf{S}_1, \mathbf{S}_2)$.

Lemma 6.13. *Let L be a relational language and let $\mathbf{S}_1$ and $\mathbf{S}_2$ be two L-structures. Suppose $\mathbf{S}_1 \sim_{n+1} \mathbf{S}_2$.*

(a) *For any $s_1 \in S_1$, there is an $s_2 \in S_2$ such that*

$$(\mathbf{S}_1, s_1) \sim_n (\mathbf{S}_2, s_2),$$

and vice-versa. The two structures $(\mathbf{S}_1, s_1)$ and $(\mathbf{S}_2, s_2)$ are considered structures for an augmented language $\mathsf{L} \cup \{c\}$, where c is a new constant symbol, and the interpretation of c in $\mathbf{S}_i$ is s_i, for $i = 1, 2$.

(b) *For any $Q_1 \subseteq S_1$, there is a $Q_2 \subseteq S_2$ such that*

$$(\mathbf{S}_1, Q_1) \sim_n (\mathbf{S}_2, Q_2),$$

and vice-versa. The two structures $(\mathbf{S}_1, Q_1)$ and $(\mathbf{S}_2, Q_2)$ are considered structures for an augmented language $\mathsf{L} \cup \{R\}$, where R is a new unary relation symbol, and the interpretation of R in $\mathbf{S}_i$ is Q_i, for $i = 1, 2$.

PROOF. (a): Let Player I play his first move as s_1. Then let Player II play his winning strategy for the game $\Gamma_{n+1}(\mathbf{S}_1, \mathbf{S}_2)$ and choose s_2 from S_2. Then no matter how Player I plays his next n moves Player II has a sequence of moves that yields a partial isomorphism after each round. But this means that, by starting with the structures $(\mathbf{S}_1, s_1)$ and $(\mathbf{S}_2, s_2)$, the correspondence between the constants is a partial isomorphism, that is, $(\mathbf{S}_1, s_1) \sim_0 (\mathbf{S}_2, s_2)$, and Player II can win the game $\Gamma_n\big((\mathbf{S}_1, s_1), (\mathbf{S}_2, s_2)\big)$. Thus $(\mathbf{S}_1, s_1) \sim_n (\mathbf{S}_2, s_2)$.

(b): Let Player I play his first move as Q_1. Then let Player II play his winning strategy for the game $\Gamma_{n+1}(\mathbf{S}_1, \mathbf{S}_2)$ and choose a subset Q_2 of S_2. Then no matter how Player I plays his next n moves Player II has a sequence of moves that yields a partial isomorphism after each round. But this means that, by starting with the structures $(\mathbf{S}_1, Q_1)$ and $(\mathbf{S}_2, Q_2)$, the correspondence between the constants is a partial isomorphism, that is, $(\mathbf{S}_1, Q_1) \sim_0 (\mathbf{S}_2, Q_2)$, and Player II can win the game $\Gamma_n\big((\mathbf{S}_1, Q_1), (\mathbf{S}_2, Q_2)\big)$. Thus $(\mathbf{S}_1, Q_1) \sim_n (\mathbf{S}_2, Q_2)$. $\square$

The converse of the above lemma also holds.

Lemma 6.14. *Let* L *be a relational language and let* $\mathbf{S}_1$ *and* $\mathbf{S}_2$ *be two* L*-structures.*

(a) *Suppose that for every* $s_1 \in S_1$ *there is an* $s_2 \in S_2$ *such that*

$$(\mathbf{S}_1, s_1) \ \sim_n \ (\mathbf{S}_2, s_2);$$

and suppose the same assertion holds with $\mathbf{S}_1$ *and* $\mathbf{S}_2$ *interchanged.*

(b) *Furthermore, suppose that for every* $Q_1 \subseteq S_1$ *there is a* $Q_2 \subseteq S_2$ *such that*

$$(\mathbf{S}_1, Q_1) \ \sim_n \ (\mathbf{S}_2, Q_2);$$

and suppose the same assertion holds with $\mathbf{S}_1$ *and* $\mathbf{S}_2$ *interchanged.*

Then

$$\mathbf{S}_1 \sim_{n+1} \mathbf{S}_2.$$

PROOF. By (a), for every choice of a first move in the game $\Gamma_{n+1}(\mathbf{S}_1, \mathbf{S}_2)$ by Player I that consists of choosing an element s_i, there is a move by Player II such that, with the pair (s_1, s_2) resulting from the first round, one has $(\mathbf{S}_1, s_1) \sim_n (\mathbf{S}_2, s_2)$. Player II has a winning strategy for the next n moves by playing the winning strategy for the game $\Gamma_n\big((\mathbf{S}_1, s_1), (\mathbf{S}_2, s_2)\big)$.

Likewise, by (b), for every choice of a first move by Player I that consists of choosing a subset Q_i, there is a move by Player II such that, with the pair (Q_1, Q_2) resulting from the first round, one has $(\mathbf{S}_1, Q_1) \sim_n (\mathbf{S}_2, Q_2)$. Player II has a winning strategy for the next n moves by playing the winning strategy for the game $\Gamma_n\big((\mathbf{S}_1, Q_1), (\mathbf{S}_2, Q_2)\big)$.

Thus Player II has a winning strategy for the game $\Gamma_{n+1}(\mathbf{S}_1, \mathbf{S}_2)$. $\square$

Lemma 6.15. *Let* L *be a relational language and let* $\mathbf{S}_1$ *and* $\mathbf{S}_2$ *be two* L*-structures. Then*

$$\mathbf{S}_1 \sim_n \mathbf{S}_2 \quad \textit{implies} \quad \mathbf{S}_1 \equiv_n \mathbf{S}_2.$$

PROOF. This will be proved by induction over n.

The Ground Case is $n = 0$. Let L be a relational language and let $\mathbf{S}_1$ and $\mathbf{S}_2$ be L-structures with $\mathbf{S}_1 \sim_0 \mathbf{S}_2$. This says that the set G_0 of pairs $(c^{\mathbf{S}_1}, c^{\mathbf{S}_2})$ obtained as the interpretation of the constant symbols c in $\mathbf{S}_1$, respectively $\mathbf{S}_2$, gives a partial isomorphism. Now, let φ be a sentence with $\mathrm{qr}(\varphi) = 0$, that is, a sentence with no quantifiers. Then this sentence is equivalent to a Boolean combination of atomic sentences. By the partial isomorphism G_0 it is seen that $\mathbf{S}_1$ and $\mathbf{S}_2$ satisfy the same atomic sentences, and thus they both satisfy φ or they both satisfy $\neg\,\varphi$. Thus $\mathbf{S}_1 \equiv_0 \mathbf{S}_2$.

The Induction Hypothesis is that, for any relational language L and any L-structures $\mathbf{S}_1, \mathbf{S}_2$, the condition $\mathbf{S}_1 \sim_n \mathbf{S}_2$ implies $\mathbf{S}_1 \equiv_n \mathbf{S}_2$.

The Induction Step is to assume that a relational language L and two L-structures $\mathbf{S}_1$ and $\mathbf{S}_2$ are given with $\mathbf{S}_1 \sim_{n+1} \mathbf{S}_2$. The goal now is to show that if φ is any L-sentence with $\mathrm{qr}(\varphi) = n+1$ then either both or neither of the structures satisfy φ. Since a universal quantifier $\forall$ can be replaced by $\neg\exists\neg$, φ is equivalent to a Boolean combination of L-sentences of quantifier rank at most $n+1$, sentences that are either of quantifier-rank at most n, or of quantifier rank $n+1$ and of the form $\exists\mu\,\psi(\mu)$, where μ is a variable. The two structures agree on L-sentences of quantifier rank at most n by the induction hypothesis, as $\mathbf{S}_1 \sim_{n+1} \mathbf{S}_2$ implies $\mathbf{S}_1 \sim_n \mathbf{S}_2$. So it suffices to show that they both agree on L-sentences of the form $\exists\mu\,\psi(\mu)$ that are of quantifier rank $n+1$. Let $\varphi = \exists\mu\,\psi(\mu)$ be an L-sentence with quantifier rank $n+1$, and assume $\mathbf{S}_1 \models \varphi$.

If $\mu = x$, a variable for elements, choose an element $s_1 \in S_1$ such that $\mathbf{S}_1 \models \psi(s_1)$. Then

$$(\mathbf{S}_1, s_1) \models \psi(s_1),$$

and the sentence on the right (in the language L augmented by a new constant symbol) has quantifier rank n. By Lemma 6.13 (a) there is an element $s_2 \in S_2$ such that

$$(\mathbf{S}_1, s_1) \sim_n (\mathbf{S}_2, s_2).$$

By the induction hypothesis, $(\mathbf{S}_1, s_1) \equiv_n (\mathbf{S}_2, s_2)$, and thus

$$(\mathbf{S}_2, s_2) \models \psi(s_2).$$

But this means that $\mathbf{S}_2 \models \exists x\,\psi(x)$, that is, $\mathbf{S}_2 \models \varphi$. Likewise, $\mathbf{S}_2 \models \varphi$ implies $\mathbf{S}_1 \models \varphi$.

If $\mu = U$, a variable for unary relations, choose $Q_1 \subseteq S_1$ such that $\mathbf{S}_1 \models \psi(Q_1)$. Then

$$(\mathbf{S}_1, Q_1) \models \psi(Q_1),$$

and the sentence on the right (in the language L augmented by a new unary relation symbol) has quantifier rank n. By Lemma 6.13 (b) there is $Q_2 \subseteq S_2$ such that

$$(\mathbf{S}_1, Q_1) \sim_n (\mathbf{S}_2, Q_2).$$

By the induction hypothesis, $(\mathbf{S}_1, Q_1) \equiv_n (\mathbf{S}_2, Q_2)$, and thus

$$(\mathbf{S}_2, Q_2) \models \psi(Q_2).$$

But this means that $\mathbf{S}_2 \models \exists U\,\psi(U)$, that is, $\mathbf{S}_2 \models \varphi$. Likewise, $\mathbf{S}_2 \models \varphi$ implies $\mathbf{S}_1 \models \varphi$.

Combining the above gives $\mathbf{S}_1 \equiv_{n+1} \mathbf{S}_2$. $\square$

The next lemma uses the observation that if $\varphi(x)$ and $\psi(x)$ are two L-formulas and c is a new constant symbol, then $\varphi(x) \equiv \psi(x)$ holds iff $\varphi(c) \equiv \psi(c)$. Likewise, if $\varphi(U)$ and $\psi(U)$ are two L-formulas and R is a new unary predicate symbol, then $\varphi(U) \equiv \psi(U)$ iff $\varphi(R) \equiv \psi(R)$.

Lemma 6.16. *Let L be a finite relational language and let $\mathbf{S}_1$ and $\mathbf{S}_2$ be L-structures. Then, for any n,*

$$\mathbf{S}_1 \equiv_n \mathbf{S}_2 \quad \textit{implies} \quad \mathbf{S}_1 \sim_n \mathbf{S}_2.$$

PROOF. The proof is by induction on n.

The Ground Case is $n = 0$. Suppose L is a finite relational language and $\mathbf{S}_1$ and $\mathbf{S}_2$ are two L-structures such that $\mathbf{S}_1 \equiv_0 \mathbf{S}_2$. This means that, for any quantifier-free sentence φ, either φ is true for both $\mathbf{S}_1$ and $\mathbf{S}_2$, or false for both. Thus the substructures of $\mathbf{S}_i$, $i = 1, 2$, whose universes are the interpretations of the constant symbols, are isomorphic. This means $\mathbf{S}_1 \sim_0 \mathbf{S}_2$.

The Induction Hypothesis is that, for any finite relational language L, and any two L-structures $\mathbf{S}_1$ and $\mathbf{S}_2$,

$$\mathbf{S}_1 \equiv_n \mathbf{S}_2 \quad \text{implies} \quad \mathbf{S}_1 \sim_n \mathbf{S}_2.$$

The Induction Step is to assume that a finite relational language L and two L-structures $\mathbf{S}_1$ and $\mathbf{S}_2$ are given that satisfy

$$\mathbf{S}_1 \equiv_{n+1} \mathbf{S}_2.$$

First choose any element $s_1 \in S_1$. By Lemma 6.9 there are only finitely many equivalence classes of L-formulas in $\Phi_n \cap \Phi(x)$, so select a representative $\varphi_i(x)$ from each class. Now, let $\varphi_1(x), \ldots, \varphi_k(x)$ be those representatives such that

$$\mathbf{S}_1 \models \varphi_i(s_1).$$

Then $\mathbf{S}_1 \models \exists x \Big(\varphi_1(x) \wedge \cdots \wedge \varphi_k(x)\Big)$. Since $\mathbf{S}_1 \equiv_{n+1} \mathbf{S}_2$ it follows that

$$\mathbf{S}_2 \models \exists x \Big(\varphi_1(x) \wedge \cdots \wedge \varphi_k(x)\Big),$$

so choose $s_2 \in S_2$ such that $\mathbf{S}_2 \models \varphi_1(s_2) \wedge \cdots \wedge \varphi_k(s_2)$. Then

$$(\mathbf{S}_1, s_1) \equiv_n (\mathbf{S}_2, s_2)$$

since, for $\varphi(x)$ an L-formula of quantifier rank at most n, $(\mathbf{S}_1, s_1) \models \varphi(s_1)$ iff $\varphi(x)$ is equivalent to one of the $\varphi_i(x)$, and then $(\mathbf{S}_2, s_2) \models \varphi(s_2)$; furthermore $(\mathbf{S}_1, s_1) \not\models \varphi(s_1)$ iff $\neg\varphi(x)$ is equivalent to one of the $\varphi_i(x)$, and then $(\mathbf{S}_2, s_2) \not\models \varphi(s_2)$. Thus the $(\mathbf{S}_i, s_i)$ satisfy the same sentences of quantifier rank at most n, as claimed. From the induction hypothesis follows

$$(\mathbf{S}_1, s_1) \sim_n (\mathbf{S}_2, s_2).$$

Likewise, for $s_2 \in S_2$ there is an $s_1 \in S_1$ that makes this last assertion true.

Next choose any subset $Q_1 \subseteq S_1$. By Lemma 6.9 there are only finitely many equivalence classes of L-formulas in $\Phi_n \cap \Phi(U)$, so select a representative $\varphi_i(U)$ from each class. Now, let $\varphi_1(U), \ldots, \varphi_k(U)$ be those representatives such that

$$\mathbf{S}_1 \models \varphi_i(Q_1).$$

Then $\mathbf{S}_1 \models \exists U \left(\varphi_1(U) \wedge \cdots \wedge \varphi_k(U)\right)$. Since $\mathbf{S}_1 \equiv_{n+1} \mathbf{S}_2$ it follows that

$$\mathbf{S}_2 \models \exists U \left(\varphi_1(U) \wedge \cdots \wedge \varphi_k(U)\right),$$

so choose $Q_2 \subseteq S_2$ such that $\mathbf{S}_2 \models \varphi_1(Q_2) \wedge \cdots \wedge \varphi_k(Q_2)$. Then

$$(\mathbf{S}_1, Q_1) \equiv_n (\mathbf{S}_2, Q_2)$$

since, for $\varphi(U)$ an L-formula of quantifier rank at most n, $(\mathbf{S}_1, Q_1) \models \varphi(Q_1)$ iff $\varphi(U)$ is equivalent to one of the $\varphi_i(U)$, and then $(\mathbf{S}_2, Q_2) \models \varphi(Q_2)$; furthermore $(\mathbf{S}_1, Q_1) \not\models \varphi(Q_1)$ iff $\neg\varphi(U)$ is equivalent to one of the $\varphi_i(U)$, and then $(\mathbf{S}_2, s_2) \not\models \varphi(Q_2)$. Thus the $(\mathbf{S}_i, Q_i)$ satisfy the same sentences of quantifier rank at most n, as claimed. From the induction hypothesis,

$$(\mathbf{S}_1, Q_1) \sim_n (\mathbf{S}_2, Q_2).$$

Likewise, for $Q_2 \subseteq S_2$ there is a $Q_1 \subseteq S_1$ that makes this last assertion true.

By Lemma 6.14 it follows that $\mathbf{S}_1 \sim_{n+1} \mathbf{S}_2$. □

Combining the two preceding lemmas gives the basic tool that arises from Ehrenfeucht-Fraissé games.

Theorem 6.17. *Let* L *be a finite relational language and let* $\mathbf{S}_1$ *and* $\mathbf{S}_2$ *be* L*-structures. Then*

$$\mathbf{S}_1 \sim_n \mathbf{S}_2 \quad \textit{iff} \quad \mathbf{S}_1 \equiv_n \mathbf{S}_2.$$

By using this theorem the properties of $\equiv_n$ that are needed for the asymptotic density results will be proved. First a simple corollary.

Corollary 6.18. *Given a finite relational language* L *there are, for each* n*, only finitely many* $\sim_n$ *equivalence classes of* L*-structures.*

PROOF. This is immediate from Lemma 6.12 and Theorem 6.17. □

The notion of the disjoint union of two (purely relational) L-structures is intuitively clear—namely, replace the structures by isomorphic copies that have no common elements in their universes, and then take the union of the two universes, with the new relations being the union of the individual relations. One can give a formal definition such as the following.

Definition 6.19. Let L be a purely relational language, and let $\mathbf{S}_1$ and $\mathbf{S}_2$ be two L-structures. Define the *disjoint union* $\mathbf{S} = \mathbf{S}_1 \sqcup \mathbf{S}_2$ by first defining the universe of $\mathbf{S}$ as

$$S_1 \sqcup S_2 = \{(s_1, 1) : s_1 \in S_1\} \cup \{(s_2, 2) : s_2 \in S_2\}.$$

Let π be the mapping from $S_1 \sqcup S_2$ defined by

$$\pi : (s_1, 1) \mapsto s_1$$
$$\pi : (s_2, 2) \mapsto s_2.$$

Then, for R an n-ary relation symbol of L with interpretations Q_i in $\mathbf{S}_i$, let the interpretation of R in $\mathbf{S}$ be given by

$$Q_1 \sqcup Q_2 \;=\; \Big(\pi^{-1}(Q_1) \cap (S_1 \times \{1\})^n\Big) \;\cup\; \Big(\pi^{-1}(Q_2) \cap (S_2 \times \{2\})^n\Big).$$

However, in the discussion below, the more common practice of assuming that $\mathbf{S}_1 \sqcup \mathbf{S}_2$ has been constructed from $\mathbf{S}_1$ and $\mathbf{S}_2$ that have disjoint universes, will be followed. Then the disjoint union is defined to be the union, for example,

$$(S_1, Q_1) \;\sqcup\; (S_2, Q_2) \;=\; (S_1 \cup S_2,\, Q_1 \cup Q_2).$$

Lemma 6.20. *Given a purely relational language* L *and* L*-structures* $\mathbf{S}, \mathbf{S}^\star, \mathbf{S}_i, \mathbf{S}_i^\star$, *the following hold:*

(a) $\mathbf{S} \;\cong\; \mathbf{S}^\star$ *implies* $\mathbf{S} \;\sim_n\; \mathbf{S}^\star$

(b) $\mathbf{S} \sqcup \mathbf{S}^\star \;\sim_n\; \mathbf{S}^\star \sqcup \mathbf{S}$

(c) $\mathbf{S}_1 \sqcup (\mathbf{S}_2 \sqcup \mathbf{S}_3) \;\sim_n\; (\mathbf{S}_1 \sqcup \mathbf{S}_2) \sqcup \mathbf{S}_3$

(d) $\mathbf{S}_i \;\sim_n\; \mathbf{S}_i^\star$, *for* $i = 1, 2$, *implies*

$$\mathbf{S}_1 \sqcup \mathbf{S}_2 \;\sim_n\; \mathbf{S}_1^\star \sqcup \mathbf{S}_2^\star$$

(e) $(\mathbf{S}_1, s) \;\sim_n\; (\mathbf{S}_1^\star, s^\star)$ *and* $\mathbf{S}_2 \;\sim_n\; \mathbf{S}_2^\star$ *imply*

$$\big(\mathbf{S}_1 \sqcup \mathbf{S}_2, s\big) \;\sim_n\; \big(\mathbf{S}_1^\star \sqcup \mathbf{S}_2^\star,\, s^\star\big).$$

PROOF. (a) is clear since Player II just uses an isomorphism $\sigma : \mathbf{S} \to \mathbf{S}^\star$ to mimic the moves of Player I. Then (b) and (c) follow from (a).

For (d), Player II has a winning strategy by playing $\Gamma_n\big(\mathbf{S}_1 \sqcup \mathbf{S}_2, \mathbf{S}_1^\star \sqcup \mathbf{S}_2^\star\big)$ as two separate games.

If Player I chooses an element from either S_1 or $S_1^\star$ then Player II uses his winning strategy from $\Gamma_n(\mathbf{S}_1, \mathbf{S}_1^\star)$. Likewise, if Player I chooses an element from either S_2 or $S_2^\star$ then Player II uses his winning strategy from $\Gamma_n(\mathbf{S}_2, \mathbf{S}_2^\star)$.

If Player I chooses $Q \subseteq S_1 \sqcup S_2$ let $Q_i = Q \cap S_i$. Then Player II uses his winning strategy for the two games $\Gamma_n(\mathbf{S}_i, \mathbf{S}_i^\star)$, assuming Player II has just played Q_i, to choose $Q_i^\star \subseteq S_i^\star$, and then chooses $Q^\star = Q_1^\star \sqcup Q_2^\star$. Likewise, if Player I chooses $Q^\star \subseteq S_1^\star \sqcup S_2^\star$ Player II will use this two-game strategy to choose a $Q \subseteq S_1 \sqcup S_2$.

The proof of (e) is essentially the same as for (d). □

Now comes the main step, to show that, for any given n, any disjoint union of enough structures $\mathbf{S}_i$ is $\sim_n$ equivalent to any other disjoint union of enough structures $\mathbf{S}_j^\star$, provided all the structures under consideration are $\sim_n$ equivalent to one another. The proof of this particular step is transparent in the case of first-order logic, because enough just means n structures, but, for monadic second-order logic, more effort is needed. The first step is to define what is meant by 'enough structures'.

Definition 6.21. Given a finite purely relational language L let $\mathsf{L}^{(k)}$ be the language obtained by adding k new unary relation symbols to L. Then let $E_{\mathsf{L}}(k,n)$ be the number of equivalence classes of $\mathsf{L}^{(k)}$-structures using the equivalence relation $\sim_n$. This is finite by Corollary 6.18.

From this definition, a recursive definition of $N_{\mathsf{L}}(k,n)$ is given which provides the 'enough' discussed above.

Definition 6.22. Let L be a purely relational language, and define $N_{\mathsf{L}}(k,n)$ by

$$\begin{aligned} N_{\mathsf{L}}(k,0) &= 1 \\ N_{\mathsf{L}}(k,n) &= \big(E_{\mathsf{L}}(k+1,n-1)\ +\ 1\big)\cdot N_{\mathsf{L}}(k+1,n-1), \quad \text{for } k\geq 0,\ n\geq 1. \end{aligned}$$

Thus

$$N_{\mathsf{L}}(k,n) = \big(E_{\mathsf{L}}(k+1,n-1)\ +\ 1\big)\ \cdots\ \big(E_{\mathsf{L}}(k+n,0)+1\big).$$

Now comes a key result, noting that

$$\begin{aligned} E_{\mathsf{L}^{(1)}}(k,n) &= E_{\mathsf{L}}(k+1,n) \\ N_{\mathsf{L}^{(1)}}(k,n) &= N_{\mathsf{L}}(k+1,n). \end{aligned}$$

Proposition 6.23 (Compton, 1989). *Let L be a purely relational language, and let $\mathbf{S}_i$, $i\in I$, and $\mathbf{S}^\star_j$, $j\in J$, be L-structures such that, for a given n,*

$$\mathbf{S}_i \sim_n \mathbf{S}^\star_j, \textit{ for } i\in I,\ j\in J,$$

that is, all the given structures are $\sim_n$ equivalent. If $|I|,|J|\geq N_{\mathsf{L}}(0,n)$ then

$$\bigsqcup_{i\in I}\mathbf{S}_i \sim_n \bigsqcup_{j\in J}\mathbf{S}^\star_j.$$

PROOF. The assertion, that will be proved by induction on n, is that, for each n, the proposition holds for all purely relational languages L.

For the Ground Case, $n=0$, assume that $\mathbf{S}_i\sim_0\mathbf{S}^\star_j$ for $i\in I$ and $j\in J$, and that $|I|,|J|\geq N_{\mathsf{L}}(0,0)$ $(=1)$. The assertion

$$\bigsqcup_{i\in I}\mathbf{S}_i \sim_0 \bigsqcup_{j\in J}\mathbf{S}^\star_j$$

holds trivially since there are no constants in the language.

The Induction Hypothesis is that the assertion holds for n.

Then, for the Induction Step, it is assumed that $\mathbf{S}_i\sim_{n+1}\mathbf{S}^\star_j$, for $i\in I$ and $j\in J$, and that $|I|,|J|\geq N_{\mathsf{L}}(0,n+1)$. The strategy is to apply Lemma 6.14, and this leads to consideration of two cases of augmenting $\bigsqcup_{i\in I}\mathbf{S}_i$, first by an element, and second by a unary relation, as well as the symmetric cases of augmenting $\bigsqcup_{j\in J}\mathbf{S}^\star_j$.

Claim 1: Let $s \in \bigsqcup_{i\in I} S_i$. Then there is an $s^\star \in \bigsqcup_{j\in J} S_j^\star$ such that

$$\Big(\bigsqcup_{i\in I} \mathbf{S}_i, s\Big) \sim_n \Big(\bigsqcup_{j\in J} \mathbf{S}_j^\star, s^\star\Big).$$

To see this let $i_0 \in I$ be such that $s \in S_{i_0}$. Choose any $j_0 \in J$, and then choose $s^\star \in S_{j_0}^\star$ such that

$$(6.23.1) \qquad \big(\mathbf{S}_{i_0}, s\big) \sim_n \big(\mathbf{S}_{j_0}^\star, s^\star\big).$$

Such an element $s^\star$ exists by Lemma 6.13 (a). Let $I_0 = I \smallsetminus \{i_0\}$ and let $J_0 = J \smallsetminus \{j_0\}$. As

$$N_{\mathsf{L}}(0, n+1) > N_{\mathsf{L}}(0, n)$$

it follows that

$$|I_0|, |J_0| \geq N_{\mathsf{L}}(0, n).$$

Clearly, $\mathbf{S}_i \sim_n \mathbf{S}_j^\star$, for all i, j, since these structures are $\sim_{n+1}$ equivalent. This means that the L-structures $\mathbf{S}_i$, $i \in I_0$, and $\mathbf{S}_j^\star$, $j \in J_0$, satisfy the induction hypothesis, so (see Figure 3)

$$(6.23.2) \qquad \bigsqcup_{i\in I_0} \mathbf{S}_i \sim_n \bigsqcup_{j\in J_0} \mathbf{S}_j^\star.$$

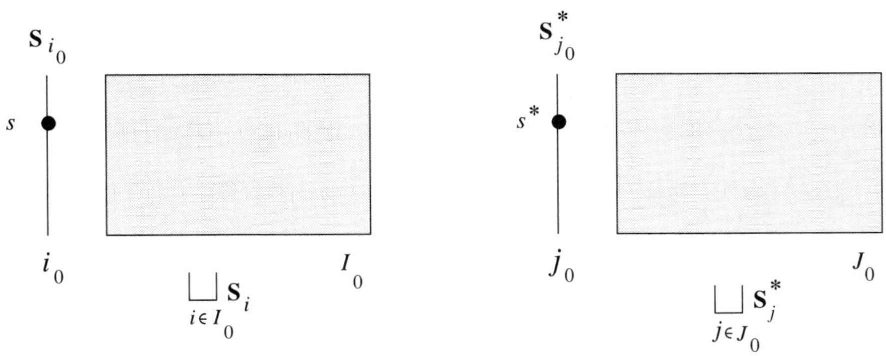

FIGURE 3

By Lemma 6.20 (e) applied to (6.23.1) and (6.23.2) it follows that

$$\Big(\bigsqcup_{i\in I} \mathbf{S}_i, s\Big) \sim_n \Big(\bigsqcup_{j\in J} \mathbf{S}_j^\star, s^\star\Big).$$

This proves Claim 1. The same argument applies with the roles of $\mathbf{S}_i$ and $\mathbf{S}_j^\star$ interchanged.

Claim 2: Let $Q \subseteq \bigsqcup_{i\in I} S_i$. Then there is a $Q^\star \subseteq \bigsqcup_{j\in J} S_j^\star$ such that

$$\Big(\bigsqcup_{i\in I} \mathbf{S}_i, Q\Big) \sim_n \Big(\bigsqcup_{j\in J} \mathbf{S}_j^\star, Q^\star\Big).$$

The first step is to consider the unary relations Q_i that are induced on the $\mathbf{S}_i$ by Q. Then $Q = \bigsqcup_{i \in I} Q_i$. Each structure $(\mathbf{S}_i, Q_i)$ is considered as an $\mathsf{L}^{(1)}$-structure. Let $I_1, \ldots, I_m$ be the partition of I determined by the $\sim_n$ equivalence relation on the $(\mathbf{S}_i, Q_i)$, that is, two elements $i_1, i_2 \in I$ are in the same partition block I_k iff

$$(\mathbf{S}_{i_1}, Q_{i_1}) \ \sim_n \ (\mathbf{S}_{i_2}, Q_{i_2}).$$

Clearly,

$$m \ \leq \ E_{\mathsf{L}^{(1)}}(0, n) \ = \ E_{\mathsf{L}}(1, n),$$

and thus

$$\begin{aligned} |J| \ &\geq \ N_{\mathsf{L}}(0, n+1) \\ &= \ \big(E_{\mathsf{L}}(1, n) \ + \ 1\big) \cdot N_{\mathsf{L}}(1, n) \\ &\geq \ m \cdot N_{\mathsf{L}}(1, n). \end{aligned}$$

This means that J can be partitioned into m blocks $J_1, \ldots, J_m$ such that, for $1 \leq k \leq m$,

$$\begin{array}{lllll} |J_k| & = & |I_k| & \text{if} & |I_k| \ < \ N_{\mathsf{L}}(1, n) \\ |J_k| & \geq & N_{\mathsf{L}}(1, n) & \text{if} & |I_k| \ \geq \ N_{\mathsf{L}}(1, n). \end{array}$$

Now, for $j \in J_k$, apply Lemma 6.13 (b) to choose a subset $Q_j^\star$ of $S_j^\star$ such that

$$(\mathbf{S}_i, Q_i) \ \sim_n \ (\mathbf{S}_j^\star, Q_j^\star),$$

for $i \in I_k$. Then, for each k such that $|I_k| = |J_k|$,

$$\Big(\bigsqcup_{i \in I_k} \mathbf{S}_i, \bigsqcup_{i \in I_k} Q_i\Big) \ \sim_n \ \Big(\bigsqcup_{j \in J_k} \mathbf{S}_j^\star, \bigsqcup_{j \in J_k} Q_j^\star\Big) \tag{6.23.3}$$

by Lemma 6.20 (d); and for the other k, $|I_k|, |J_k| \geq N_{\mathsf{L}}(1, n) = N_{\mathsf{L}^{(1)}}(0, n)$, so, by the induction hypothesis, the assertion (6.23.3) also holds for such k. (See Figure 4.)

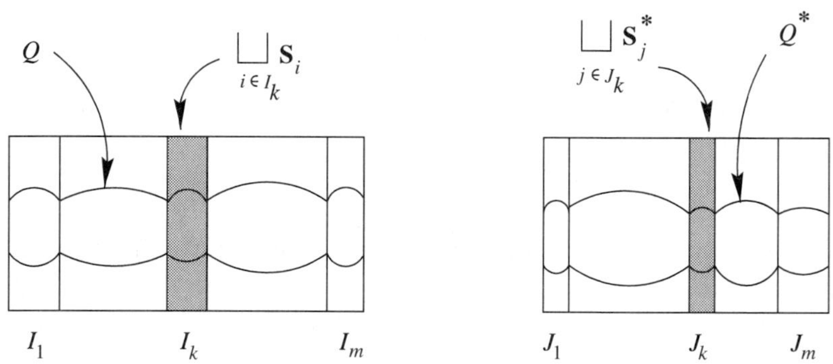

FIGURE 4

Now Lemma 6.20 (d) is applied again to obtain:

$$\bigsqcup_{1\le k\le m}\Big(\bigsqcup_{i\in I_k}\mathbf{S}_i,\bigsqcup_{i\in I_k}Q_i\Big)\sim_n\bigsqcup_{1\le k\le m}\Big(\bigsqcup_{j\in J_k}\mathbf{S}_j^\star,\bigsqcup_{j\in J_k}Q_j^\star\Big),$$

which gives

$$\Big(\bigsqcup_{i\in I}\mathbf{S}_i,\bigsqcup_{i\in I}Q_i\Big)\sim_n\Big(\bigsqcup_{j\in J}\mathbf{S}_j^\star,\bigsqcup_{j\in J}Q_j^\star\Big).$$

Letting $Q^\star=\bigsqcup_{j\in J}Q_j^\star$ the above proves Claim 2. Again the same argument applies with the roles of $\mathbf{S}_i$ and $\mathbf{S}_j^\star$ interchanged.

Now Lemma 6.14 is applied to the two claims to obtain:

$$\bigsqcup_{i\in I}\mathbf{S}_i\sim_{n+1}\bigsqcup_{j\in J}\mathbf{S}_j^\star,$$

completing the proof of the induction step. □

Remark 6.24. If one only wants to prove the above proposition for first-order logic then the proof is much simpler. First one replaces the recursively defined number $N_{\mathsf{L}}(k,n)$ by n. Then the strategy of Player II is as follows: if Player I plays in a new structure $\mathbf{S}_i$ then Player II chooses a new structure $\mathbf{S}_j^\star$ and subsequently plays his winning strategy for $\Gamma_n(\mathbf{S}_i,\mathbf{S}_j^\star)$ whenever Player I plays in either of these structures. Likewise, if Player I plays in a new structure $\mathbf{S}_j^\star$ then Player II chooses a new structure $\mathbf{S}_i$ and subsequently plays his winning strategy for $\Gamma_n(\mathbf{S}_i,\mathbf{S}_j^\star)$ whenever Player I plays in either of these structures.

The result that is really desired is the following.

Corollary 6.25. *Let* L *be a finite purely relational language, and let* $\mathbf{S}_i$, $i\in I$, *and* $\mathbf{S}_j^\star$, $j\in J$, *be* L*-structures such that, for a given* n,

$$\mathbf{S}_i\equiv_n\mathbf{S}_j^\star,\quad \textit{for } i\in I,\ j\in J,$$

that is, all the given structures are $\equiv_n$ *equivalent. If* $|I|,|J|\ge N_{\mathsf{L}}(0,n)$ *then*

$$\bigsqcup_{i\in I}\mathbf{S}_i\equiv_n\bigsqcup_{j\in J}\mathbf{S}_j^\star.$$

PROOF. This is immediate from Theorem 6.17 and Proposition 6.23. □

6.6. Adequate classes of structures

To apply the results of the previous section, one needs a class $\mathcal{K}$ of finite relational structures in a finite purely relational language L that naturally leads to a generalized additive number system. Compton realized that certain simple properties of classes of relational structures play a key role in combinatorial asymptotics, in particular he noted that *disjoint union* plays the role of addition.

Definition 6.26. A finite structure $\mathbf{S}$, with $S \neq \varnothing$, in a class $\mathcal{K}$ of relational structures, is said to be *$\mathcal{K}$-indecomposable* if it is not isomorphic to a disjoint union of two or more smaller members of $\mathcal{K}$.

Given a first-order language L of relational structures, a class $\mathcal{K}$ of finite L-structures that is closed under isomorphism is said to be *adequate* if:

(i) $\mathcal{K}$ is closed under disjoint union,
(ii) members of $\mathcal{K}$ with nonempty universe can be uniquely decomposed into (a disjoint union of) $\mathcal{K}$-indecomposable structures, and
(iii) the L-structure $\mathbf{O}$ on the empty set is in $\mathcal{K}$.

Compton actually used a generalized notion of *component* of a relational structure rather than the yet more general notion of $\mathcal{K}$-indecomposable. His version of condition (ii) was that $\mathcal{K}$ would be closed under components.

For $\mathcal{K}$ an adequate class of structures, let $\mathcal{P}$ be the subclass of $\mathcal{K}$-indecomposables. It can be easily seen, for adequate classes $\mathcal{K}$, that one has a generalized additive number system $\mathcal{A}_{\mathcal{K}}$ on $\widetilde{\mathcal{K}}$, the set of isomorphism types of $\mathcal{K}$, by defining:

$$\begin{aligned} \|\widetilde{\mathbf{S}}\| &= |S| \\ \widetilde{\mathbf{S}_1} + \widetilde{\mathbf{S}_2} &= \widetilde{\mathbf{S}_1 \sqcup \mathbf{S}_2}\widetilde{\mathbf{S}}. \end{aligned}$$

The indecomposable elements of $\mathcal{A}_{\mathcal{K}}$ are the elements of $\widetilde{\mathcal{P}}$.

The classes $(\leq m_i)\mathcal{P}_i$, $m_i\mathcal{P}_i$, $(\geq m_i)\mathcal{P}_i$, and their disjoint unions, are defined in the obvious manner. For example, $(\geq 3)\mathcal{P}_1 \sqcup 5\mathcal{P}_2$ is the class of structures that can be obtained as a disjoint union of at least 3 structures from $\mathcal{P}_1$ and exactly five structures from $\mathcal{P}_2$.

Definition 6.27. A *partition* $\mathcal{P}_1, \ldots, \mathcal{P}_k$ of $\mathcal{P}$ is assumed to have the property that each $\mathcal{P}_i$ is closed under isomorphism. A *partition class* of $\mathcal{K}$ is a class of the form $\bigsqcup \gamma_i \mathcal{P}_i$, where the $\mathcal{P}_i$ partition $\mathcal{P}$.

Proposition 6.28 (Compton, 1989). *For L a finite purely relational language, let $\mathcal{K}$ be an adequate class of L-structures. For φ a first-order, or more generally a monadic second-order, L-sentence there is an integer n and a partition $\mathcal{P}_1, \ldots, \mathcal{P}_k$ of the class $\mathcal{P}$ of indecomposables of $\mathcal{K}$ such that $\mathcal{K}_\varphi$ is a union of disjoint partition classes $\gamma_1\mathcal{P}_1 \sqcup \cdots \sqcup \gamma_k\mathcal{P}_k$, where each γ_i is one of $0, 1, \ldots, n-1, (\geq n)$.*

PROOF. Let r be the quantifier rank of φ, and define n to be the number $N_{\mathsf{L}}(0, r)$ as given in Definition 6.22. Let $\mathcal{P}_1, \ldots, \mathcal{P}_k$ be the partition of the class $\mathcal{P}$ of indecomposables of $\mathcal{K}$ by $\equiv_r$.

For γ_i one of $0, 1, \ldots, n-1, (\geq n)$, one can use Lemma 6.20 (d) and Corollary 6.25 to show that all structures in each $\gamma_i\mathcal{P}_i$ are $\equiv_r$ equivalent (using the lemma if $\gamma_i = m_i$, and the corollary if $\gamma_i = (\geq n)$). Then, by Lemma 6.20 (d), it follows that all structures in the partition class $\mathcal{B} = \gamma_1\mathcal{P}_1 \sqcup \cdots \sqcup \gamma_k\mathcal{P}_k$ are $\equiv_r$ equivalent. Thus all members of $\mathcal{B}$ satisfy φ, or else none of the structures in $\mathcal{B}$ satisfy φ. Since $\mathcal{K}$ is the union of the various

$\gamma_1\mathcal{P}_1 \sqcup \cdots \sqcup \gamma_k\mathcal{P}_k$, for the γ_i as described in the proposition, it follows that $\mathcal{K}_\varphi$ is the union of the $\gamma_1\mathcal{P}_1 \sqcup \cdots \sqcup \gamma_k\mathcal{P}_k$ whose members satisfy φ.

Note that, for distinct choices of the sequence γ_i, the corresponding partition sets $\gamma_1\mathcal{P}_1 \sqcup \cdots \sqcup \gamma_k\mathcal{P}_k$ are disjoint. □

Theorem 6.29 (Compton, 1989). *Let $\mathcal{K}$ be an adequate class of finite* L*-structures, for a finite purely relational language* L*. If all partition sets of the additive number system $\mathcal{A}_\mathcal{K}$ have local [global] asymptotic density in $\mathcal{A}_\mathcal{K}$ then, for any monadic second-order* L*-sentence φ, the class $\mathcal{K}_\varphi$ has local [global] asymptotic density in $\mathcal{K}$.*

Definition 6.30. Let $a(n), p(n)$ be the counting functions of $\mathcal{A}_\mathcal{K}$. $a(n)$ is called the *local spectrum* of $\mathcal{K}$, and $A(n)$ the *global spectrum* of $\mathcal{K}$.

Proposition 6.31. *Let $\mathcal{K}$ be an adequate class of finite* L*-structures, for a finite purely relational language* L*. Then, assuming $\mathcal{A}_\mathcal{K}$ is reduced, that is, $a(n)$ is eventually positive, the following hold:*

(a) $a(n) \in \mathsf{RT}_1$ *implies $\mathcal{K}$ has a local monadic second-order 0–1 law.*
(b) $A(n) \in \mathsf{RT}_1$ *implies $\mathcal{K}$ has a global monadic second-order 0–1 law.*
(c) $a(n) \in \mathsf{RT}_\rho$, *with* $0 < \rho < 1$ *and* $\dfrac{a(n-k)}{a(n)} \le C\rho^{-k}$, *for any choice of k, n satisfying $K \le k \le n$, with C and K both positive, implies $\mathcal{K}$ has a local monadic second-order law.*
(d) $a(n) \sim c\beta^n$, *with $c > 0$ and $\beta \ge 1$, implies $\mathcal{K}$ has a local monadic second-order law.*
(e) $a(n) \sim C\beta^n/n^\mu$ *and $p(n) = O(\beta^n/n)$ with $\beta > 1$, $\mu < 1$, and $C > 0$, implies $\mathcal{K}$ has a local monadic second-order law.*
(f) $p(n) = \mathrm{O}(n^c)$ *implies $\mathcal{K}$ has a local monadic second-order 0–1 law.*
(g) $p(n) \in \mathsf{RT}_1$ *implies $\mathcal{K}$ has a local monadic second-order 0–1 law.*
(h) $p(n) \in \mathsf{RT}_\rho$ *and $\liminf_n np(n)\rho^n > 1$ implies $\mathcal{K}$ has a local monadic second-order limit law.*

PROOF. These follow from Theorem 6.29 and: Theorem 4.2 for (a), Theorem 4.3 for (b), Theorem 5.13 for (c), Corollary 5.15 for (d), and Theorem E.2 for (e), [**6**] for (f), [**8**] for (g), (h). □

Remark 6.32. The above also apply even when $a(n)$ is not eventually positive—simply replace $a(n)$ and $p(n)$ by $a^\star(n)$ and $p^\star(n)$, the counting functions for the reduced form $\mathcal{A}^\star_\mathcal{K}$ of $\mathcal{A}_\mathcal{K}$.

Such results have been called *spectrally determined* logical limit laws in [**12**] since they depend only on the growth rate of the counting functions (that is, of the spectra) of $\mathcal{K}$, given that $\mathcal{K}$ is adequate. For a given adequate $\mathcal{K}$, only a portion of the partition classes of $\mathcal{K}$ can be defined by monadic second-order sentences—after all, there are continuum many partition classes if $\mathcal{P}$ is infinite, but only a countable number of sentences when the language is countable. However, it is necessary that *all* partition classes have asymptotic density in order to deduce a logical limit law from the counting functions alone.

Proposition 6.33. *Suppose $\mathcal{K}$ is adequate, but not all partition classes have asymptotic density. Then one can construct another adequate class $\widehat{\mathcal{K}}$, with the same counting functions as $\mathcal{K}$, by expanding the language of $\mathcal{K}$ by finitely many unary predicates and a single binary predicate, with the property that there is a first-order sentence φ such that $\widehat{\mathcal{K}}_\varphi$ does not have asymptotic density in $\widehat{\mathcal{K}}$.*

PROOF. Suppose

$$\mathcal{B} = \gamma_1 \mathcal{P}_1 \sqcup \cdots \sqcup \gamma_k \mathcal{P}_k$$

is a partition class of $\mathcal{K}$ that does not have asymptotic density.

Let $\mathcal{Q} \subseteq \mathcal{P}$ consist of one representative from each isomorphism class of indecomposables in $\mathcal{P}$. Add a binary relation $\leq$ to each indecomposable in $\mathcal{Q}$ that is a linear order, and add unary predicates $U_1, \ldots, U_k$ and require that U_i hold for all elements of members of $\mathcal{Q} \cap \mathcal{P}_i$, and for no elements of other members of $\mathcal{Q}$. Now close this expansion of $\mathcal{Q}$ under finite disjoint union, and then close this class under isomorphism to obtain an expansion $\widehat{\mathcal{K}}$ of $\mathcal{K}$, with the new relations $\leq, U_1, \ldots, U_k$, but with the same count functions.

Each γ_i (in the expression for $\mathcal{B}$) is in one of the forms $(\leq m_i)$, m_i, $(\geq m_i)$, so construct a first-order sentence φ that says:

- there are m_i, but not $m_i + 1$, pairwise incomparable elements that satisfy U_i, for $\gamma_i = m_i$;
- there are at most m_i pairwise incomparable elements that satisfy U_i, for $\gamma_i = (\leq m_i)$;
- there are m_i pairwise incomparable elements that satisfy U_i, for $\gamma_i = (\geq m_i)$.

Then such a sentence is satisfied by precisely the members of $\widehat{\mathcal{B}}$, the expansion of $\mathcal{B}$, so $\widehat{\mathcal{K}}_\varphi$ does not have asymptotic density. □

Compton gives the following examples in **[20]** as applications of the Bateman-Erdös and Meinardus Theorems to his work. They have a monadic second-order 0–1 law.

- *Permutations*, as binary relations. Then $p(n) = 1$, for all n.
- *Oriented Forests of Height One.* (See Figure 5.) Then $p(n) = 1$, for all n.

FIGURE 5. Indecomposable Oriented Forests of Height One

- *Linear Forests (as graphs or as posets).* Then $p(n) = 1$, for all n.
- Any adequate $\mathcal{K}$ *with finitely many indecomposables.*
- *Equivalence Relations.* Then $p(n) = 1$, for all n.
- *Partitions with Select Subsets.* Then $p(n) = n$, for all n.

Using (d) of Proposition 6.31, and the asymptotics of Knopfmacher, et al, many examples with monadic second-order limit laws can be found:

- *Linear Forests (as posets) with k Unary Predicates.* Then one has $p(n) = (2^k)^n$.
- *k-Colored Linear Forests (as posets).* Then $p(n) = k^n$.

(From Appendix E) Woods had the following example in mind in his work:

- *A Partial [or Total] Unary Operation with k-Unary Predicates* (see Figure 6) defines a class that has a monadic second-order limit law. The operation is to be treated as a binary relation.

FIGURE 6. Indecomposable Partial Unary Operations With one Unary Predicate

Part 2

Multiplicative Number Systems

CHAPTER 7

Background from Analysis

This chapter collects together elementary facts about Dirichlet series,[1] the Euler product, and regular variation at infinity, facts that will be needed in Part 2. The reader familiar with this material can skip over it. However, look at Definition 7.12 for the meaning of $\mathcal{DS}_c$.

7.1. Series and infinite products

This section gives the basic definitions and properties.

7.1.1. Dirichlet series basics.

Definition 7.1. A series $\mathbf{S}(x) = \sum_{n\geq 1} s(n)n^{-x}$ is called a *Dirichlet series.* $s(n)$ is the *coefficient* of n^{-x} in $\mathbf{S}(x)$, also written as $[n^{-x}]\mathbf{S}(x)$.

Dirichlet series are designated by upper case boldface letters, with the corresponding lower case italic letters used for the coefficients, and the corresponding upper case italic letters used for partial sums of the coefficients.

Dirichlet series	Partial sums of coefficients
$\mathbf{A}(x) = \sum_n a(n)n^{-x}$	$A(x) = \sum_{n\leq x} a(n)$
$\vdots$	$\vdots$
$\mathbf{S}(x) = \sum_n s(n)n^{-x}$	$S(x) = \sum_{n\leq x} s(n)$
$\vdots$	$\vdots$

The logical limit laws of Part 2 are based on making a connection between logic and Dirichlet series.

[1]The study of Dirichlet series has remarkable parallels to the study of power series, and the reader will see this in the similarities between the results of Part 1 and Part 2. One of the big differences in the two subjects is that Dirichlet series are not considered an elementary, or even essential, topic in an undergraduate mathematics curriculum. This is largely due to the fact that the functions that can be represented by Dirichlet series are not the ones usually encountered in physics and calculus courses. Dirichlet series are primarily useful to study the counting functions of multiplicative number systems. For a substantial introduction to the subject see Apostol [**1**], Chapter 11, or Titchmarsh [**54**], Chapter 9.

Definition 7.2. The *Dirichlet product (convolution)* $\mathbf{S}(x) * \mathbf{T}(x)$ of two Dirichlet series $\mathbf{S}(x)$ and $\mathbf{T}(x)$ is defined by

$$\mathbf{S}(x) * \mathbf{T}(x) = \sum_{n\geq 1}\Big(\sum_{i\cdot j=n} s(i)\cdot t(j)\Big)n^{-x}.$$

The Dirichlet product is associative, and for $\mathbf{S}_k(x) = \sum_n s_k(n)n^{-x}$, $1 \leq k \leq m$, the Dirichlet product is given by:

$$\begin{aligned}\prod^{*}_{1\leq k\leq m} \mathbf{S}_k(x) &= \mathbf{S}_1(x) * \cdots * \mathbf{S}_m(x)\\ &= \sum_n \Big(\sum_{k_1\cdots k_m=n} s_1(k_1)\cdots s_m(k_m)\Big)n^{-x}.\end{aligned}$$

Proposition 7.3. *Let $\mathbf{S}(x)$ and $\mathbf{T}(x)$ be Dirichlet series with real coefficients.*

(a) *There are α^- and α in $[-\infty,\infty]$ such that*

$$\begin{array}{rll} x > \alpha & \textit{implies} \ \ \mathbf{S}(x) & \textit{converges absolutely}\\ \alpha^- < x < \alpha & \textit{implies} \ \ \mathbf{S}(x) & \textit{converges conditionally}\\ x < \alpha^- & \textit{implies} \ \ \mathbf{S}(x) & \textit{diverges.}\end{array}$$

α^- *is called the* abscissa of convergence *of* $\mathbf{S}(x)$, *and* α *the* abscissa of absolute convergence *of* $\mathbf{S}(x)$.

If the $s(n)$ are eventually nonnegative then $\alpha^- = \alpha$. The abscissa of absolute convergence α of $\mathbf{S}(x)$ will usually be written α_s.

(b) *Suppose $\mathbf{S}(x)$ does not converge at $x = 0$, that is, the sequence $S(n)$ does not converge. Then*

$$\alpha_s^- = \limsup_{n\to\infty}\frac{\log|S(n)|}{\log n}.$$

If, furthermore, the $s(n)$ are eventually nonnegative then

$$\alpha_s = \limsup_{n\to\infty}\frac{\log|S(n)|}{\log n}.$$

(c) *If all $s(n) \geq 0$ then $\mathbf{S}(x)$ defines a function on (α_s,∞) that is nonnegative and nonincreasing.*

(d) *If $\mathbf{S}(x)$ converges absolutely at $x = c$ then $\mathbf{S}(x)$ converges uniformly on $[c,\infty)$.*

(e) *If $\mathbf{S}(x)$ converges absolutely at $x = c$ then $\mathbf{S}(x)$ is continuous on $[c,\infty)$.*

(f) *If $\mathbf{S}(x)$ and $\mathbf{T}(x)$ converge absolutely at $x = c$ then*

$$\mathbf{S}(x)\cdot\mathbf{T}(x) = \mathbf{S}(x) * \mathbf{T}(x)$$

holds on $[c,\infty)$.

(g) *One can differentiate a Dirichlet series term-by-term, i.e.,*

$$\frac{d^k}{dx^k}\mathbf{S}(x) = \sum s(n)(-\log n)^k n^{-x}, \qquad \textit{for } x > \alpha_s.$$

PROOF. These are standard results in Dirichlet series. For example, (d) follows from the Weierstrass M-test since $|s(n)n^{-x}| \le |s(n)n^{-c}|$. Then (e) follows from uniform convergence.

However, the formula in (b) for the abscissa of convergence is omitted from some basic references. Since this is a formula that we will rely on, a proof is given here.

Let

$$\lambda = \limsup_{n\to\infty} \frac{\log\bigl|S(n)\bigr|}{\log n}.$$

From the assumption that $\mathbf{S}(0)$ diverges it follows that $\lambda \ge 0$. To see this, note that, since $S(n)$ does not converge to 0, there is a $c > 0$ such that $\bigl|S(n)\bigr| > c$ holds infinitely often. This implies that infinitely often

$$\frac{\log\bigl|S(n)\bigr|}{\log n} > \frac{\log c}{\log n},$$

and the right side tends to 0.

Define $\mathbf{S}|_0(x) = 0$, and let $\mathbf{S}|_m(x)$ be the sum of the first m terms of $\mathbf{S}(x)$:

$$\mathbf{S}|_m(x) = s(1)1^{-x} + \cdots + s(m)m^{-x}.$$

Suppose x is such that $\mathbf{S}(x)$ converges. From the assumption that $\mathbf{S}(0)$ diverges, it follows that $x > 0$, by item (a). To show that $x \ge \lambda$, first note that

$$\begin{aligned} S(n) &= \sum_{k=1}^{n} \bigl(s(k)k^{-x}\bigr)\cdot k^{x} \\ &= \sum_{k=1}^{n} \bigl(\mathbf{S}|_k(x) - \mathbf{S}|_{k-1}(x)\bigr)\cdot k^{x} \\ &= \sum_{k=1}^{n-1} \mathbf{S}|_k(x)\bigl(k^{x} - (k+1)^{x}\bigr) + \mathbf{S}|_n(x)n^{x}. \end{aligned}$$

Let B be a bound on the absolute values of the convergent sequence $\mathbf{S}|_n(x)$. Then, since $x > 0$,

$$\bigl|S(n)\bigr| \le B\sum_{k=1}^{n-1}\bigl((k+1)^{x} - k^{x}\bigr) + Bn^{x} \le 2Bn^{x},$$

hence $\log\bigl|S(n)\bigr| < x\log n + \log 2B$. This leads to

$$\frac{\log\bigl|S(n)\bigr|}{\log n} < x + \frac{\log 2B}{\log n},$$

and thus $\lambda \le x$. This means $\lambda \le \alpha_s^{-}$, the abscissa of convergence of $\mathbf{S}(x)$.

To show that λ is the abscissa of convergence, fix an $x > \lambda$ (assuming $\lambda \neq \infty$). The objective is to show that $\mathbf{S}(x)$ converges. Choose $\varepsilon > 0$ so that $\lambda + \varepsilon < x$. From the definition of λ one has, for some $C > 0$,

$$\left|S(n)\right| \leq Cn^{\lambda+\varepsilon}, \quad \text{for } n \geq 0.$$

Observe that, since $x > 0$, for $k \geq 1$,

$$0 < k^{-x} - (k+1)^{-x} = x\int_k^{k+1} \frac{du}{u^{x+1}} < xk^{-x-1}. \tag{7.3.1}$$

Using these facts one has, with $\mu = \lambda + \epsilon - x < 0$,

$$\begin{aligned}
\Big|\sum_{k=m+1}^{n} s(k)k^{-x}\Big| &= \Big|\sum_{k=m+1}^{n} \big(S(k) - S(k-1)\big)k^{-x}\Big| \\
&\leq \sum_{k=m+1}^{n-1} \left|S(k)\right|\Big(\frac{1}{k^x} - \frac{1}{(k+1)^x}\Big) + \frac{\left|S(n)\right|}{n^x} + \frac{\left|S(m)\right|}{(m+1)^x} \\
&\leq Cx\sum_{k=m+1}^{n-1} k^{\lambda+\varepsilon}\cdot k^{-x-1} + \frac{Cn^{\lambda+\varepsilon}}{n^x} + \frac{Cm^{\lambda+\varepsilon}}{m^x} \\
&= Cx\sum_{k=m+1}^{n-1} k^{\mu-1} + Cn^{\mu} + Cm^{\mu} \\
&\leq Cx\frac{m^{\mu}}{|\mu|} + 2Cm^{\mu} \\
&\to 0 \quad \text{as } m \to \infty, \text{ since } \mu < 0.
\end{aligned}$$

This says that $\mathbf{S}(x)$ converges for $x > \lambda$. Thus $\lambda = \alpha_s^-$.

If the $s(n)$ are eventually nonnegative then, by item (a) above, one has $\alpha_s = \alpha_s^-$. □

Remark 7.4. The formula for α_s^- in Proposition 7.3 (b) indeed depends on the fact that $S(n)$ does not converge. For example, if $s(n)$ is eventually 0 then $\alpha_s^- = -\infty$, whereas the formula would give $\alpha_s^- = 0$ (assuming $\mathbf{S}(x)$ is not the zero Dirichlet series). The correct formula in the case that $\mathbf{S}(0)$ converges is

$$\alpha_s^- = \limsup_{n\to\infty} \frac{\log\left|\sum_{k=n}^{\infty} s(k)\right|}{\log n}.$$

See [**2**], p. 165 or [**54**], p. 292.

7.1.2. Infinite Dirichlet products. Underlying the density results of Part 2 are the Dirichlet series that can be expressed as products of geometric Dirichlet series. In this section a Dirichlet product of infinitely many geometric Dirichlet series is defined.

Let $\mathbb{P}'$ be the set $\{2, 3, 4, \dots\}$ of integers greater than 1.

Definition 7.5. Given $g : \mathbb{N} \to \mathbb{P}'$ with the property $g(n) \to \infty$, let

$$\begin{aligned} \mathbf{S}_n^g(x) &= 1 + g(n)^{-x} + g(n)^{-2x} + \cdots + g(n)^{-kx} + \cdots \\ \mathbf{T}_n^g(x) &= \mathop{\circledast\!\!\prod}_{j \le n} \mathbf{S}_j^g(x) = \Big(\sum_{k\ge 0} g(0)^{-kx}\Big) * \cdots * \Big(\sum_{k \ge 0} g(n)^{-kx}\Big). \end{aligned}$$

$\mathbf{S}_n^g(x)$ is a geometric Dirichlet series with increment factor $g(n)^{-x}$, and $\mathbf{T}_n^g(x)$ is the product of the first $n+1$ of these geometric series. A remarkable property of the coefficients of the $\mathbf{T}_n^g(x)$ is that they *stabilize*. This means, for any fixed m, that the sequence of coefficients of m^{-x},

$$\big[m^{-x}\big]\mathbf{T}_0^g(x), \quad \big[m^{-x}\big]\mathbf{T}_1^g(x), \quad \ldots, \quad \big[m^{-x}\big]\mathbf{T}_n^g(x), \quad \ldots,$$

eventually becomes constant. The reason is simple—if $g(k) > m$ then the Dirichlet series

$$\mathbf{S}_k^g(x) = 1 + g(k)^{-x} + g(k)^{-2x} + \cdots$$

does not contribute to the coefficient of m^{-x} in $\mathbf{T}_n^g(x)$.

Definition 7.6. Suppose $g : \mathbb{N} \to \mathbb{P}'$ with $g(n) \to \infty$. The *infinite Dirichlet product* $\circledast\!\!\prod_{n\ge 0} \mathbf{S}_n^g(x)$ is the Dirichlet series $\mathbf{T}^g(x)$ whose coefficients are given by

$$\big[m^{-x}\big]\mathbf{T}^g(x) = \lim_{n\to\infty} \big[m^{-x}\big]\mathbf{T}_n^g(x),$$

that is, the mth coefficient of the infinite Dirichlet product is the stable value of the mth coefficients of the $\mathbf{T}_n^g(x)$.

Clearly this defines a Dirichlet series with nonnegative integer coefficients. However, work is needed to show that there is an analytic connection between this Dirichlet series and $\lim_{n\to\infty} \mathbf{T}_n^g(x)$.

7.1.3. Convergence of infinite products.

The basic facts from §1.1.4 about $\prod_n \big(1 + r(n)\big)$, where $r(n) \ge 0$, will be used.

Definition 7.7. Given a Dirichlet series $\mathbf{R}(x) = \sum_n r(n)n^{-x}$, define the *mth truncation* $\mathbf{R}|_m(x)$ of $\mathbf{R}(x)$ to be the *Dirichlet polynomial*

$$\mathbf{R}|_m(x) = r(1) + r(2)2^{-x} + \cdots + r(m)m^{-x}.$$

Lemma 7.8. *If* $\mathbf{R}(x)$ *has nonnegative coefficients then, for* $x \ge 0$,

$$\lim_{m\to\infty} \mathbf{R}|_m(x) = \mathbf{R}(x).$$

PROOF. This is simply a consequence of the definition of the value of a Dirichlet series $\mathbf{R}(x)$ at a real number x, namely

$$\mathbf{R}(x) = \lim_{m\to\infty} \mathbf{R}|_m(x),$$

provided the right side exists. If the coefficients are nonnegative then the limit exists. □

For $g : \mathbb{N} \to \mathbb{P}'$, and, for $x \geq 0$, the $\mathbf{S}_n^g(x)$ take values in $[1, \infty]$, so the infinite product $\prod_{n \geq 0} \mathbf{S}_n^g(x)$ exists, possibly with the value ∞. Since $g(n) \to \infty$, this infinite product has the same value as the infinite Dirichlet product.

Lemma 7.9. *If $g : \mathbb{N} \to \mathbb{P}'$ with $g(n) \to \infty$ then, for $x \geq 0$,*

$$\mathbf{T}^g(x) \;=\; \mathop{\boxed{\ast}}_{n \geq 0} \mathbf{S}_n^g(x) \;=\; \prod_{n \geq 0} \mathbf{S}_n^g(x).$$

Proof. Given m, choose N such that $n \geq N$ implies $g(n) > m$. Then, making use of the fact that the Dirichlet product of finitely many Dirichlet polynomials is the product of the Dirichlet polynomials,

$$\begin{aligned} \Big(\mathop{\boxed{\ast}}_{n \geq 0} \mathbf{S}_n^g\Big)\Big|_m(x) \;&\leq\; \mathop{\boxed{\ast}}_{n \leq N} \mathbf{S}_n^g\big|_m(x) \;=\; \prod_{n \leq N} \mathbf{S}_n^g\big|_m(x) \\ &\leq\; \prod_{n \leq N} \mathbf{S}_n^g(x) \;\leq\; \prod_{n \geq 0} \mathbf{S}_n^g(x), \end{aligned}$$

so

$$\mathop{\boxed{\ast}}_{n \geq 0} \mathbf{S}_n^g(x) \;\leq\; \prod_{n \geq 0} \mathbf{S}_n^g(x).$$

For the other inequality,

$$\prod_{n \leq M} \mathbf{S}_n^g\Big|_m(x) \;=\; \mathop{\boxed{\ast}}_{n \leq M} \mathbf{S}_n^g\Big|_m(x) \;\leq\; \mathop{\boxed{\ast}}_{n \geq 0} \mathbf{S}_n^g(x),$$

so

$$\prod_{n \geq 0} \mathbf{S}_n^g(x) \;\leq\; \mathop{\boxed{\ast}}_{n \geq 0} \mathbf{S}_n^g(x).$$

□

The content of this lemma can be expressed as 'One can take the infinite Dirichlet product of the $\mathbf{S}_n^g$ and then evaluate at x, or one can evaluate each $\mathbf{S}_n^g$ at x, and then take the infinite product of their values. The results are the same.'

This gives a key tool when working with fundamental identities.

Proposition 7.10. *If $g : \mathbb{N} \to \mathbb{P}'$ with $g(n) \to \infty$ then, for $x > 0$,*

$$\mathop{\boxed{\ast}}_{n \geq 0} \sum_{k \geq 0} g(n)^{-kx} \;=\; \prod_{n \geq 0} \sum_{k \geq 0} g(n)^{-kx} \;=\; \prod_{n \geq 0} \left(1 - g(n)^{-x}\right)^{-1}.$$

Proof. This follows from Lemma 7.9, noting that, for $x > 0$,

$$\mathbf{S}_n^g(x) \;=\; \sum_{k \geq 0} g(n)^{-kx} \;=\; \left(1 - g(n)^{-x}\right)^{-1}.$$

□

Lemma 7.11. *Let $f : \{2, 3, \dots\} \to \mathbb{N}$ be given. Then, for $x > 0$,*

$$\prod_{n \geq 2} (1 - n^{-x})^{-f(n)} \text{ converges iff } \sum_{n \geq 2} f(n) n^{-x} \text{ converges.}$$

PROOF. Since

$$(1 - n^{-x})^{-f(n)} = \left(1 + \frac{n^{-x}}{1 - n^{-x}}\right)^{f(n)},$$

Lemma 1.14 shows that $\prod_n (1 - n^{-x})^{-f(n)}$ converges iff $\sum_n f(n) \frac{n^{-x}}{1 - n^{-x}}$ converges. For $x > 0$, and, for sufficiently large n,

$$n^{-x} \leq \frac{n^{-x}}{1 - n^{-x}} \leq 2n^{-x},$$

so, by the comparison test, $\sum_n f(n) \frac{n^{-x}}{1 - n^{-x}}$ converges iff $\sum_n f(n) n^{-x}$ converges. □

7.2. Dirichlet series expansions

Definition 7.12. A Dirichlet series $\mathbf{S}(x)$ is a *Dirichlet series expansion* of a function $f(x)$ if, for some $c \in \mathbb{R}$, one has $f(x) = \mathbf{S}(x)$ on (c, ∞). Then one says that $f(x)$ has a *Dirichlet series expansion* $\mathbf{S}(x)$ *on* (c, ∞). The expression $f(x) \in \mathcal{DS}_c$ means that $f(x)$ has a Dirichlet series expansion on (c, ∞) that converges absolutely at c.

If $f(x)$ has a Dirichlet series expansion $\mathbf{S}_1(x)$ on (c_1, ∞), and a Dirichlet series expansion $\mathbf{S}_2(x)$ on (c_2, ∞), then the two Dirichlet series are identical. Thus, if $f(x)$ has *a* Dirichlet series expansion then one can speak of *the* Dirichlet series expansion of $f(x)$.

For $\mathbf{S}(x)$ a Dirichlet series, note that $\mathbf{S}(x) \in \mathcal{DS}_c$ holds iff $\mathbf{S}(c)$ converges absolutely.

Lemma 7.13. *Suppose $f(x) \in \mathcal{DS}_c$ has the Dirichlet series expansion $\mathbf{S}(x)$. Then*

$$\lim_{x \to c^+} f(x) = \mathbf{S}(c).$$

PROOF. $f(x) \in \mathcal{DS}_c$ means that $\mathbf{S}(x)$ converges absolutely at c, so $\mathbf{S}(x)$ is continuous on $[c, \infty)$. Then

$$\lim_{x \to c^+} \mathbf{S}(x) = \mathbf{S}(c).$$

Since $f(x) = \mathbf{S}(x)$ on (c, ∞), the lemma follows. □

The next lemma says that $\mathcal{DS}_c$ is an algebra of functions, in view of the obvious fact that constant functions are in $\mathcal{DS}_c$.

Lemma 7.14. *If $f, g \in \mathcal{DS}_c$ and $\lambda \in \mathbb{R}$ then*

$$\begin{aligned}\lambda f &\in \mathcal{DS}_c\\ f+g &\in \mathcal{DS}_c\\ f\cdot g &\in \mathcal{DS}_c.\end{aligned}$$

The next lemma shows how to find the Dirichlet series expansions corresponding to these operations.

Lemma 7.15. *Suppose $\mathbf{S}(x)$, $\mathbf{T}(x) \in \mathcal{DS}_c$, and $\lambda \in \mathbb{R}$. Then the following hold, for $x \geq c$:*

$$\begin{aligned}\lambda\mathbf{S}(x) &= \sum \big(\lambda s(n)\big)n^{-x}\\ \mathbf{S}(x)+\mathbf{T}(x) &= \sum \big(s(n)+t(n)\big)n^{-x}\\ \mathbf{S}(x)\cdot\mathbf{T}(x) &= \mathbf{S}(x)*\mathbf{T}(x).\end{aligned}$$

Furthermore, if the Dirichlet series $\mathbf{S}(x)$ and $\mathbf{T}(x)$ have nonnegative coefficients, and $\lambda > 0$, then the above hold for any $x \geq 0$.

PROOF. The above results are standard except, perhaps, for the last statement. In the case that the coefficients are nonnegative, the first two equations are straightforward to verify, for $x \geq 0$. For the third, let

$$\mathbf{R}(x) = \mathbf{S}(x)*\mathbf{T}(x).$$

Then, for $x \geq 0$,

$$\mathbf{S}|_M(x)\cdot\mathbf{T}|_M(x) \leq \mathbf{R}|_{M^2}(x) \leq \mathbf{S}|_{M^2}(x)\cdot\mathbf{T}|_{M^2}(x)$$

leads to the desired conclusion. □

Lemma 7.16. *Suppose $f_k(x) \in \mathcal{DS}_c$, say $f_k(x)$ has the Dirichlet series expansion $\sum_n s_k(n)n^{-x}$, and suppose the double series $\sum_k\sum_n |s_k(n)n^{-c}|$ converges. Then $\sum_k f_k(x) \in \mathcal{DS}_c$, and the Dirichlet series expansion of $\sum_k f_k(x)$ is given by $\sum_n \big(\sum_k s_k(n)\big)n^{-x}$.*

PROOF. For $x \geq c$, the comparison test shows that $\sum_k\sum_n s_k(n)n^{-x}$, a double series, converges absolutely. It follows that

$$\sum_k f_k(x) = \sum_k\sum_n s_k(n)n^{-x},$$

with both sides convergent for $x > c$. By Lemma 1.5, for $x \geq c$,

$$\begin{aligned}\sum_k\sum_n s_k(n)n^{-x} &= \sum_n\sum_k s_k(n)n^{-x}\\ &= \sum_n\Big(\sum_k s_k(n)\Big)n^{-x}\end{aligned}$$

with the absolute convergence of all sums mentioned. Thus $\sum_k f_k(x) \in \mathcal{DS}_c$, and the Dirichlet series expansion of $\sum_k f_k(x)$ is as described. □

7.2.1. Exponentiation. As with additive number systems, exponentiation plays a big role in understanding asymptotic density.

Lemma 7.17. *If* $f(x) \in \mathcal{DS}_c$ *then* $e^{f(x)} \in \mathcal{DS}_c$.

PROOF. If $f(x) \in \mathcal{DS}_c$ then there is a Dirichlet series $\mathbf{S}(x) = \sum_n s(n)n^{-x}$ such that $f(x) = \mathbf{S}(x)$ on (c, ∞), and $\sum_n |s(n)|n^{-c}$ converges. Thus, for $x \in (c, \infty)$,

$$\begin{aligned} e^{f(x)} &= \sum_k \frac{1}{k!}\mathbf{S}(x)^k \\ &= \sum_k \frac{1}{k!}\sum_n \sum_{n_1\cdots n_k=n} s(n_1)\cdots s(n_k)n^{-x} \\ &= \sum_k \sum_n \Big(\frac{1}{k!}\sum_{n_1\cdots n_k=n} s(n_1)\cdots s(n_k)\Big)n^{-x}. \qquad (7.17.1) \end{aligned}$$

Since, by Lemma 7.14, $\mathbf{S}(x)^k/k! \in \mathcal{DS}_c$, if the doubly infinite series (7.17.1) is absolutely convergent at $x = c$ then, by Lemma 7.16, $e^{f(x)} \in \mathcal{DS}_c$.

To show (7.17.1) is absolutely convergent at $x = c$ consider the Dirichlet series $\mathbf{T}(x) = \sum_n |s(n)|n^{-x}$. As noted above, this series is (absolutely) convergent at $x = c$. Thus

$$\begin{aligned} e^{\mathbf{T}(c)} &= \sum_k \frac{1}{k!}\mathbf{T}(c)^k \\ &= \sum_k \frac{1}{k!}\sum_n \sum_{n_1\cdots n_k=n} |s(n_1)|\cdots |s(n_k)|n^{-c} \\ &= \sum_k \sum_n \Big(\frac{1}{k!}\sum_{n_1\cdots n_k=n} |s(n_1)|\cdots |s(n_k)|\Big)n^{-c}. \qquad (7.17.2) \end{aligned}$$

So, the double series (7.17.2) is convergent (to $e^{\mathbf{T}(c)}$). As

$$\Big|\sum_{n_1\cdots n_k=n} s(n_1)\cdots s(n_k)\Big| \;\le\; \sum_{n_1\cdots n_k=n} |s(n_1)|\cdots |s(n_k)|,$$

the comparison test shows that (7.17.1) is indeed absolutely convergent at $x = c$. This finishes the proof. □

This section concludes with a look at the abscissa of absolute convergence of the Dirichlet series expansion of $e^{\mathbf{S}(x)}$, given that $\mathbf{S}(x)$ is a Dirichlet series with abscissa of absolute convergence $\alpha_s < \infty$. For $c > \alpha_s$ one has $\mathbf{S}(x) \in \mathcal{DS}_c$, so, by Lemma 7.17, $e^{\mathbf{S}(x)} \in \mathcal{DS}_c$ This means that the abscissa of absolute convergence $\widehat{\alpha}$ of the Dirichlet series expansion of $e^{\mathbf{S}(x)}$ is at most α_s. There is an interesting setting where $\widehat{\alpha} = \alpha_s$, and this will be important to find asymptotic formulas (see Proposition 9.57).

Lemma 7.18. *Suppose* $\mathbf{S}(x)$ *is a Dirichlet series with nonnegative coefficients and abscissa of (absolute) convergence* $\alpha_s < \infty$. *Then* $e^{\mathbf{S}(x)}$ *also has a Dirichlet series expansion with nonnegative coefficients and abscissa of (absolute) convergence* α_s.

PROOF. From the above, $\widehat{\alpha} \le \alpha_s$. Let $e^{\mathbf{S}(x)}$ have the Dirichlet series expansion $\mathbf{T}(x)$. Since the $s(n)$ are nonnegative it follows that the $t(n)$ are nonnegative, since $t(n)$ is the coefficient of n^{-x} in the Dirichlet series expansion of

$$\sum_j \frac{1}{j!}\Big(\sum_{k\ge 1} s(k)k^{-x}\Big)^j \;=\; 1 \;+\; \big(s(1)+\cdots+s(n)n^{-x}+\cdots\big) \;+\; \cdots.$$

Indeed $t(n) \ge s(n)$, for all n, and since $\sum_n s(n)n^{-x}$ diverges for $x < \alpha$, it follows that $\sum_n t(n)n^{-x}$ also diverges for $x < \alpha$. Thus $\widehat{\alpha} \ge \alpha_s$, so from the above, $\widehat{\alpha} = \alpha_s$. □

7.3. The abscissa of convergence and RV_α

Fortunately, the analog of the ratio test that is needed to study density in multiplicative number systems is an established concept in analysis called *regular variation at infinity*, a concept introduced by Karamata in the 1930s.

7.3.1. Regular variation at infinity.

Definition 7.19. Suppose $f(x)$ is eventually defined on the reals, and eventually positive. Given $\alpha \in \mathbb{R}$, $f(x)$ has *regular variation at infinity of index* α, abbreviated as $f(x) \in \mathsf{RV}_\alpha$, if

$$\lim_{t\to\infty} \frac{f(tx)}{f(t)} \;=\; x^\alpha, \qquad \text{for } x > 0. \tag{7.19.1}$$

If $f(x) \in \mathsf{RV}_0$ then $f(x)$ is also said to be *slowly varying at infinity.*

The simplest example of an $f(x) \in \mathsf{RV}_\alpha$ is surely x^α. This example occurs in a back and forth mapping between RV_0 and RV_α.

Proposition 7.20. *Suppose* $f(x)$ *is eventually defined on* $\mathbb{R}$, *and eventually positive. Then, for* $\alpha \in \mathbb{R}$,

$$f(x) \in \mathsf{RV}_\alpha \quad \textit{iff} \quad f(x)/x^\alpha \in \mathsf{RV}_0.$$

PROOF.

$$\begin{aligned} f(x) \in \mathsf{RV}_\alpha \quad &\Leftrightarrow \quad \lim_{t\to\infty} \frac{f(tx)}{f(t)} = x^\alpha \\ &\Leftrightarrow \quad \lim_{t\to\infty} \frac{f(tx)/(tx)^\alpha}{f(t)/t^\alpha} = 1 \\ &\Leftrightarrow \quad f(x)/x^\alpha \in \mathsf{RV}_0. \end{aligned}$$

□

Here are three of the simplest examples from RV_0:

$$\begin{array}{rclll} f(x) & = & c & & \text{for } c > 0 \\ f(x) & = & (\log x)^c & & \text{for } c \in \mathbb{R} \\ f(x) & = & d^{(\log x)^c} & & \text{for } c < 1 < d. \end{array}$$

These examples, combined with the following, lead to a substantial collection of $f(x)$ in RV_0.

Proposition 7.21.

(a) *If* $f(x) \sim g(x) \in \mathsf{RV}_\alpha$ *then* $f(x) \in \mathsf{RV}_\alpha$.

(b) *If* $f(x) \in \mathsf{RV}_\alpha$ *and* $g(x) \in \mathsf{RV}_\beta$ *then the reciprocal* $1/f(x) \in \mathsf{RV}_{-\alpha}$, *and the product* $f(x) \cdot g(x) \in \mathsf{RV}_{\alpha+\beta}$.

(c) RV_0 *is closed under multiplication, division, and asymptotic equality.*

PROOF. (a) Suppose $f(x) \sim g(x) \in \mathsf{RV}_\alpha$. Then $f(x)$ must be eventually positive, and

$$\frac{f(tx)}{f(t)} = \frac{f(tx)/g(tx)}{f(t)/g(t)} \cdot \frac{g(tx)}{g(t)} \rightarrow x^\alpha \qquad \text{as } t \rightarrow \infty,$$

so $f \in \mathsf{RV}_\alpha$.

(b) Suppose $f(x) \in \mathsf{RV}_\alpha$ and $g(x) \in \mathsf{RV}_\beta$. Then

$$\frac{1/f(tx)}{1/f(t)} = \frac{f(t)}{f(tx)} \rightarrow 1/x^\alpha \qquad \text{as } t \rightarrow \infty,$$

so $1/f(x) \in \mathsf{RV}_{-\alpha}$, and

$$\frac{f(tx)g(tx)}{f(t)g(t)} = \frac{f(tx)}{f(t)} \cdot \frac{g(tx)}{g(t)} \rightarrow x^\alpha x^\beta = x^{\alpha+\beta} \qquad \text{as } t \rightarrow \infty,$$

so $f(x) \cdot g(x) \in \mathsf{RV}_{\alpha+\beta}$.

(c) This follows from parts (a) and (b). □

With the above it is easily seen, for example, with $a, c > 0$ and $\alpha, b \in \mathbb{R}$, if $f(x) \sim a(\log x)^b e^{c\sqrt{\log x}} x^\alpha$ then $f(x) \in \mathsf{RV}_\alpha$. Such an example occurs with the asymptotics of §11.5.

The reader may have noticed a strong parallel between the introduction to RV_α above and that of RT_ρ in §1.4. The relation between the two goes even further.

Proposition 7.22. *Let* $f : \mathbb{R} \rightarrow \mathbb{R}$ *be [asymptotic to] an eventually positive and nondecreasing function. Then*

$$f(n) \in \mathsf{RT}_1 \quad \textit{iff} \quad f(\log x) \in \mathsf{RV}_0.$$

PROOF. ($\Rightarrow$) First note that, given $f(n) \in \mathsf{RT}_1$, $f(n+k)/f(n) \rightarrow 1$ as $n \rightarrow \infty$, for any given $k \in \mathbb{P}$. So, given an $x > 0$, for t sufficiently large,

$$\frac{f(\lfloor \log tx \rfloor)}{f(\lfloor \log t \rfloor + 1)} \leq \frac{f(\log tx)}{f(\log t)} \leq \frac{f(\lfloor \log tx \rfloor + 1)}{f(\lfloor \log t \rfloor)}.$$

Since the ratios on the two ends tend to 1 as $t \rightarrow \infty$, it follows that $f \in \mathsf{RV}_0$.

($\Leftarrow$) Note that, for $g \in \mathsf{RV}_0$ eventually nondecreasing, and for t suitably large, $g(t/2) \le g(t-1) \le g(t)$, so $g(t-1)/g(t) \to 1$ as $t \to \infty$.

Now, with $f(\log x) \in \mathsf{RV}_0$, if $\log = \log_d$ then

$$\frac{f(n-1)}{f(n)} = \frac{f\big(\log d^n - 1\big)}{f\big(\log d^n\big)} \to 1 \qquad \text{as } n \to \infty.$$

Thus $f(n) \in \mathsf{RT}_1$. □

The examples above of $f(x) \in \mathsf{RV}_0$ can be obtained via this proposition from the examples of $f(n) \in \mathsf{RT}_1$ listed in §1.4.

7.3.2. Regular variation and the abscissa of convergence. This section, which ties the index of regular variation to the abscissa of convergence, gives a very useful piece of information in the study of density in multiplicative number systems. (The following is based on the paper of Bateman and Diamond [**3**], p. 166.)

Proposition 7.23. *Suppose $f(x) \in \mathsf{RV}_\alpha$ is eventually nondecreasing and diverges (to infinity). Then*

$$\alpha = \lim_{n\to\infty} \frac{\log f(n)}{\log n}.$$

Proof. Since $f(2x)/f(x) \to 2^\alpha$ as $x \to \infty$ it follows, from the eventual monotonicity of $f(x)$, that $\alpha \ge 0$.

Given $\varepsilon > 0$, it is claimed that there are positive constants C_1, C_2 and M such that

$$C_1 x^{\alpha-\varepsilon} \le f(x) \le C_2 x^{\alpha+\varepsilon}, \qquad \text{for } x \ge M.$$

This will suffice to prove the proposition.

From $f(x) \in \mathsf{RV}_\alpha$ follows

$$\lim_{x\to\infty} \frac{f(2x)}{f(x)} = 2^\alpha,$$

so

$$\lim_{x\to\infty} \frac{f(2x)/(2x)^{\alpha-\varepsilon}}{f(x)/x^{\alpha-\varepsilon}} = 2^\varepsilon. \tag{7.23.1}$$

Choose $M_1 \ge 1$ such that $f(M_1) > 1$, $f(x)$ is nondecreasing on $[M_1, \infty)$ and

$$\left|\frac{f(2x)/(2x)^{\alpha-\varepsilon}}{f(x)/x^{\alpha-\varepsilon}} - 2^\varepsilon\right| < \frac{2^\varepsilon - 1}{2}, \qquad \text{for } x \ge M_1.$$

This implies, since $\varepsilon > 0$, that

$$\frac{f(2x)/(2x)^{\alpha-\varepsilon}}{f(x)/x^{\alpha-\varepsilon}} > \frac{2^\varepsilon + 1}{2} > 1,$$

and thus

$$f(2x)/(2x)^{\alpha-\varepsilon} > f(x)/x^{\alpha-\varepsilon}, \qquad \text{for } x \ge M_1. \tag{7.23.2}$$

For $x \geq M_1$, choose $k \in \mathbb{N}$ such that

$$2^k M_1 \;\leq\; x \;<\; 2^{k+1} M_1.$$

If $\alpha - \varepsilon \leq 0$ clearly $x^{\alpha-\varepsilon} \leq f(x)$, for $x \geq M_1$.

So assume $\alpha - \varepsilon > 0$. Then

$$\begin{aligned}
f(x)/x^{\alpha-\varepsilon} &= \frac{f(x)/x^{\alpha-\varepsilon}}{f(M_1)/M_1^{\alpha-\varepsilon}} \cdot f(M_1)/M_1^{\alpha-\varepsilon} \\
&\geq \frac{f(x)/x^{\alpha-\varepsilon}}{f(M_1)/M_1^{\alpha-\varepsilon}} \cdot 1/M_1^{\alpha} \\
&\geq \frac{f(x)/x^{\alpha-\varepsilon}}{f(2^k M_1)/(2^k M_1)^{\alpha-\varepsilon}} \cdot 1/M_1^{\alpha} \qquad \text{by (7.23.2)} \\
&= \frac{f(x)}{f(2^k M_1)} \cdot \frac{1}{2^{\alpha-\varepsilon}} \cdot \left(\frac{2^{k+1} M_1}{x}\right)^{\alpha-\varepsilon} \cdot 1/M_1^{\alpha} \\
&\geq 1 \cdot 1/2^{\alpha} \cdot 1 \cdot 1/M_1^{\alpha} \\
&= 1/(2M_1)^{\alpha}.
\end{aligned}$$

This establishes the lower bound $C_1 x^{\alpha-\varepsilon}$ of $f(x)$, with $C_1 = 1/(2M_1)^{\alpha}$.

For the upper bound, proceed in a similar fashion. It suffices to consider $\varepsilon \in (0,1)$. First modify (7.23.1) to

$$\lim_{x\to\infty} \frac{f(2x)/(2x)^{\alpha+\varepsilon}}{f(x)/x^{\alpha+\varepsilon}} \;=\; 2^{-\varepsilon},$$

and then choose an $M_2 \geq 1$ such that $f(x)$ is positive and nondecreasing on $[M_2, \infty)$, and

$$\left|\frac{f(2x)/(2x)^{\alpha+\varepsilon}}{f(x)/x^{\alpha+\varepsilon}} - 2^{-\varepsilon}\right| \;<\; \frac{1-2^{-\varepsilon}}{2} \qquad \text{for } x \geq M_2.$$

Then

$$\text{(7.23.3)} \qquad \frac{f(2x)/(2x)^{\alpha+\varepsilon}}{f(x)/x^{\alpha+\varepsilon}} \;<\; \frac{1+2^{-\varepsilon}}{2} \;<\; 1,$$

so,

$$\text{(7.23.4)} \qquad f(2x)/(2x)^{\alpha+\varepsilon} \;<\; f(x)/x^{\alpha+\varepsilon}, \qquad \text{for } x \geq M_2.$$

Choose $k \in \mathbb{N}$ such that, for $x \geq M_2$,

$$2^k M_2 \;\leq\; x \;<\; 2^{k+1} M_2.$$

Then, for $x \geq M_2$,

$$\begin{aligned}
f(x)/x^{\alpha+\varepsilon} &= \frac{f(x)/x^{\alpha+\varepsilon}}{f(M_2)/M_2^{\alpha+\varepsilon}} \cdot f(M_2)/M_2^{\alpha+\varepsilon} \\
&\leq \frac{f(x)/x^{\alpha+\varepsilon}}{f(M_2)/M_2^{\alpha+\varepsilon}} \cdot f(M_2)
\end{aligned}$$

$$
\begin{aligned}
&\le \frac{f(x)/x^{\alpha+\varepsilon}}{f(2^{k+1}M_2)/(2^{k+1}M_2)^{\alpha+\varepsilon}} \cdot f(M_2) \qquad \text{by (7.23.4)}\\
&= \frac{f(x)}{f(2^{k+1}M_2)} \cdot 2^{\alpha+\varepsilon} \cdot \left(\frac{2^k M_2}{x}\right)^{\alpha+\varepsilon} \cdot f(M_2)\\
&\le 1 \cdot 2^{\alpha+1} \cdot 1 \cdot f(M_2)\\
&= 2^{\alpha+1} f(M_2).
\end{aligned}
$$

This establishes the upper bound $C_2 x^{\alpha+\varepsilon}$ of $f(x)$, with $C_2 = 2^{\alpha+1} f(M_2)$.

With $M = \max(M_1, M_2)$ the proposition is proved. □

Corollary 7.24. *Suppose* $\mathbf{S}(x)$ *is a Dirichlet series such that one has* $S(x) \in \mathsf{RV}_\alpha$. *If* $S(x)$ *is eventually nondecreasing and diverges (to infinity) then* α *is the abscissa of (absolute) convergence of* $\mathbf{S}(x)$.

Proof. This is immediate from Proposition 7.23 in view of Proposition 7.3 (b). □

7.3.3. A simple test for $f \in \mathsf{RV}_0$. Knowing $\lim_{t\to\infty} \frac{f(tx)}{f(t)}$, for even a single $x \neq 1$, can be very informative.

Lemma 7.25 ([**17**], 1997). *Suppose* $f(x)$ *is eventually positive and non-decreasing, and, for some* $c \neq 1$ *in* $(0, \infty)$, $\lim_{t\to\infty} \frac{f(tc)}{f(t)} = 1$. *Then* $f(x) \in \mathsf{RV}_0$.

Proof. For such a c, $\lim_{t\to\infty} \frac{f(tc^m)}{f(t)} = 1$ holds, for any integer m, so assume $c > 1$. For any x in $(0, \infty)$, choose an integer m such that

$$c^m \le x \le c^{m+1}.$$

Then, for t sufficiently large,

$$\frac{f(tc^m)}{f(t)} \le \frac{f(tx)}{f(t)} \le \frac{f(tc^{m+1})}{f(t)}.$$

By the observation on powers of c, $\lim_{t\to\infty} \frac{f(tx)}{f(t)} = 1$. □

7.4. Growth rate of coefficients

Definition 7.26. For $f : \mathbb{P} \to \mathbb{R}$, let α_f be the abscissa of absolute convergence of $\sum f(n) n^{-x}$.

Definition 7.27. Given a function $f : \mathbb{P} \to \mathbb{R}$,

(a) f is *polynomially bounded* if

$$f(n) = \mathrm{O}(n^c), \qquad \text{for some } c > 0;$$

(b) f is *subpolynomial* if

$$f(n) = \mathrm{O}(n^c), \qquad \text{for every } c > 0.$$

Lemma 7.28. *Let $f : \mathbb{P} \to \mathbb{N}$. Then*
(a) $\alpha_f < \infty$ *iff* $f(n)$ *is polynomially bounded*
(b) $\alpha_f \leq 0$ *iff* $F(n)$ *is subpolynomial*
(c) $\alpha_f < 0$ *iff* $\sum f(n) < \infty$ *iff* $\alpha_f = -\infty$.

PROOF. For (a), it is clear that if $\sum_n f(n) < \infty$ then $f(n) \geq 0$ is eventually 0, so $f(n)$ is polynomially bounded and $\alpha_f = -\infty$. Thus only the case $\sum_n f(n) = \infty$ needs to be considered. Then, by Proposition 7.3 (b),

$$\begin{aligned}
\alpha_f < \infty \quad &\text{iff} \quad \limsup_{n\to\infty} \frac{\log F(n)}{\log n} < \infty \\
&\text{iff} \quad (\exists c > 0)\Big(\frac{\log F(n)}{\log n} < c\Big) \\
&\text{iff} \quad (\exists c > 0)\Big(F(n) < n^c\Big) \\
&\text{iff} \quad (\exists c > 0)\Big(f(n) = \mathrm{O}(n^c)\Big) \\
&\text{iff} \quad f \text{ is polynomially bounded .}
\end{aligned}$$

For (b), note that if $\sum_n f(n) < \infty$ then $\alpha_f = -\infty$ and $F(n)$ is eventually constant, and hence subpolynomial. If $\sum_n f(n) = \infty$ then $\alpha_f \geq 0$, and thus it suffices to show that $\alpha_f = 0$ iff $F(n)$ is subpolynomial. This follows from

$$\begin{aligned}
\alpha_f = 0 \quad &\text{iff} \quad \limsup_{n\to\infty} \frac{\log F(n)}{\log n} = 0 \\
&\text{iff} \quad (\forall c > 0)(\exists M > 0)\Big(\frac{\log F(n)}{\log n} < c, \quad \text{for } n > M\Big) \\
&\text{iff} \quad (\forall c > 0)(\exists M > 0)\Big(F(n) < n^c, \quad \text{for } n > M\Big) \\
&\text{iff} \quad (\forall c > 0)\big(F(n) = \mathrm{O}(n^c)\big) \\
&\text{iff} \quad F(n) \text{ is subpolynomial.}
\end{aligned}$$

(c) If $\lim_{n\to\infty} F(n) = \infty$ then $\alpha_f \geq 0$ by Proposition 7.3 (b). Thus $\alpha_f < 0$ implies $\sum_n f(n) < \infty$. Now if $\sum_n f(n) < \infty$ then $\sum_n f(n)n^{-x}$ is a finite sum, so $\alpha_f = -\infty$, and this in turn clearly implies $\alpha_f < 0$. □

Remark 7.29. If $F(n)$ is subpolynomial then it is clear that $f(n)$ is subpolynomial, but the converse does not hold, for example, when $f(n) = 1$, for all n.

CHAPTER 8

Counting Functions and Fundamental Identities

One goal of Part 2 is, where possible, to parallel the results of Part 1 in the more complex setting of multiplicative number systems derived from classes of algebraic structures.

8.1. Defining multiplicative number systems

Multiplicative number systems are generalizations of the usual multiplicative number system of positive integers. In Part 2 the multiplicative notation $\mathcal{M} = (M, \cdot, 1)$ is used for commutative monoids. For $a \in M$, one defines $a^0 = 1$, and a^n is the product of n copies of a, for $n \geq 1$. The *factorization* of an element of a free commutative monoid into indecomposables is expressed in product notation by

$$\prod p^{n_p},$$

where $p \in P$, $n_p \in \mathbb{N}$, and all but finitely many of the n_p are 0.

Definition 8.1. A *multiplicative norm* $\| \; \|$ on a commutative monoid $\mathcal{M}$ is a mapping from M to the positive integers such that

- $\|a\| \;=\; 1 \;\text{ iff }\; a \;=\; 1$
- $\|a \cdot b\| \;=\; \|a\| \cdot \|b\|.$

If $\mathcal{M}$ is a free commutative monoid, freely generated by P, then any multiplicative norm on M is completely determined by its restriction to P. Furthermore, any mapping $\| \; \|$ from P to the set of positive integers greater than 1 extends (uniquely) to a multiplicative norm on $\mathcal{M}$ via

$$\left\| \prod p^{n_p} \right\| \;=\; \prod \|p\|^{n_p}.$$

Definition 8.2. A *multiplicative number system*

$$\mathcal{A} = \bigl(\mathsf{A}, \mathsf{P}, \cdot, 1, \| \; \|\bigr)$$

consists of a countable free commutative monoid $(\mathsf{A}, \cdot, 1)$, with P being the set of *indecomposable elements*, also called *primes*, and with $\| \; \|$ a multiplicative norm that satisfies, for every $n \in \mathbb{P}$,

$$\{\mathsf{a} \in \mathsf{A} \;:\; \|\mathsf{a}\| = n\} \;\text{ is finite,}$$

that is, the norm is a finite-to-one mapping. The *rank* of $\mathcal{A}$ is $|\mathsf{P}|$.

Throughout Part 2 the letter $\mathcal{A}$ will denote a multiplicative number system whose rank is positive, that is, $\mathsf{P} \neq \varnothing$.

Definition 8.3. A multiplicative number system $\mathcal{B} = (\mathsf{B}, \mathsf{Q}, \cdot, 1, \| \ \|)$ is a *subsystem* of $\mathcal{A} = (\mathsf{A}, \mathsf{P}, \cdot, 1, \| \ \|)$ if $\mathsf{Q} \subseteq \mathsf{P}$, and if the operation $\cdot$, the constant 1, and the norm $\| \ \|$ of $\mathcal{B}$ agree with the corresponding items of $\mathcal{A}$. In this case one writes $\mathcal{B} \leq \mathcal{A}$, and says that $\mathcal{B}$ is the *subsystem generated by* Q. To indicate that $\mathcal{B}$ is a *proper* subsystem, that is, $\mathsf{B} \neq \mathsf{A}$, the expression $\mathcal{B} < \mathcal{A}$ is used. A subsystem $\mathcal{B}$ is *trivial* if $\mathsf{Q} = \varnothing$.

The concept of two multiplicative number systems being 'essentially the same' is captured in the next definition.

Definition 8.4. Let $\mathcal{A}_1$ and $\mathcal{A}_2$ be two multiplicative number systems. $\mathcal{A}_1$ is *isomorphic to* $\mathcal{A}_2$, written $\mathcal{A}_1 \cong \mathcal{A}_2$, if there is a bijection $\mu : \mathsf{A}_1 \longrightarrow \mathsf{A}_2$ that preserves multiplication, one, and norm, that is,

$$\begin{aligned} \mu(\mathsf{a} \cdot_1 \mathsf{b}) &= \mu(\mathsf{a}) \cdot_2 \mu(\mathsf{b}) && \text{for } \mathsf{a}, \mathsf{b} \in \mathsf{A}_1 \\ \mu(1_1) &= 1_2 && \\ \|\mu(\mathsf{a})\|_2 &= \|\mathsf{a}\|_1 && \text{for } \mathsf{a} \in \mathsf{A}_1. \end{aligned}$$

Such a mapping μ is called an *isomorphism.*

As is common practice in algebra, the distinct notations $\cdot_i$, 1_i, and $\| \ \|_i$ are replaced by $\cdot$, 1, and $\| \ \|$.

Isomorphism is an equivalence relation on the class of multiplicative number systems. Note that the indecomposables were not mentioned in the definition of isomorphism. This is because the above definition of isomorphism ensures that indecomposables map to indecomposables.

Lemma 8.5. *If* $\mu : \mathcal{A}_1 \longrightarrow \mathcal{A}_2$ *is an isomorphism then* $\mu(\mathsf{P}_1) = \mathsf{P}_2$.

PROOF. An element $\mathsf{a} \in \mathsf{A}_1$ is indecomposable iff $\mathsf{a} = \mathsf{b} \cdot \mathsf{c}$ implies either $\mathsf{b} = 1$ or $\mathsf{c} = 1$. Now $\mathsf{a} = \mathsf{b} \cdot \mathsf{c}$ iff $\mu(\mathsf{a}) = \mu(\mathsf{b}) \cdot \mu(\mathsf{c})$, $\mathsf{b} = 1$ iff $\mu(\mathsf{b}) = 1$, and $\mathsf{c} = 1$ iff $\mu(\mathsf{c}) = 1$. Thus a is indecomposable iff $\mu(\mathsf{a})$ is indecomposable. □

8.2. Examples of multiplicative number systems

By far the most famous multiplicative number system is the set $\mathbb{P}$ of positive integers with the usual multiplication. In this case the indecomposables are the prime numbers, so the rank is infinite. In the 19th century there was considerable interest in the multiplicative number systems, like the Gaussian integers, obtained from algebraic number fields. Here one has to be careful since the unique factorization property can fail. Also, one has to consider the possibility of many units, giving many elements of norm 1, which violates the definition of a multiplicative number system. The cure, for both problems, is to go to the number system of ideals. There has been a great deal of research on these systems, including Landau's prime number theorem. This program is continued in the more general context of multiplicative number systems in John Knopfmacher's books [**29**] and [**30**].

Multiplicative number systems derived from classes of finite structures, especially algebraic structures, provide the connection with first-order limit laws. Several examples are introduced in this section. As in the additive case, the idea behind a derived multiplicative number system is simply to consider structures up to isomorphism. The operation of multiplication will be the direct product, and the norm the number of elements in the underlying set of the structure. One has to be careful with finite algebraic systems since unique factorization is not automatic. However, any algebraic structure with an underlying group, or ring, or lattice, or even semilattice, will have unique factorization. (For an in-depth discussion of unique factorization see [**11**] and [**35**].)

8.2.1. Boolean algebras. Boolean algebras $\mathbf{B} = (B, \vee, \wedge, ', 0, 1)$ are defined by the equations that govern union ($\cup$), intersection($\cap$), and complement ($'$) with regard to a universe of sets, with 0 corresponding to the empty set, and 1 to the universe. The collection of finite Boolean algebras give a multiplicative number system in which the only indecomposable is the famous two-element Boolean algebra. If the norm is $\log_2$ of the size then one has an additive number system.

8.2.2. Abelian p-groups. Finite Abelian p-groups, where p is a fixed prime number, give a multiplicative number system in which the indecomposables are the cyclic p-groups, the $\mathbf{Z}_{p^k}$. If $\log_p$ of the size is used for the norm, then again one has an additive number system.

8.2.3. $x^m = x$ rings. The finite rings that satisfy an equation $x^m = x$, for a fixed m, give a multiplicative number system in which the indecomposables are the finite fields $\mathbf{F}_q$ with q a power of a prime number, and $q-1 \big| m-1$. Thus there are only finitely many indecomposables. In general, there is no way to change the norm so that one has an additive number system.

8.2.4. The equational class $\mathcal{HA}_3$. This is the equational class of Heyting algebras (see [**16**]) generated by the three-element chain. The finite members of this class give a multiplicative number system with exactly one indecomposable of size $2^n + 1$, for each $n \geq 1$.

8.2.5. Abelian groups. The finite Abelian groups form a multiplicative number system with the indecomposables being the finite cyclic groups, the $\mathbf{Z}_{p^n}$, where p is any prime number. Thus there is exactly one indecomposable of size p^n, for each prime number p and positive integer n.

8.3. Counting functions, fundamental identities

Definition 8.6. For $\mathcal{A} = (\mathsf{A}, \mathsf{P}, \cdot, 1, \| \ \|)$ a multiplicative number system and $\mathsf{B} \subseteq \mathsf{A}$, define, just as with additive number systems, the (*local*) *counting function* $b(n)$ of B by

$$b(n) \quad = \quad \big|\{\mathsf{b} \in \mathsf{B} : \|\mathsf{b}\| = n\}\big|,$$

that is, $b(n)$ gives the number of elements in B with norm n.

In some situations, particularly when the set B has a more complicated name than just a single letter, the notation $|\mathsf{B}|_n$ is used to denote $b(n)$. The corresponding Dirichlet series $\mathbf{B}(x)$, called the *(Dirichlet) generating series* of B, is defined by:[1]

$$\mathbf{B}(x) = \sum_{n\geq 1} b(n)n^{-x}.$$

$\mathbf{B}(x)$ can also be expressed by

$$\mathbf{B}(x) = \sum_{\mathsf{b}\in\mathsf{B}} \|\mathsf{b}\|^{-x}.$$

The two most significant counting functions are $a(n)$ and $p(n)$, for A and P, called the *(local) counting functions of* $\mathcal{A}$. Note that $a(1) = 1$ and $p(1) = 0$. So the generating series of A and P can be expressed as:

$$\mathbf{A}(x) = \sum_{n\geq 1} a(n)n^{-x}$$

$$\mathbf{P}(x) = \sum_{n\geq 2} p(n)n^{-x}.$$

A multiplicative number system is completely determined, up to isomorphism, by either of its counting functions $a(n)$ or $p(n)$.

Proposition 8.7. *Let $\mathcal{A}_i$, $i = 1, 2$, be two multiplicative number systems, and let $a_i(n), p_i(n)$ be their counting functions. Then*

$$\mathcal{A}_1 \cong \mathcal{A}_2 \quad \textit{iff} \quad p_1 = p_2 \quad \textit{iff} \quad a_1 = a_2.$$

PROOF. If $\mu : \mathcal{A}_1 \to \mathcal{A}_2$ is an isomorphism then, by Lemma 8.5, $\mathsf{a} \in \mathsf{A}_1$ is an indecomposable iff $\mu(\mathsf{a})$ is an indecomposable in A_2. As $\|\mathsf{a}\| = \|\mu(\mathsf{a})\|$, it follows that a is an indecomposable of norm n iff $\mu(\mathsf{a})$ is an indecomposable of norm n. Thus $p_1(n) = p_2(n)$.

If $p_1 = p_2$ then let $\mu : \mathsf{P}_1 \to \mathsf{P}_2$ be a bijection that preserves the norm. Now extend this to A_1 by

$$\mu\Big(\prod \mathsf{p}^{n_\mathsf{p}}\Big) = \prod \mu(\mathsf{p})^{n_\mathsf{p}}.$$

It is easily seen that this defines an isomorphism, so $\mathcal{A}_1 \cong \mathcal{A}_2$.

For any multiplicative number system $\mathcal{A}$, one can determine $a(n)$ from $p(n)$ as the elements of norm n are uniquely expressible as products of indecomposables of size at most n. Also, one can recursively find $p(n)$ from $a(n)$ by noting that $p(2) = a(2)$, and, having determined $p(m)$ for $m < n$,

[1]For the usual multiplicative number system $(\mathsf{A}, \mathsf{P}, \cdot, 1, \|\ \|)$, where A is the set $\mathbb{P}$ of positive integers, the function $\mathbf{A}(x)$ is the Riemann zeta function $\zeta(x)$, and, for this reason, notations similar to $\zeta_\mathsf{B}(x)$, instead of $\mathbf{B}(x)$, are used by some authors. The simpler notation $\mathbf{B}(x)$ is preferred because of the modified names of sets B, for example, $\mathsf{B}^{\leq n}$, that will be used. This convention reduces the levels of sub/superscripting needed by one.

compute the number $b(n)$ of elements of norm n that can be expressed as products of indecomposables of size less than n. Then $p(n) = a(n) - b(n)$. □

Definition 8.8. The *fundamental identity* of $\mathcal{A}$ is

$$\sum_{n\geq 1} a(n)n^{-x} = \prod_{n\geq 2}(1-n^{-x})^{-p(n)}. \tag{8.8.1}$$

The right side is called the *Euler product* of $\mathcal{A}$.[2]

A detailed discussion of how to understand this, as describing a method to compute the values of $a(n)$ from $p(n)$, is given in §8.3.7, as well as a proof that it holds as an identity between two functions, for $x > 0$.

8.3.1. Boolean algebras.

The fundamental identity of the multiplicative number system of finite Boolean algebras is

$$\sum_{n\geq 1} a(n)n^{-x} = \left(1-2^{-x}\right)^{-1},$$

where $a(n) = 1$ if n is a power of 2, otherwise $a(n) = 0$, since there is exactly one finite Boolean algebra, up to isomorphism, of size 2^n, for $n \geq 0$, and no others. In the multiplicative setting the primary concern is with global results. However, since the number system derived from Boolean algebras can also be considered as an additive number system, simply by changing the norm, a stronger result holds—namely, since $p(n) \leq 1$ there is a *local* first-order 0–1 law.[3]

8.3.2. Abelian p-groups.

The fundamental identity of the multiplicative number system of finite Abelian p-groups is

$$\sum_{n\geq 1} a(n)n^{-x} = \prod_{n\geq 1}\left(1-(p^n)^{-x}\right)^{-1}.$$

Here $a(p^k)$ is equal to the number of ways to partition the number k as a sum of positive integers; and otherwise $a(n) = 0$. However, since this number system can also be considered as an additive number system with $p(n) \leq 1$ simply by changing the norm, there is a local first-order 0–1 law. Thus it will be shown, for example, that the local probability that a randomly chosen finite Abelian p-group has at least k copies of an indecomposable $\mathbf{Z}_{p^m}$ in its factorization (into indecomposables) is 1, for any given k and m.

---

[2]Used by Euler with the usual multiplicative number system of positive integers.

[3]This means, for any given sentence φ, that the probability of φ holding is the limit as n goes to infinity of the proportion of structures of size n that satisfy φ, considering only those n for which structures of size n exist.

8.3.3. $x^m = x$ rings. The fundamental identity of the multiplicative number system of finite rings satisfying $x^m = x$ is

$$\sum_{n\geq 1} a(n)n^{-x} = \prod_{\substack{p^n-1|m-1 \\ p \text{ prime}, n\geq 1}} \left(1-(p^n)^{-x}\right)^{-1}.$$

Since this number system is finitely generated, $x^m = x$ rings must have a global first-order 0–1 law. (But generally they do not have a local first-order 0–1 law.) Thus it will be proved, for example, that the global probability that a randomly chosen finite ring satisfying $x^m = x$ has at least k copies of a finite field $\mathbf{F}_q$ in its factorization (into indecomposables) is 1, for any given k, and prime power q such that $q-1|m-1$.

8.3.4. The equational class $\mathcal{HA}_3$. The fundamental identity of the multiplicative number system of the class $\mathcal{HA}_3$ of Heyting algebras is

$$\sum_{n\geq 1} a(n)n^{-x} = \prod_{n\geq 1} \left(1-(2^n+1)^{-x}\right)^{-1}.$$

Since $P(n) \leq \log_2 n$, from §10.3 it follows that $\mathcal{HA}_3$ must have a global first-order 0–1 law (but not a local first-order 0–1 law). The global probability that a given indecomposable appears at least k times in the factorization of a member of $\mathcal{HA}_3$ is 1.

8.3.5. Abelian groups. The fundamental identity of the multiplicative number system of finite Abelian groups is

$$\sum_{n\geq 1} a(n)n^{-x} = \prod_{\substack{p \text{ prime} \\ n\geq 1}} \left(1-(p^n)^{-x}\right)^{-1}.$$

In this case there is no 0–1 law. Instead, deeper results lead to a global first-order law, simply because $a(n) \sim cn$, for a suitable choice of c. It will be proved, by letting $\mathbf{B}(x)$ be the generating function of the collection of finite Abelian groups with an element of order 2, that the global probability of a randomly chosen finite Abelian group having an element of order 2 is given by

$$\lim_{x\to 1} \frac{\mathbf{B}(x)}{\mathbf{A}(x)} = 1 - \prod_{n\geq 1} \left(1-2^{-n}\right) \approx 0.71,$$

or a probability of about 71%.

8.3.6. A Dirichlet product. As a prelude to analyzing the fundamental identity, the Dirichlet product

$$\prod^{*}_{\mathsf{p}\in\mathsf{P}} \left(1+\|\mathsf{p}\|^{-x}+\|\mathsf{p}\|^{-2x}+\cdots\right)$$

will be discussed. The case that P is infinite is examined in this section. The discussion when P is finite is essentially the same, but easier, since one is not concerned with the convergence of infinite products.

So, given that P is infinite, let p_n be an ordering of the elements of P, and let $g(n) = \|\mathsf{p}_n\|$. Then $g : \mathbb{N} \to \mathbb{P}'$, and $g(n) \to \infty$. In the notation of §7.1.2,

$$\mathbf{S}_n^g(x) \;=\; 1 + \|\mathsf{p}_n\|^{-x} + \|\mathsf{p}_n\|^{-2x} + \cdots,$$

so

$$\prod^{*}_{n\geq 0} \left(1 + \|\mathsf{p}_n\|^{-x} + \|\mathsf{p}_n\|^{-2x} + \cdots\right)$$

is the infinite Dirichlet product $\prod^{*}_{n\geq 0} \mathbf{S}_n^g(x)$ discussed in §7.1.2.

Lemma 8.9. *For $x \geq 0$,*

$$\prod^{*}_{n\geq 0} \left(1 + \|\mathsf{p}_n\|^{-x} + \|\mathsf{p}_n\|^{-2x} + \cdots\right) \;=\; \prod_{n\geq 0} \left(1 + \|\mathsf{p}_n\|^{-x} + \|\mathsf{p}_n\|^{-2x} + \cdots\right),$$

and the products do not depend on the ordering of the factors.

Proof. This follows from Lemmas 1.9 and 7.9. □

Since the ordering of the factors is unimportant, it is preferable, when possible, not to refer to any particular ordering. Then the above equation can be written more simply as

$$\prod^{*}_{\mathsf{p}\in\mathsf{P}} \left(1 + \|\mathsf{p}\|^{-x} + \|\mathsf{p}\|^{-2x} + \cdots\right) \;=\; \prod_{\mathsf{p}\in\mathsf{P}} \left(1 + \|\mathsf{p}\|^{-x} + \|\mathsf{p}\|^{-2x} + \cdots\right).$$

Lemma 8.10. *For $x \geq 0$,*

$$\prod_{\mathsf{p}\in\mathsf{P}} \left(1 + \|\mathsf{p}\|^{-x} + \|\mathsf{p}\|^{-2x} + \cdots\right) \;=\; \prod_{n\geq 2} \left(1 + n^{-x} + n^{-2x} + \cdots\right)^{p(n)}.$$

Proof. Since the left side has $p(n)$ factors involving p with $\|\mathsf{p}\| = n$, the right side is just a particular arrangement of the left side. □

Proposition 8.11. *For $x > 0$,*

$$\begin{aligned}
\prod_{n\geq 2} \left(1 - n^{-x}\right)^{-p(n)} \;&=\; \prod_{\mathsf{p}\in\mathsf{P}} \left(1 - \|\mathsf{p}\|^{-x}\right)^{-1} \\
&=\; \prod_{\mathsf{p}\in\mathsf{P}} \left(1 + \|\mathsf{p}\|^{-x} + \|\mathsf{p}\|^{-2x} + \cdots\right) \\
&=\; \prod^{*}_{\mathsf{p}\in\mathsf{P}} \left(1 + \|\mathsf{p}\|^{-x} + \|\mathsf{p}\|^{-2x} + \cdots\right).
\end{aligned}$$

Proof. This follows from Proposition 7.10 and the previous two lemmas. □

8.3.7. A(x) as a Dirichlet product. The Dirichlet product described in the last section is $\mathbf{A}(x)$. This gives a combinatorial connection between the counting functions $a(n)$ and $p(n)$ that includes the fundamental identity.

Proposition 8.12. *For any multiplicative number system $\mathcal{A}$,*

$$\mathbf{A}(x) \;=\; \prod^{*}_{\mathsf{p}\in\mathsf{P}} \left(1+\|\mathsf{p}\|^{-x}+\|\mathsf{p}\|^{-2x}+\cdots\right).$$

PROOF. The case that P is infinite will be discussed—the case that P is finite requiring a similar but simpler analysis. The equation asserts that $\mathbf{A}(x)$ is the indicated infinite Dirichlet product, so $a(m)$, the number of elements of norm m in A, is the stable value of the coefficient of m^{-x} in the initial products of the infinite Dirichlet product:

$$a(m) \;=\; \left[m^{-x}\right] \prod^{*}_{\substack{\mathsf{p}\in\mathsf{P}\\ \|\mathsf{p}\|\le n}} \left(1+\|\mathsf{p}\|^{-x}+\|\mathsf{p}\|^{-2x}+\cdots\right),$$

for any $n \ge m$.

To see that this is actually true, proceed as follows. Let $\mathsf{p}_0, \mathsf{p}_1, \ldots$ be a listing of the indecomposable elements of A. There is a natural one-to-one correspondence between the elements of A and the sequences (m_i) of nonnegative integers that have only finitely many nonzero entries, namely, (m_i) corresponds to the element a such that m_i is the exponent of p_i in the factorization of a into indecomposables. Thus the sequences $\left(\|\mathsf{p}_i\|^{-m_i x}\right)$ of monomials, where almost all monomials are 1, are also naturally in a one-to-one correspondence with the elements of A.

Now, $a(m)$ is the number of sequences (m_i) of nonnegative integers such that $m = \prod \|\mathsf{p}_i\|^{m_i}$. This is because each element of A with norm m can be expressed as a product $\prod \mathsf{p}_i^{m_i}$ with $\prod \|\mathsf{p}_i\|^{m_i} = m$, and distinct sequences (m_i) give rise to distinct elements $\prod \mathsf{p}_i^{m_i}$ of A. Note that $m = \prod \|\mathsf{p}_i\|^{m_i}$ iff $m^{-x} = \prod \|\mathsf{p}_i\|^{-m_i x}$.

Thus $a(m)$ is the number of sequences (m_i) of nonnegative integers such that $m^{-x} = \prod \|\mathsf{p}_i\|^{-m_i x}$. Now consider the product

(8.12.1)
$$\left(1+\|\mathsf{p}_0\|^{-x}+\|\mathsf{p}_0\|^{-2x}+\cdots\right)\;\cdots\;\left(1+\|\mathsf{p}_n\|^{-x}+\|\mathsf{p}_n\|^{-2x}+\cdots\right),$$

where n is sufficiently large to ensure that all indecomposables of norm at most m appear in the list $\mathsf{p}_0, \ldots, \mathsf{p}_n$. Then each sequence (m_i) with $m^{-x} = \prod \|\mathsf{p}_i\|^{-m_i x}$ corresponds to a choice of one monomial from each of the sums $1+\|\mathsf{p}_i\|^{-x}+\|\mathsf{p}_i\|^{-2x}+\cdots$ above, and thus $a(m)$ is the coefficient of m^{-x} in the infinite Dirichlet product $\prod^{*}_{\mathsf{p}\in\mathsf{P}} \left(1+\|\mathsf{p}\|^{-x}+\|\mathsf{p}\|^{-2x}+\cdots\right)$. $\square$

Combining the results of this section and the previous one gives the following.

Theorem 8.13. *For $x \geq 0$,*

$$\begin{aligned}\mathbf{A}(x) &= \prod^{*}_{\mathsf{p}\in\mathsf{P}} \left(1 + \|\mathsf{p}\|^{-x} + \|\mathsf{p}\|^{-2x} + \cdots\right) \\ &= \prod_{\mathsf{p}\in\mathsf{P}} \left(1 + \|\mathsf{p}\|^{-x} + \|\mathsf{p}\|^{-2x} + \cdots\right) \\ &= \prod_{n\geq 2} \left(1 + n^{-x} + n^{-2x} + \cdots\right)^{p(n)},\end{aligned}$$

and, for $x > 0$, the following also hold:

$$\begin{aligned}\mathbf{A}(x) &= \prod_{\mathsf{p}\in\mathsf{P}} \left(1 - \|\mathsf{p}\|^{-x}\right)^{-1} \\ &= \prod_{n\geq 2} \left(1 - n^{-x}\right)^{-p(n)}.\end{aligned}$$

Remark 8.14. Requiring that the $a(n), p(n)$ in the fundamental identity be obtained from a multiplicative number system $\mathcal{A}$ is not a restriction on the possible fundamental identities. For suppose $a(n), p(n)$ are two functions with values in $\mathbb{N}$ that satisfy (8.8.1). Take a free commutative monoid of rank $r = \sum_n p(n)$ and endow it with a norm such that the number of generators with norm n will be exactly $p(n)$. For this multiplicative number system $\mathcal{A}$, the associated counting functions are precisely the given $a(n), p(n)$.

8.3.8. The abscissa of convergence of $\mathcal{A}$.

Definition 8.15. The *abscissa of convergence α of $\mathcal{A}$* is the abscissa of convergence of the generating series $\mathbf{A}(x)$, that is, $\alpha = \alpha_a$, the abscissa of convergence of $\sum_n a(n)n^{-x}$.

Lemma 8.16. *The abscissa of convergence of $\mathcal{A}$ is given by*

$$\alpha = \limsup_{n\to\infty} \frac{\log A(n)}{\log n} \geq 0.$$

Proof. Since $\mathcal{A}$ has positive rank it follows that $a(n) \geq 1$ holds infinitely often, so $\mathbf{A}(0) = \infty$. Thus Proposition 7.3 (b) applies. □

Corollary 8.17. *The fundamental identity*

$$\sum_{n\geq 1} a(n)n^{-x} = \prod_{n\geq 2}(1 - n^{-x})^{-p(n)}$$

of $\mathcal{A}$ is valid for $x > 0$. The two sides converge for $x > \alpha$. For $0 < x < \alpha$, the two sides have the value ∞.

Proof. Just add to Theorem 8.13 the observation that $\mathbf{A}(x)$ converges when $x > \alpha$, and diverges when $x < \alpha$. □

Corollary 8.18.

(a) *For* $x > 0$, $\mathbf{A}(x)$ *converges iff* $\mathbf{P}(x)$ *converges.*

(b) $\alpha \;=\; \max(0, \alpha_p)$.

(c) *The following conditions are equivalent:*

- $a(n)$ *is polynomially bounded*
- $\alpha < \infty$
- $p(n)$ *is polynomially bounded*
- $\alpha_p < \infty$.

(d) *The following conditions are equivalent:*

- $A(n)$ *is subpolynomial*
- $\alpha = 0$
- $P(n)$ *is subpolynomial*
- $\alpha_p \le 0$.

PROOF. (a) Let $x > 0$. Then

$$\begin{array}{lll} \mathbf{A}(x) \text{ converges} & \text{iff } \prod_{n\ge 2}(1-n^{-x})^{-p(n)} \text{ converges} & \text{by Proposition 8.17} \\ & \text{iff } \mathbf{P}(x) \text{ converges} & \text{by Lemma 7.11.} \end{array}$$

(b) Since $\alpha_{\mathsf{p}} \le \alpha$ use (a) and Lemma 8.16.

(c) $a(n)$ is polynomially bounded

$$\begin{array}{lll} \text{iff} & \alpha < \infty & \text{by Lemma 7.28 (a)} \\ \text{iff} & \alpha_p < \infty & \text{by part (b)} \\ \text{iff} & p(n) \text{ is polynomially bounded} & \text{by Lemma 7.28 (a).} \end{array}$$

(d) $A(n)$ is subpolynomial

$$\begin{array}{lll} \text{iff} & \alpha = 0 & \text{by Lemma 7.28 (b) and Lemma 8.16} \\ \text{iff} & \alpha_p \le 0 & \text{by part (b)} \\ \text{iff} & P(n) \text{ is subpolynomial} & \text{by Lemma 7.28 (b).} \end{array}$$

□

Remark 8.19. When working with the usual multiplicative number system of positive integers,[4] the fundamental identity is

$$\zeta(x) \;=\; \sum_{n\ge 1} n^{-x} \;=\; \prod_{p\in\text{ primes}} (1-p^{-x})^{-1}.$$

Clearly the abscissa of convergence is $\alpha = 1$. By Corollary 8.18 (a), putting $x = 1$,

$$\zeta(1) \;=\; \sum_{n\ge 1} \frac{1}{n} \text{ converges iff } \mathbf{P}(1) \;=\; \sum_{p\in\text{ primes}} \frac{1}{p} \text{ converges.}$$

Since the harmonic series $\sum_n 1/n$ diverges it follows that $\sum_p 1/p$ diverges, giving Euler's proof of the infinitude of the number of primes.

---

[4]In 1837 Dirichlet worked with $\zeta(x)$, for x real, to prove that if a, b are coprime positive integers then the arithmetic progression $a + bn$ has an infinite number of primes in it. Later, in 1859, Riemann studied $\zeta(z)$, for $z \in \mathbb{C}$. He showed that one could extend the Dirichlet series $\zeta(z)$ to a function (also called $\zeta(z)$) that is analytic everywhere except at $z = 1$. Out of this study came the famous *Riemann Hypothesis*, which says that all the zeros of $\zeta(z)$ with real part positive have the real part equal to 1/2.

For the multiplicative number systems derived from finite Boolean algebras, Abelian p-groups, $x^m = x$ rings and $\mathcal{HA}_3$, $\alpha = 0$. For the Abelian groups, $\alpha = 1$. The condition $\alpha = 0$ is needed to prove logical 0–1 laws, and Chapter 10 is devoted to this. The case $\alpha > 0$ is more difficult and is treated in Chapter 11.

Remark 8.20. One cannot extend part (a) of Corollary 8.18 to include $x = 0$ since $\mathbf{A}(0) = \infty$, but $\mathbf{P}(0)$ could be either finite or infinite, depending on whether P is finite or infinite.

Remark 8.21. If $a(n)$, equivalently $p(n)$, is not polynomially bounded, then there is no analytical information available in the fundamental identity since $\alpha = \infty$. For example, if $p(n) > n^{\log n}$ holds infinitely often, then $\alpha = \infty$.

Following the terminology of additive systems, let us call $\mathcal{A}$ *fast growing* if $\alpha = \infty$, and *slowly growing* if $\alpha < \infty$. There is very little information on density in fast growing multiplicative systems. The density results obtained in Part 2 are, as in Part 1, for slowly growing systems.

8.4. Alternate version of the fundamental identity

The fundamental identity gives a connection between the two counting functions $a(n)$ and $p(n)$ of a multiplicative number system. For the purpose of studying density of partition sets, a variation on this identity is useful.

Proposition 8.22. *The identity*

$$\mathbf{A}(x) \;=\; \exp\Big(\sum_{m\geq 1}\mathbf{P}(mx)/m\Big) \tag{8.22.1}$$

holds, for $x \geq 0$.

PROOF. For $x = 0$, both sides of (8.22.1) are clearly equal to ∞. For $x > 0$,

$$\begin{aligned}(1-n^{-x})^{-p(n)} &= \exp\big(\log\big((1-n^{-x})^{-p(n)}\big)\big)\\ &= \exp\big(-p(n)\log(1-n^{-x})\big)\\ &= \exp\Big(p(n)\sum_{m\geq 1} n^{-mx}/m\Big),\end{aligned}$$

thus,

$$\prod_{2\leq n\leq M}(1-n^{-x})^{-p(n)} \;=\; \prod_{2\leq n\leq M}\exp\Big(p(n)\sum_{m\geq 1} n^{-mx}/m\Big).$$

So, from Proposition 8.17,

$$\begin{aligned}
\mathbf{A}(x) &= \prod_{n\geq 2}\left(1-n^{-x}\right)^{-p(n)} \\
&= \lim_{M\to\infty}\prod_{2\leq n\leq M}\exp\Big(\sum_{m\geq 1}p(n)n^{-mx}/m\Big) \\
&= \lim_{M\to\infty}\exp\Big(\sum_{2\leq n\leq M}\sum_{m\geq 1}p(n)n^{-mx}/m\Big) \\
&= \exp\Big(\sum_{n\geq 2}\sum_{m\geq 1}p(n)n^{-mx}/m\Big) \\
&= \exp\Big(\sum_{m\geq 1}\sum_{n\geq 2}p(n)n^{-mx}/m\Big) \\
&= \exp\Big(\sum_{m\geq 1}\mathbf{P}(mx)/m\Big).
\end{aligned}$$

□

This shows that $\mathbf{A}(x)$ exponentially dominates $\mathbf{P}(x)$.

Corollary 8.23. *For $x \geq 0$,*

$$e^{\mathbf{P}(x)} \leq \mathbf{A}(x).$$

Definition 8.24. Let $\mathbf{H}(x) = \sum h(n)n^{-x}$ be the Dirichlet series defined by

$$h(n) = \begin{cases} 0 & \text{for } n = 1 \\ \sum_{j^m=n} p(j)/m & \text{for } n \geq 2. \end{cases}$$

Thus $h(n) \geq p(n)$, for $n \geq 1$, and $\mathbf{H}(x) = \sum_{n\geq 2} h(n)n^{-x}$.

Proposition 8.25. *The identity*

$$\mathbf{H}(x) = \sum_{m\geq 1}\mathbf{P}(mx)/m$$

holds for $x \geq 0$.

Proof. For $x \geq 0$ and $\ell \geq k \geq 0$,

$$\begin{aligned}
\sum_{n=1}^{k} h(n)n^{-x} &= \sum_{n=1}^{k}\sum_{j^m=n}\frac{p(j)}{m}n^{-x} \\
&= \sum_{1\leq j^m\leq k}\frac{p(j)}{m}j^{-mx} \\
&\leq \sum_{1\leq m\leq k}\sum_{1\leq j\leq \ell}\frac{p(j)}{m}j^{-mx}
\end{aligned}$$

$$\begin{aligned}
&\leq \sum_{1\leq j^m\leq \ell^k} \frac{p(j)}{m} j^{-mx} \\
&= \sum_{n=1}^{\ell^k} \sum_{j^m=n} \frac{p(j)}{m} j^{-mx} \\
&= \sum_{n=1}^{\ell^k} h(n)n^{-x} \\
&\leq \mathbf{H}(x),
\end{aligned}$$

so,

$$\sum_{n=1}^{k} h(n)n^{-x} \;\leq\; \sum_{1\leq m\leq k} \Big(\sum_{1\leq j\leq \ell} p(j)j^{-mx}\Big)\Big/ m \;\leq\; \mathbf{H}(x).$$

Taking the limit as $\ell \to \infty$, and then as $k \to \infty$, gives the result. □

With this the alternate version of the fundamental identity of $\mathcal{A}$ can be expressed as $\mathbf{A}(x) = e^{\mathbf{H}(x)}$, and one has the following analytic information from Proposition 8.22:

Proposition 8.26. *The identity*

$$\mathbf{A}(x) \;=\; e^{\mathbf{H}(x)} \tag{8.26.1}$$

holds, for $x \geq 0$.

Corollary 8.27. *For $x \geq 0$, $\mathbf{P}(x) \leq \mathbf{H}(x) \leq \mathbf{A}(x)$, and*

$$\mathbf{H}(x) < \infty \quad \textit{iff} \quad \mathbf{A}(x) < \infty.$$

Corollary 8.28. *Suppose $\mathbf{A}(x) \in \mathcal{DS}_c$. Then $1/\mathbf{A}(x) \in \mathcal{DS}_c$, and letting its Dirichlet series expansion be $[\mathbf{1/A}](x)$,*

$$1/\mathbf{A}(x) \;=\; [\mathbf{1/A}](x) \qquad \textit{on } [c,\infty).$$

PROOF. Since $\mathbf{A}(x) \in \mathcal{DS}_c$, Corollary 8.27 shows $\mathbf{H}(x) \in \mathcal{DS}_c$, so one has $-\mathbf{H}(x) \in \mathcal{DS}_c$, and then by Lemma 7.17

$$1/\mathbf{A}(x) \;=\; e^{-\mathbf{H}(x)} \in \mathcal{DS}_c.$$

Let $[\mathbf{1/A}](x)$ be the Dirichlet series expansion of $1/\mathbf{A}(x)$. Then one has $1/\mathbf{A}(x) = [\mathbf{1/A}](x)$ on (c,∞). Since $1/\mathbf{A}(x)$ and $[\mathbf{1/A}](x)$ are right continuous at $x = c$, they must also be equal at $x = c$. □

8.4.1. A formula for $a(n)$. The results of the last section lead to the following explicit formula to calculate $a(n)$ from $p(n)$, and then, of course, one can calculate $A(n) = a(1) + \cdots + a(n)$.

Proposition 8.29. *For $n \geq 2$,*

$$a(n) \;=\; \sum_{j=1}^{n} \frac{1}{j!} \sum_{\substack{n_1\cdots n_j=n\\ n_i\geq 2}} \sum_{a_1^{b_1}=n_1} \cdots \sum_{a_j^{b_j}=n_j} \frac{p(a_1)\cdots p(a_j)}{b_1\cdots b_j}.$$

PROOF.

$$\begin{aligned} a(n) &= [n^{-x}]e^{\mathbf{H}(x)} \\ &= \sum_{j=1}^{n} \frac{1}{j!} \sum_{\substack{n_1 \cdots n_j = n \\ n_i \ge 2}} h(n_1) \cdots h(n_j) \\ &= \sum_{j=1}^{n} \frac{1}{j!} \sum_{\substack{n_1 \cdots n_j = n \\ n_i \ge 2}} \Bigg(\sum_{a_1^{b_1} = n_1} \frac{p(a_1)}{b_1} \Bigg) \cdots \Bigg(\sum_{a_j^{b_j} = n_j} \frac{p(a_j)}{b_j} \Bigg). \end{aligned}$$

□

Thus, for example,

$$\begin{aligned} a(2) &= p(2) \\ a(3) &= p(3) \\ a(4) &= p(4) + \frac{1}{2}p(2) + \frac{1}{2}p(2)^2. \end{aligned}$$

8.5. Finitely generated multiplicative number systems

If $\mathcal{A}$ is finitely generated then, parallel to the additive case, there is a formula for the asymptotics of $A(x)$. (This has been used in the study of limit laws in Burris and Idziak [**15**].)

Theorem 8.30. *Suppose the set* P *of indecomposables of* $\mathcal{A}$ *is finite. Let* $\mathsf{supp}\, p(n) = \{d_1, \ldots, d_k\}$ *and let* $p(d_i) = m_i$. *Then, with* r *the rank of* $\mathcal{A}$, *that is,* $r = |\mathsf{P}|$,

$$A(x) \sim \frac{1}{r! \prod (\log d_i)^{m_i}} \cdot (\log x)^r.$$

PROOF. Let $\mathsf{P} = \{\mathsf{p}_1, \ldots, \mathsf{p}_r\}$. $A(x)$ is the number of $(x_1, \ldots, x_r) \in \mathbb{N}^r$ such that

$$\left\| \mathsf{p}_1^{x_1} \cdots \mathsf{p}_r^{x_r} \right\| \le x,$$

that is,

$$\|\mathsf{p}_1\|^{x_1} \cdots \|\mathsf{p}_r\|^{x_r} \le x.$$

Taking the log of both sides, and letting $\mu_i = \log\|\mathsf{p}_i\|$, it follows that $A(x)$ is the number of $(x_1, \ldots, x_r) \in \mathbb{N}^r$ such that

$$\mu_1 x_1 + \cdots + \mu_r x_r \le \log x.$$

The number of solutions is clearly asymptotic to the volume of the region in $\mathbb{R}^r$ bounded by $x_1 = 0, \ldots, x_r = 0$, and $\mu_1 x_1 + \cdots + \mu_r x_r = \log x$. Figure 1 illustrates the two-dimensional case, that is, $r = 2$. The volume computation gives the desired answer.

□

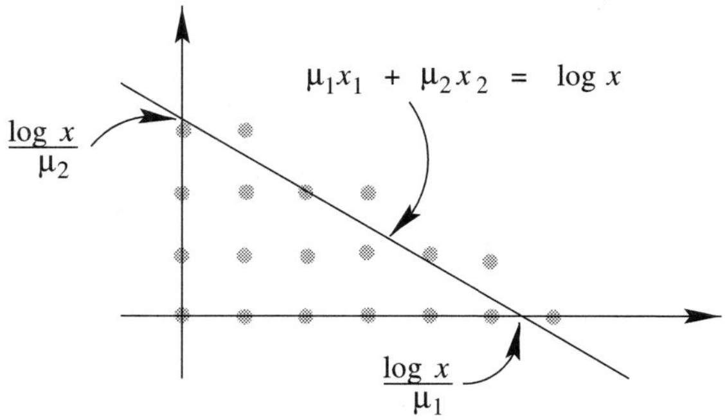

FIGURE 1. The case $k = 2$

CHAPTER 9

Density and Partition Sets

Now it is time to look at the concept of density in the context of multiplicative systems. It turns out that global density plays the dominant role. There is no analog of 'being reduced', for global density, but there is a serious split of multiplicative systems into two very different kinds: discrete and strictly multiplicative.

9.1. Asymptotic density

Let $\mathcal{A} = \big(\mathsf{A}, \mathsf{P}, \cdot, 1, \|\ \|\big)$ be a multiplicative number system with counting functions $a(n), p(n)$ and abscissa of convergence α. This section repeats key definitions from §3.1.

Definition 9.1. For $\mathsf{B} \subseteq \mathsf{A}$, the *local asymptotic density* $\delta(\mathsf{B})$ is as follows, provided the indicated limit exists:

$$\delta(\mathsf{B}) \quad = \quad \lim_{\substack{n\to\infty \\ a(n)\neq 0}} \frac{b(n)}{a(n)}.$$

The following lemma says that the collection of subsets of A that have local asymptotic density is closed under complement as well as union of disjoint members.

Lemma 9.2. *Suppose* B, C *are subsets of* A *such that* $\delta(\mathsf{B})$ *and* $\delta(\mathsf{C})$ *exist. Then*

(a) $\delta(\mathsf{A} \smallsetminus \mathsf{B}) \;=\; 1 - \delta(\mathsf{B})$

(b) $\mathsf{B} \cap \mathsf{C} = \varnothing \quad \Rightarrow \quad \delta(\mathsf{B} \cup \mathsf{C}) \;=\; \delta(\mathsf{B}) + \delta(\mathsf{C})$.

PROOF. (Same as for Lemma 3.2.) □

This notion of the density of B captures the idea of *the local probability of a randomly chosen element from* A *being in* B. It turns out that the existence of local asymptotic density for all partition sets of a multiplicative number system is a very restrictive condition (see Proposition 9.29).

The more interesting part of density in multiplicative number systems deals with global concepts. Define the global asymptotic density as in the additive case.

Definition 9.3. For $\mathsf{B} \subseteq \mathsf{A}$, the *global asymptotic density* $\Delta(\mathsf{B})$ is as follows, whenever the indicated limit exists:

$$\Delta(\mathsf{B}) \;=\; \lim_{n\to\infty} \frac{B(n)}{A(n)}.$$

This notion of the density of B captures the idea of *the global probability of a randomly chosen element from* A *being in* B.

The following lemma says that the collection of subsets of A that have global asymptotic density is closed under complement as well as union of disjoint members.

Lemma 9.4. *Suppose* B, C *are subsets of* A *such that* $\Delta(\mathsf{B})$ *and* $\Delta(\mathsf{C})$ *exist. Then*

(a) $\Delta(\mathsf{A} \setminus \mathsf{B}) \;=\; 1 - \Delta(\mathsf{B})$

(b) $\mathsf{B} \cap \mathsf{C} = \varnothing \quad\Rightarrow\quad \Delta(\mathsf{B} \cup \mathsf{C}) \;=\; \Delta(\mathsf{B}) + \Delta(\mathsf{C})$.

PROOF. (Same as for Lemma 3.5.) □

What follows is a repeat of Proposition 3.6 (use the same proof):

Proposition 9.5 ($\delta \subseteq \Delta$). *Suppose* B *is a subset of* A. *If the local asymptotic density* $\delta(\mathsf{B})$ *exists then the global asymptotic density* $\Delta(\mathsf{B})$ *also exists, and* $\Delta(\mathsf{B}) = \delta(\mathsf{B})$.

9.2. Dirichlet density

Dirichlet used what is called Dirichlet density, relative to the set of prime numbers, to show, for any given positive integer b, that the primes are evenly distributed among the $\varphi(b)$ arithmetical sequences $a + nb$, where $1 \le a < b$ and $\gcd(a, b) = 1$, and φ is the Euler phi function. For Q a subset of the primes P, let

$$\partial(\mathsf{Q}) \;=\; \lim_{x \to 1} \frac{\mathbf{Q}(x)}{\mathbf{P}(x)}.$$

In 1837 Dirichlet showed, given b, that if P_a is the set of primes congruent to a modulo b, where $\gcd(a, b) = 1$, then $\partial(\mathsf{P}_a) = 1/\varphi(b)$. It was only much later, in 1896, that de la Vallee Poussin proved that this 'evenly distributed' property held in the global asymptotic sense, that is, $\lim\limits_{n \to \infty} \dfrac{P_a(n)}{P(n)} \;=\; 1/\varphi(b)$.

Dirichlet's density is defined relative to the set of primes. A similar density is used here, defined relative to an entire number system $\mathcal{A}$. It seems appropriate to also call it Dirichlet density (as in Part 1).

Definition 9.6. Suppose $\alpha < \infty$, where α is the abscissa of convergence of $\mathcal{A}$. For B a subset of A, the *Dirichlet density* $\partial(\mathsf{B})$ of B is defined as follows, whenever the indicated limit exists:

$$\partial(\mathsf{B}) \;=\; \lim_{x \to \alpha} \frac{\mathbf{B}(x)}{\mathbf{A}(x)}.$$

Of course, $\lim\limits_{x \to \alpha}$ means the limit as x approaches α from the right, since $\mathbf{A}(x)$ is not convergent when $x < \alpha$.

The following lemma says that the collection of subsets of A that have Dirichlet density is closed under complement as well as union of disjoint members.

Lemma 9.7. *Suppose* B, C *are subsets of* A *such that* $\partial(\mathsf{B})$ *and* $\partial(\mathsf{C})$ *exist. Then*

(a) $\partial(\mathsf{A} \setminus \mathsf{B}) \;=\; 1 - \partial(\mathsf{B})$

(b) $\mathsf{B} \cap \mathsf{C} = \varnothing \quad \Rightarrow \quad \partial(\mathsf{B} \cup \mathsf{C}) \;=\; \partial(\mathsf{B}) + \partial(\mathsf{C})$.

PROOF. (Same as for Lemma 3.9 after replacing $x \to \rho$ by $x \to \alpha$.) □

If $\mathbf{A}(\alpha) < \infty$ then it is easy to see that *every* subset of A has Dirichlet density. Furthermore, the Dirichlet density of any nonempty subset is positive.

Lemma 9.8. *Assume* $\mathbf{A}(\alpha) < \infty$. *Then, for* $\mathsf{B} \subseteq \mathsf{A}$,

$$\partial(\mathsf{B}) \;=\; \frac{\mathbf{B}(\alpha)}{\mathbf{A}(\alpha)}.$$

If $\mathsf{B} \neq \varnothing$ *then* $\partial(\mathsf{B}) > 0$.

PROOF. $\mathbf{A}(\alpha) < \infty$ implies, for any $\mathsf{B} \subseteq \mathsf{A}$, $\mathbf{B}(x)$ is continuous on $[\alpha, \infty)$, so

$$\partial(\mathsf{B}) \;=\; \lim_{x \to \alpha} \frac{\mathbf{B}(x)}{\mathbf{A}(x)} \;=\; \frac{\mathbf{B}(\alpha)}{\mathbf{A}(\alpha)}.$$

If $\mathsf{B} \neq \varnothing$ then $0 < \mathbf{B}(\alpha) \leq \mathbf{A}(\alpha)$. □

Thus if $\mathbf{A}(\alpha) < \infty$ then the Dirichlet density and asymptotic density disagree on all nonempty finite subsets of A since the asymptotic density of any finite set is 0 since $A(n) \to \infty$. The condition that guarantees that Dirichlet density exists and agrees with local asymptotic density, whenever the latter exists, is $\mathbf{A}(\alpha) = \infty$. This condition always holds if $\alpha = 0$, but one can easily construct examples of multiplicative number systems, with $\alpha > 0$ and $\mathbf{A}(\alpha) < \infty$, for example,

$$p(n) \;=\; \begin{cases} \lfloor n/m^2 \rfloor & \text{if } n = 2^m \\ 0 & \text{otherwise.} \end{cases}$$

9.2.1. δ and ∂.

Proposition 9.9. *If* $\mathbf{A}(\alpha) = \infty$ *then*

$$\delta \;\subseteq\; \partial,$$

that is, for any $\mathsf{B} \subseteq \mathsf{A}$ *for which* $\delta(\mathsf{B})$ *exists,* $\partial(\mathsf{B})$ *exists and* $\partial(\mathsf{B}) = \delta(\mathsf{B})$. *Thus* ∂ *extends* δ.

PROOF. Assume $\delta(\mathsf{B})$ exists. Then, for $x > \alpha$,

$$\begin{aligned} \left|\frac{\mathbf{B}(x)}{\mathbf{A}(x)} - \delta(\mathsf{B})\right| \;&=\; \big|\mathbf{B}(x) - \delta(\mathsf{B}) \cdot \mathbf{A}(x)\big| / \mathbf{A}(x) \\ &=\; \Big|\Big(\sum_{n \geq 1} b(n) n^{-x}\Big) - \delta(\mathsf{B}) \cdot \Big(\sum_{n \geq 1} a(n) n^{-x}\Big)\Big| \Big/ \mathbf{A}(x) \end{aligned}$$

$$
\begin{aligned}
&= \left|\sum_{n\geq 1}\Big(b(n)-\delta(\mathsf{B})\cdot a(n)\Big)n^{-x}\right|\Big/\mathbf{A}(x)\\
&\leq \underbrace{\sum_{1\leq n\leq M}\left|\Big(b(n)-\delta(\mathsf{B})\cdot a(n)\Big)n^{-x}\right|\Big/\mathbf{A}(x)}_{\text{I}}\\
&\qquad + \underbrace{\sum_{n>M}\left|\Big(b(n)-\delta(\mathsf{B})\cdot a(n)\Big)n^{-x}\right|\Big/\mathbf{A}(x)}_{\text{II}}.
\end{aligned}
$$

For any choice of M, term I goes to 0 as $x \to \alpha$ since $\mathbf{A}(\alpha) = \infty$.

Given $\varepsilon > 0$ choose M such that, for $n > M$ and $a(n) \neq 0$,

$$\left|\frac{b(n)}{a(n)} - \delta(\mathsf{B})\right| \leq \varepsilon.$$

Then, for term II,

$$
\begin{aligned}
\sum_{n>M}\left|\Big(b(n)-\delta(\mathsf{B})\cdot a(n)\Big)n^{-x}\right|\Big/\mathbf{A}(x) &\leq \sum_{n>M}\varepsilon a(n)n^{-x}\Big/\mathbf{A}(x)\\
&\leq \varepsilon\mathbf{A}(x)/\mathbf{A}(x)\\
&= \varepsilon.
\end{aligned}
$$

This suffices to prove the result. □

9.2.2. Δ and ∂^+. As with additive number systems, $\mathbf{A}(\alpha) = \infty$ does not guarantee that $\delta = \partial$. Unlike the results for additive number systems, it is not known if $\Delta \subseteq \partial$ when $\mathbf{A}(\alpha) = \infty$, although $\delta \subseteq \Delta$. However, a connection between Δ and the density ∂^+ can be established.

Definition 9.10. For $\mathsf{B} \subseteq \mathsf{A}$, define

$$
\begin{aligned}
\mathbf{B}^+(x) &= \sum_{n\geq 1} B(n)n^{-x}\\
\partial^+(\mathsf{B}) &= \lim_{x\to\alpha^+}\frac{\mathbf{B}^+(x)}{\mathbf{A}^+(x)} \qquad \text{if the limit exists,}
\end{aligned}
$$

where α^+ is the abscissa of convergence of $\mathbf{A}^+(x)$.

In the additive case $\mathbf{A}^+(x)$ has the same radius of convergence as $\mathbf{A}(x)$, and any $\mathbf{B}^+(x)$ is obtained by simply multiplying $\mathbf{B}(x)$ by $(1-x)^{-1}$. These niceties do not hold in the multiplicative setting. $\mathbf{A}^+(x)$ can have a different abscissa of convergence, for example, if $\mathbf{A}(x) = \zeta(x)$ then $\mathbf{A}^+(x) = \sum_{n\geq 1} nn^{-x} = \zeta(x-1)$, so $\alpha^+ = 2$ whereas $\alpha = 1$. Also, it is not known if $\partial^+ = \partial$, as happens in the additive case.

Proposition 9.11. *If* $\mathbf{A}^+(\alpha^+) = \infty$ *then*

$$\Delta \subseteq \partial^+,$$

that is, for any $\mathsf{B} \subseteq \mathsf{A}$ *for which* $\Delta(\mathsf{B})$ *exists, one has* $\partial^+(\mathsf{B})$ *exists and* $\partial^+(\mathsf{B}) = \Delta(\mathsf{B})$*. Thus* ∂^+ *extends* Δ*.*

PROOF. Assume $\Delta(\mathsf{B})$ exists. Then

$$\begin{aligned}
\left|\frac{\mathbf{B}^+(x)}{\mathbf{A}^+(x)} - \Delta(\mathsf{B})\right| &= \left|\mathbf{B}^+(x) - \Delta(\mathsf{B}) \cdot \mathbf{A}^+(x)\right| / \mathbf{A}^+(x) \\
&= \left|\Big(\sum_{n\geq 1} B(n)n^{-x}\Big) - \Delta(\mathsf{B}) \cdot \Big(\sum_{n\geq 1} A(n)n^{-x}\Big)\right| \Big/ \mathbf{A}^+(x) \\
&= \left|\sum_{n\geq 1}\Big(B(n) - \Delta(\mathsf{B}) \cdot A(n)\Big)n^{-x}\right| \Big/ \mathbf{A}^+(x) \\
&\leq \underbrace{\sum_{1\leq n\leq M}\left|\Big(B(n) - \Delta(\mathsf{B}) \cdot A(n)\Big)n^{-x}\right| \Big/ \mathbf{A}^+(x)}_{\text{I}} \\
&\qquad + \underbrace{\sum_{n>M}\left|\Big(B(n) - \Delta(\mathsf{B}) \cdot A(n)\Big)n^{-x}\right| \Big/ \mathbf{A}^+(x)}_{\text{II}}.
\end{aligned}$$

For any choice of M, term I goes to 0 as $x \to \alpha^+$ since $\mathbf{A}^+(\alpha^+) = \infty$.

Given $\varepsilon > 0$ choose M such that, for $n > M$,

$$\left|\frac{B(n)}{A(n)} - \Delta(\mathsf{B})\right| < \varepsilon.$$

Then

$$\begin{aligned}
\sum_{n>M}\left|\Big(B(n) - \Delta(\mathsf{B}) \cdot A(n)\Big)n^{-x}\right| \Big/ \mathbf{A}^+(x) &\leq \sum_{n>M} \varepsilon A(n)n^{-x} \Big/ \mathbf{A}^+(x) \\
&\leq \varepsilon \mathbf{A}^+(x)/\mathbf{A}^+(x) \\
&= \varepsilon.
\end{aligned}$$

This suffices to prove the result. □

Although a clear connection has been established between ∂^+ and Δ, the density results in Part 2 are based on studying the connection between ∂ and Δ.

Global asymptotic density, beyond the elementary remarks in this chapter, remains a mystery in the case $\mathbf{A}(\alpha) < \infty$. Also, the connection between $\mathbf{A}(x)$ and $\mathbf{A}^+(x)$ provides opportunity for future investigation.

9.3. The standard assumption

The **standard assumption** for the remainder of Part 2 is:

> $\mathcal{A} = (\mathsf{A}, \mathsf{P}, \cdot, 1, \| \ \|)$ is a multiplicative number system with local counting functions $a(n), p(n)$, global counting functions $A(x), P(x)$, generating series $\mathbf{A}(x), \mathbf{P}(x)$, and abscissa of convergence $\alpha < \infty$.

9.4. The set of multiples of an element

Definition 9.12. For $\mathsf{b} \in \mathsf{A}$ let

$$\mathsf{bA} \;=\; \{\mathsf{ba} : \mathsf{a} \in \mathsf{A}\},$$

the set of multiples of b.

Such sets always have Dirichlet density, and there is a simple formula for this density. If $\mathcal{A} \in \mathsf{RV}_\alpha$ then such sets also have global asymptotic density, and it agrees with the Dirichlet density.

Lemma 9.13. *For* $\mathsf{b} \in \mathsf{A}$,

$$|\mathsf{bA}|_{\leq n} \;=\; A(n/\|\mathsf{b}\|).$$

PROOF. The number of elements of bA with norm at most n is the number of $\mathsf{a} \in \mathsf{A}$ with $\|\mathsf{a}\|$ at most $n/\|\mathsf{b}\|$. □

Proposition 9.14. *Every set* bA *has Dirichlet density and*

$$\partial(\mathsf{bA}) \;=\; \|\mathsf{b}\|^{-\alpha}.$$

PROOF. For $x > \alpha$,

$$\begin{aligned}[\mathsf{b}\mathbf{A}](x) &= \sum_{\mathsf{a}\in\mathsf{A}} \|\mathsf{ab}\|^{-x} \\ &= \|\mathsf{b}\|^{-x} \sum_{\mathsf{a}\in\mathsf{A}} \|\mathsf{a}\|^{-x} \\ &= \|\mathsf{b}\|^{-x}\mathbf{A}(x),\end{aligned}$$

so $\partial(\mathsf{bA}) = \|\mathsf{b}\|^{-\alpha}$. □

Lemma 9.15. *A set of multiples* bA *has global asymptotic density iff*

$$\lim_{n\to\infty} \frac{A(n/\|\mathsf{b}\|)}{A(n)} \qquad \textit{exists,}$$

in which case

$$\Delta(\mathsf{bA}) \;=\; \lim_{n\to\infty} \frac{A(n/\|\mathsf{b}\|)}{A(n)}$$

$$(9.15.1) \qquad\qquad = \lim_{t\to\infty} \frac{A(t/\|\mathsf{b}\|)}{A(t)}.$$

PROOF. The claims, except for the last equality, follow from Lemma 9.13. For the last equality, note that

$$\begin{aligned} A(t) &= A(\lfloor t \rfloor) \\ A(t/\|\mathsf{b}\|) &= A(\lfloor t \rfloor/\|\mathsf{b}\|), \end{aligned}$$

the last equation following from the fact that if

$$\lfloor t \rfloor/\|\mathsf{b}\| \;\leq\; m \;\leq\; t/\|\mathsf{b}\|,$$

where $m \in \mathbb{N}$, then $\lfloor t \rfloor/\|\mathsf{b}\| = m$.

□

The expression (9.15.1) reminds one of the definition of RV_α, an important concept in the study of density.

Definition 9.16. $\mathcal{A} \in \mathsf{RV}_\alpha$ means $A(x) \in \mathsf{RV}_\alpha$, that is,

$$\lim_{t\to\infty} \frac{A(tx)}{A(t)} \;=\; x^\alpha \quad \text{for } x > 0.$$

Proposition 9.17 ([**17**], 1997). *If $\mathcal{A} \in \mathsf{RV}_\alpha$ then all sets of multiples* bA *have global asymptotic density that agrees with the Dirichlet density, that is,*

$$\Delta(\mathsf{bA}) \;=\; \partial(\mathsf{bA}) \;=\; \|\mathsf{b}\|^{-\alpha}.$$

PROOF. If $\mathcal{A} \in \mathsf{RV}_\alpha$ then

$$\lim_{t\to\infty} \frac{A(t/\|\mathsf{b}\|)}{A(t)} \;=\; \|\mathsf{b}\|^{-\alpha},$$

so, from Lemma 9.15, $\Delta(\mathsf{bA}) \;=\; \|\mathsf{b}\|^{-\alpha}$. Now use Proposition 9.14. □

The following is needed for the classification of multiplicative number systems in §9.9.3. (Recall that the standard assumption includes $\alpha < \infty$.)

Proposition 9.18. *If* $\Delta(\mathsf{bA})$ *exists then* $\Delta(\mathsf{bA}) > 0$.

PROOF. Suppose $\Delta(\mathsf{bA}) = 0$. By Lemma 9.15 one has

$$\lim_{n\to\infty} \frac{A(n/\|\mathsf{b}\|)}{A(n)} \;=\; 0.$$

Letting $k = \|\mathsf{b}\|$, and substituting nk for n, gives $\lim\limits_{n\to\infty} \dfrac{A(n)}{A(nk)} = 0$, so

$$\lim_{n\to\infty} \frac{A(nk)}{A(n)} \;=\; \infty.$$

Thus, for any $R > 0$, there is an $M > 0$ such that

$$\frac{A(nk)}{A(n)} \;>\; R, \qquad \text{for } n \geq M,$$

and thus

$$\frac{A(Mk^j)}{A(Mk^{j-1})} > R, \qquad \text{for } j \geq 1.$$

This implies

$$A(Mk^\ell) \;>\; R^\ell, \qquad \text{for } \ell \geq 1.$$

Thus, for $\ell \geq 1$,

$$\frac{\log A(Mk^\ell)}{\log(Mk^\ell)} \;>\; \frac{\log(R^\ell)}{\log(Mk^\ell)} \;=\; \frac{\ell \log R}{\log M + \ell \log k}.$$

Taking the $\limsup_{\ell\to\infty}$, and referring to Lemma 8.16, gives

$$\alpha \;\geq\; \limsup_{\ell\to\infty} \frac{\log A(Mk^\ell)}{\log(Mk^\ell)} \;\geq\; \frac{\log R}{\log k}.$$

Since R can be arbitrarily large it follows that $\alpha = \infty$, and this contradicts the standard assumption. □

Remark 9.19. More generally one can look at *the multiples of a set.* This means, for a given $\mathsf{B} \subseteq \mathsf{A}$, the set BA of all multiples of the elements of B. Such sets appear to be good candidates to have a nice distribution of elements, and they might well have global asymptotic density. However, Besicovitch (see [**38**], p. 265) constructed a set B of positive integers such that $\mathsf{B}\mathbb{P}$ does not have global asymptotic density in the positive integers $\mathbb{P}$.

There is much more to say about the density of bA, and the topic will be resumed in §9.9.

9.5. Partition sets

It is easy to construct sets $\mathsf{B} \subseteq \mathsf{A}$ such that B does not have global asymptotic density, by appropriately choosing all or none of the elements of each size n from A. One finds interesting candidates for asymptotic density in the collection of *partition sets* defined below.

Definition 9.20. For $\mathsf{B}_1, \ldots, \mathsf{B}_k \subseteq \mathsf{A}$, let

$$\mathsf{B}_1 \cdots \mathsf{B}_k \;=\; \{\mathsf{b}_1 \cdots \mathsf{b}_k \;:\; \mathsf{b}_i \in \mathsf{B}_i\}.$$

The following simple estimate for the generating series of the product of two subsets of A will play a key role in the development.

Lemma 9.21. *Suppose* $\mathsf{B}_1, \mathsf{B}_2$ *are two nonempty subsets of* A. *Then*

$$[\mathbf{B}_1 \cdot \mathbf{B}_2](x) \;\leq\; \mathbf{B}_1(x) \cdot \mathbf{B}_2(x), \quad \text{for } x \geq 0.$$

PROOF. This follows from observing, for all $n \in \mathbb{P}$, that

$$\left[n^{-x}\right]\left([\mathbf{B}_1 \cdot \mathbf{B}_2](x)\right) \;\le\; \left[n^{-x}\right]\left(\mathbf{B}_1(x) \cdot \mathbf{B}_2(x)\right),$$

that is,

$$\left|\left\{\mathsf{b}_1 \cdot \mathsf{b}_2 : \mathsf{b}_i \in \mathsf{B}_i,\ \|\mathsf{b}_1 \cdot \mathsf{b}_2\| = n\right\}\right| \;\le\; \sum_{jk=n} b_1(j) \cdot b_2(k).$$

□

Definition 9.22. For $\mathsf{B} \subseteq \mathsf{A}$,

$$\begin{aligned} \mathsf{B}^0 &= \{1\} \\ \mathsf{B}^m &= \underbrace{\mathsf{B} \cdots \mathsf{B}}_{m \text{ times}}, \qquad \text{for } m > 0 \\ \mathsf{B}^{\le m} &= \mathsf{B}^0 \cup \cdots \cup \mathsf{B}^m, \qquad \text{for } m \ge 0 \\ \mathsf{B}^{\ge m} &= \bigcup_{n \ge m} \mathsf{B}^n, \qquad \text{for } m \ge 0. \end{aligned}$$

The exponents m and $\le m$ are said to be *small*, and $\ge m$ is *large*.

Definition 9.23. $\mathsf{P}_1, \ldots, \mathsf{P}_k$ is a *partition* of P if each P_i is not empty, $\mathsf{P}_i \cap \mathsf{P}_j = \varnothing$, for $i \ne j$, and $\mathsf{P} = \bigcup_{1 \le i \le k} \mathsf{P}_i$.

Definition 9.24. A subset B of A is a *partition set* if B can be written in the form

$$\mathsf{B} \;=\; \mathsf{P}_1^{\gamma_1} \cdots \mathsf{P}_k^{\gamma_k},$$

where $\mathsf{P}_1, \ldots, \mathsf{P}_k$ is a partition of P, and each γ_i is in one of the forms m, $\le m$, $\ge m$.

So, for example,

$$\mathsf{B} \;=\; \mathsf{P}_1^{\le 5}\, \mathsf{P}_2^{\ge 4}\, \mathsf{P}_3^2$$

says that B is the subset of elements of A which can be expressed as the product of at most 5 elements from P_1 times the product of at least 4 elements from P_2 times the product of exactly 2 elements from P_3. There is one exception to this straightforward reading of the exponents, namely if they are all 0—for note that $\mathsf{P}_1^0 \cdots \mathsf{P}_k^0 = \{1\}$. The exponents $<m$ and $>m$ will be used as shorthand for $\le m-1$, respectively $\ge m+1$. As in the additive case, the definition of partition set does not include the empty set. The above definition of B^0 is needed to make sure Corollary 9.31 holds, for all choices of exponents.

Definition 9.25. ω denotes the exponent ≥ 0.

Lemma 9.26. *Partition sets include the sets of multiples of single elements.*

PROOF. Given $\mathsf{b} \in \mathsf{A}$ let $\mathsf{b} = \mathsf{p}_1^{n_1} \cdots \mathsf{p}_k^{n_k}$, where the p_i are indecomposable. Define a partition of P by $\mathsf{P}_1 = \{\mathsf{p}_1\}, \ldots, \mathsf{P}_k = \{\mathsf{p}_k\}$, and $\mathsf{P}_{k+1} = \mathsf{P} \setminus \{\mathsf{p}_1, \ldots, \mathsf{p}_k\}$. Then

$$\mathsf{bA} \;=\; \mathsf{P}_1^{\geq n_1} \cdots \mathsf{P}_k^{\geq n_k} \mathsf{P}_{k+1}^{\omega}.$$

□

Next Definition 3.27 from Part 1 is repeated.

Definition 9.27. For B a subset of A and n a positive integer, let

$$\begin{aligned} \mathsf{B}_{(\leq n)} &= \{\mathsf{a} \in \mathsf{B} : \|\mathsf{a}\| \leq n\} \\ \mathsf{B}_{(n)} &= \{\mathsf{a} \in \mathsf{B} : \|\mathsf{a}\| = n\} \\ \mathsf{B}_{(\geq n)} &= \{\mathsf{a} \in \mathsf{B} : \|\mathsf{a}\| \geq n\}. \end{aligned}$$

It is easy to see that every nonempty subset Q of P is a partition set. P itself is a partition set, and, for $\mathsf{Q} \neq \mathsf{P}$, one has a partition $\mathsf{Q}, \mathsf{P} \setminus \mathsf{Q}$ of P, and $\mathsf{Q} = \mathsf{Q} \cdot (\mathsf{P} \setminus \mathsf{Q})^0$. From this the following proposition is easily proved.

Proposition 9.28 ([**17**], 1997). *Suppose every partition set of $\mathcal{A}$ has global asymptotic density. Then*

$$\Delta(\mathsf{P}) \;=\; 0.$$

PROOF. Since P is a partition set, by assumption $\Delta(\mathsf{P})$ exists. If $\Delta(\mathsf{P})$ is not 0, say $\Delta(\mathsf{P}) = c > 0$, then define $\mathsf{Q} \subset \mathsf{P}$ by considering the sequence $\mathsf{P}_{(1)}, \mathsf{P}_{(2)}, \ldots$ and alternately taking all elements of the $\mathsf{P}_{(n)}$ until one has $\dfrac{Q(n)}{A(n)} > \frac{2}{3}c$ and then taking no elements from the $\mathsf{P}_{(n)}$ until $\dfrac{Q(n)}{A(n)} < \frac{1}{3}c$. Then (see Figure 1),

$$\limsup_{n\to\infty} \frac{Q(n)}{A(n)} \geq \tfrac{2}{3}c \qquad \text{and} \qquad \liminf_{n\to\infty} \frac{Q(n)}{A(n)} \leq \tfrac{1}{3}c.$$

But this means Q does not have global asymptotic density, which contradicts the assumption that all partition sets have such a density.

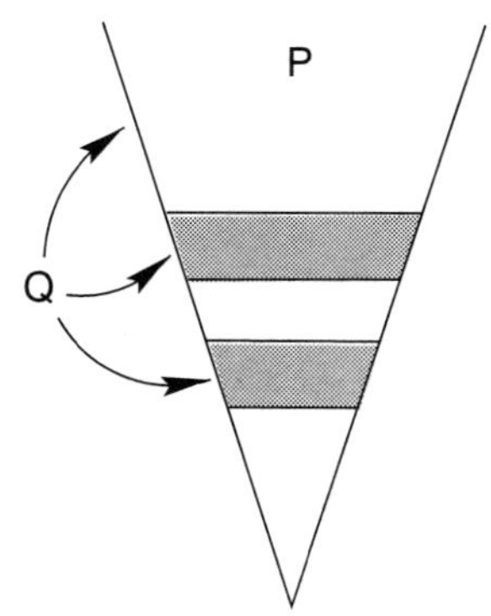

FIGURE 1

□

At the beginning of Chapter 9 it was mentioned that requiring that all partition sets have local asymptotic density is a very restrictive condition. Here is the meaning of that assertion.

Proposition 9.29. *If all partition sets of $\mathcal{A}$ have local asymptotic density then there is a finite set $\{q_1, \ldots, q_n\}$ of prime numbers such that, for any $\mathsf{a} \in \mathsf{A} \setminus \{1\}$, the prime divisors of $\|\mathsf{a}\|$ are precisely the q_i.*

PROOF. Suppose all partition sets of $\mathcal{A}$ have local asymptotic density. Let q be a prime number that divides the norm of some member of A. Let Q be the set of indecomposables whose norms are divisible by q. Then the partition set

$$\mathsf{B} \;=\; \mathsf{Q}^{\geq 1}(\mathsf{P} \setminus \mathsf{Q})^{\omega}$$

gives all the elements of $\mathcal{A}$ that have a norm divisible by q. Since this is a partition set, by assumption $\delta(\mathsf{B})$ exists. The value of this density can easily be determined since

$$b(n) \;=\; \begin{cases} a(n) & \text{if } q|n \\ 0 & \text{otherwise.} \end{cases}$$

But this implies $\dfrac{b(n)}{a(n)}$ is infinitely often 1, so $\delta(\mathsf{B}) = 1$.

Now suppose there is an element $\mathsf{p} \in \mathsf{P}$ such that q does not divide $\|\mathsf{p}\|$. This means that there are elements of size $\|\mathsf{p}\|^n$, for every n, and none of these elements are in B. But this, by the same reasoning as before, would force $\delta(\mathsf{B}) = 0$. Thus it follows that all members of A must have sizes that are divisible by the same (finite) set $\{q_1, \ldots, q_n\}$ of prime numbers. □

9.6. Generating series of partition sets

One of the beautiful facts about Dirichlet generating series is that one has the following product theorem.

Proposition 9.30. *Let $\mathsf{P}_1, \cdots, \mathsf{P}_k$ be a partition of P. Then, for any nonempty $\mathsf{B}_i \subseteq \mathsf{P}_i^{\omega}$, and, for $x \geq 0$, the generating series $[\mathbf{B}_1 \cdots \mathbf{B}_k](x)$ converges iff each $\mathbf{B}_i(x)$ converges; and, for $x \geq 0$,*

$$[\mathbf{B}_1 \cdots \mathbf{B}_k](x) \;=\; \mathbf{B}_1(x) \;\cdots\; \mathbf{B}_k(x).$$

PROOF. Let $\mathsf{B} = \mathsf{B}_1 \cdots \mathsf{B}_k$. Then, for $\mathsf{b} \in \mathsf{B}$, there are unique $\mathsf{b}_i \in \mathsf{B}_i$ such that $\mathsf{b} = \mathsf{b}_1 \cdots \mathsf{b}_k$ (by the unique decomposition property of $\mathcal{A}$), so

$$b(n) \;=\; \sum_{n_1 \cdots n_k = n} b_1(n_1) \;\cdots\; b_k(n_k).$$

Thus $b(n)$ is the coefficient of n^{-x} in the Dirichlet product

$$\Big(\sum_{m \geq 1} b_1(m) m^{-x}\Big) * \cdots * \Big(\sum_{m \geq 1} b_k(m) m^{-x}\Big).$$

For any $x \geq 0$,

$$b(1)1^{-x} + \cdots + b(n)n^{-x} \leq \Big(\sum_{m=1}^{n} b_1(m)m^{-x}\Big) \cdots \Big(\sum_{m=1}^{n} b_k(m)m^{-x}\Big) \leq \mathbf{B}(x),$$

so $\mathbf{B}(x)$ converges iff all $\mathbf{B}_i(x)$ converge; if so, $\mathbf{B}(x) = \mathbf{B}_1(x)\cdots\mathbf{B}_k(x)$. □

Corollary 9.31. *Let $\mathsf{P}_1, \cdots, \mathsf{P}_k$ be a partition of P. Then, for any partition set $\mathsf{B} = \mathsf{P}_1^{\gamma_1}\cdots\mathsf{P}_k^{\gamma_k}$, and, for $x \geq 0$,*

$$\begin{aligned}\mathbf{B}(x) &= \mathbf{P}_1^{\gamma_1}(x)\cdots\mathbf{P}_k^{\gamma_k}(x)\\ &= \mathbf{P}_1^{\gamma_1}(x) * \cdots * \mathbf{P}_k^{\gamma_k}(x).\end{aligned}$$

Two special cases of Dirichlet generating series are given in the following lemma, where the simpler notation of $\mathbf{p}^m(x)$, for $\{\mathbf{p}\}^m(x)$, and $\mathbf{p}^\omega(x)$, for $\{\mathbf{p}\}^\omega(x)$, has been adopted.

Lemma 9.32. *For $\mathsf{p} \in \mathsf{P}$ and $m \geq 0$,*

$$\begin{aligned}\mathbf{p}^m(x) &= \|\mathsf{p}\|^{-mx}\\ \mathbf{p}^\omega(x) &= \sum_{n\geq 0}\|\mathsf{p}\|^{-nx}.\end{aligned}$$

PROOF. This follows immediately from

$$\begin{aligned}\{\mathsf{p}\}^m &= \{\mathsf{p}^m\}\\ \{\mathsf{p}\}^\omega &= \{1, \mathsf{p}, \mathsf{p}^2, \cdots\},\end{aligned}$$

using the fact that $\|\mathsf{p}^k\| = \|\mathsf{p}\|^k$. □

9.7. Partition sets have Dirichlet density

In this section the fact that 'sets of multiples of an element have Dirichlet density' is strengthened to the result that 'all partition sets have Dirichlet density'. Further hypotheses are needed to ensure the existence of global asymptotic density.

9.7.1. The Dirichlet density of P^γ. Showing that the Dirichlet density of P^γ exists will give the key idea used in §9.7.2.[1]

Corollary 9.33. *Given $m \in \mathbb{N}$, for $x \geq 0$,*

$$\mathbf{P}^m(x) \leq \mathbf{P}(x)^m.$$

PROOF. This follows immediately from Lemma 9.21. □

Lemma 9.34. *For γ an exponent,*

$$\partial(\mathsf{P}^\gamma) = \begin{cases}\dfrac{\mathbf{P}^\gamma(\alpha)}{\mathbf{A}(\alpha)} > 0 & \textit{if } \mathbf{A}(\alpha) < \infty\\ 0 & \textit{if } \mathbf{A}(\alpha) = \infty \textit{ and } \gamma \textit{ is small}\\ 1 & \textit{if } \mathbf{A}(\alpha) = \infty \textit{ and } \gamma \textit{ is large.}\end{cases}$$

[1]For a detailed discussion of the generating series for P^m see Appendix B.

PROOF. If $\mathbf{A}(\alpha) < \infty$ then apply Lemma 9.8.

The interesting case is what happens when $\mathbf{A}(\alpha) = \infty$. Assuming this, the case that $\gamma = m$ will be considered first.

If P is finite then $\alpha = 0$, and $\mathbf{P}^m(x)$ is a Dirichlet polynomial since P^m is finite. Then $\mathbf{A}(0) = \infty$, but $\mathbf{P}^m(0) < \infty$, so clearly the Dirichlet density is 0.

The situation when P is not finite remains. In this case, from $\mathbf{A}(\alpha) = \infty$ follows $\mathbf{P}(\alpha) = \infty$ when $\alpha > 0$, and if $\alpha = 0$ clearly $\mathbf{P}(\alpha) = \infty$, so,

$$\begin{aligned} \partial(\mathsf{P}^m) &= \lim_{x \to \alpha} \frac{\mathbf{P}^m(x)}{\mathbf{A}(x)} \\ &\leq \lim_{x \to \alpha} \frac{\mathbf{P}(x)^m}{e^{\mathbf{P}(x)}} \qquad \text{by Corollaries 8.23 and 9.33} \\ &= 0. \end{aligned}$$

This finishes the case $\gamma = m$.

From the unique factorization property of $\mathcal{A}$, the sets $\mathsf{P}^0, \ldots, \mathsf{P}^m$ are pairwise disjoint, and

$$\begin{aligned} \mathsf{P}^{\leq m} &= \mathsf{P}^0 \cup \mathsf{P} \cup \cdots \cup \mathsf{P}^m \\ \mathsf{P}^{\geq m} &= \mathsf{A} \setminus \mathsf{P}^{<m}. \end{aligned}$$

Thus, by Lemma 9.7,

$$\begin{aligned} \partial\big(\mathsf{P}^{\leq m}\big) &= \partial(1) + \partial(\mathsf{P}) + \cdots + \partial(\mathsf{P}^m) \\ \partial\big(\mathsf{P}^{\geq m}\big) &= 1 - \partial\big(\mathsf{P}^{<m}\big). \end{aligned}$$

This proves the lemma. □

The asymptotic density of P^γ is discussed in §9.14, under the assumptions $\mathcal{A} \in \mathsf{RV}_\alpha$ and $\mathbf{A}(\alpha) = \infty$.

9.7.2. General partition sets. The strategy, to show the existence of $\lim_{x \to \alpha} \frac{\mathbf{B}(x)}{\mathbf{A}(x)}$, is to note that, for $\mathsf{B} = \mathsf{P}_1^{\gamma_1} \cdots \mathsf{P}_k^{\gamma_k}$,

$$\frac{\mathbf{B}(x)}{\mathbf{A}(x)} = \prod_{1 \leq j \leq m} \frac{\mathbf{P}_j^{\gamma_j}(x)}{\mathbf{P}_j^{\omega}(x)},$$

for then $\partial(\mathsf{B})$ exists if each $\lim_{x \to \alpha} \frac{\mathbf{P}_j^{\gamma_j}(x)}{\mathbf{P}_j^{\omega}(x)}$ exists.

Definition 9.35. For Q a nonempty subset of P, let $\mathsf{A}_\mathsf{Q} = \mathsf{Q}^\omega$ and, for any exponent γ, let

$$\partial_\mathsf{Q}(\mathsf{Q}^\gamma) = \lim_{x \to \alpha} \frac{\mathbf{Q}^\gamma(x)}{\mathbf{A}_\mathsf{Q}(x)},$$

whenever the (right-) limit on the right side exists.

Thus $\mathsf{A_Q}$ is the universe of the subsystem $\mathcal{A}_\mathsf{Q} = \big(\mathsf{A_Q}, \mathsf{Q}, \cdot, 1, \|\ \|\big)$ of $\mathcal{A}$ generated by the elements of Q. However, the definition of ∂_Q is not necessarily the Dirichlet density of $\mathcal{A}_\mathsf{Q}$ since the abscissa of convergence of $\mathcal{A}_\mathsf{Q}$ may be less than that of $\mathcal{A}$.

Lemma 9.36. *For $\mathsf{Q} \subseteq \mathsf{P}$, $\partial_\mathsf{Q}(\mathsf{Q}^\gamma)$ exists, for any exponent γ, and*

$$\partial_\mathsf{Q}(\mathsf{Q}^\gamma) \;=\; \begin{cases} \dfrac{\mathbf{Q}^\gamma(\alpha)}{\mathbf{A}_\mathsf{Q}(\alpha)} > 0 & \textit{if } \mathbf{A}_\mathsf{Q}(\alpha) < \infty \\ 0 & \textit{if } \mathbf{A}_\mathsf{Q}(\alpha) = \infty \textit{ and } \gamma \textit{ is small} \\ 1 & \textit{if } \mathbf{A}_\mathsf{Q}(\alpha) = \infty \textit{ and } \gamma \textit{ is large.} \end{cases}$$

PROOF. Just repeat the proof of Lemma 9.34, replacing P by Q, and $\mathbf{A}(x)$ by $\mathbf{A}_\mathsf{Q}(x)$. □

Theorem 9.37. *All partition sets of $\mathcal{A}$ have Dirichlet density. If B is $\mathsf{P}_1^{\gamma_1} \cdots \mathsf{P}_k^{\gamma_k}$ then*

$$\partial(\mathsf{B}) \;=\; \partial_{\mathsf{P}_1}(\mathsf{P}_1^{\gamma_1}) \cdots \partial_{\mathsf{P}_k}(\mathsf{P}_k^{\gamma_k}).$$

PROOF. From Corollary 9.31, for $x \geq 0$,

$$\begin{aligned} \mathbf{B}(x) &= \mathbf{P}_1^{\gamma_1}(x) \cdots \mathbf{P}_k^{\gamma_k}(x) \\ \mathbf{A}(x) &= \mathbf{A}_{\mathsf{P}_1}(x) \cdots \mathbf{A}_{\mathsf{P}_k}(x), \end{aligned}$$

so,

$$\frac{\mathbf{B}(x)}{\mathbf{A}(x)} \;=\; \prod_{1 \leq j \leq k} \frac{\mathbf{P}_j^{\gamma_j}(x)}{\mathbf{A}_{\mathsf{P}_j}(x)}, \qquad \text{for } x > \alpha. \tag{9.37.1}$$

By Lemma 9.36 each $\partial_{\mathsf{P}_i}(\mathsf{P}_i^{\gamma_i})$ exists, so the limit as $x \to \alpha$ of the right side of (9.37.1) exists, and thus equals the limit of the left side. This proves the theorem. □

9.7.3. Dirichlet density of subsystems. For each $\mathsf{Q} \subseteq \mathsf{P}$, there is a subsystem $\mathcal{A}_\mathsf{Q} = (\mathsf{A_Q}, \mathsf{Q}, \cdot, 1, \|\ \|)$ of $\mathcal{A}$, where $\mathsf{A_Q} = \mathsf{Q}^\omega$. Since Q^ω is a partition set it follows that $\partial(\mathsf{A_Q})$ must exist—the following gives a useful formula for this density. (Set $0^0 = 1$ if it appears in (9.38.1).)

Proposition 9.38. *For $\mathsf{Q} \subseteq \mathsf{P}$,*

$$\partial(\mathsf{A_Q}) \;=\; \prod_{n \geq 2} (1 - n^{-\alpha})^{p(n) - q(n)}, \tag{9.38.1}$$

where $q(n)$ is the local counting function of Q.

PROOF. Since $\mathsf{A_Q} = \mathsf{Q}^\omega \cdot (\mathsf{P} \setminus \mathsf{Q})^0$ apply Theorem 9.37 to obtain

$$\begin{aligned} \partial(\mathsf{A_Q}) &= \partial_\mathsf{Q}(\mathsf{A_Q}) \cdot \partial_{\mathsf{P} \setminus \mathsf{Q}}(\{1\}) \\ &= 1 \cdot \lim_{x \to \alpha} \mathbf{1}(x) / \mathbf{A}_{\mathsf{P} \setminus \mathsf{Q}}(x) \\ &= \lim_{x \to \alpha} \Big(1 \Big/ \prod_{n \geq 2} (1 - n^{-x})^{-\left(p(n) - q(n)\right)} \Big) \end{aligned}$$

$$= \lim_{x\to\alpha} \prod_{n\geq 2} (1 - n^{-x})^{p(n)-q(n)}$$

$$= \prod_{n\geq 2} (1 - n^{-\alpha})^{p(n)-q(n)}.$$

The last step follows from noting that, for $x \geq 0$, the partial products

$$\prod_{n=2}^{N} \left(1 - n^{-x}\right)^{p(n)-q(n)}$$

are nonnegative, and nonincreasing in N, and nondecreasing in x, so, for $x \geq 0$,

$$\begin{aligned} \prod_{n=2}^{N} \left(1 - n^{-x}\right)^{p(n)-q(n)} &\geq \prod_{n\geq 2} \left(1 - n^{-x}\right)^{p(n)-q(n)} \\ &\geq \prod_{n\geq 2} \left(1 - n^{-\alpha}\right)^{p(n)-q(n)} \end{aligned}$$

leads to

$$\begin{aligned} \prod_{n=2}^{N} \left(1 - n^{-\alpha}\right)^{p(n)-q(n)} &\geq \lim_{x\to\alpha} \prod_{n\geq 2} \left(1 - n^{-x}\right)^{p(n)-q(n)} \\ &\geq \prod_{n\geq 2} \left(1 - n^{-\alpha}\right)^{p(n)-q(n)}. \end{aligned}$$

Now take the limit as $N \to \infty$. □

9.7.4. Abelian groups. Let Q be the collection of cyclic Abelian groups $\mathbf{Z}_{p^n}$ with p a prime number greater than 2. Then, since $\alpha = 1$, from (9.38.1)

$$\begin{aligned} \partial(\mathsf{Q}^\omega) &= \prod_{\substack{p \text{ prime} \\ n\geq 1}} \left(1 - (p^n)^{-1}\right)^{p(n)-q(n)} \\ &= \prod_{n\geq 1} \left(1 - (2^n)^{-1}\right) \\ &= \prod_{n\geq 1} \left(1 - 2^{-n}\right) \approx 0.29. \end{aligned}$$

Thus, for $\mathsf{B} = \mathsf{A} \setminus \mathsf{Q}^\omega$, the collection of finite Abelian groups with an element of order 2, one has

$$\begin{aligned} \partial(\mathsf{B}) &= 1 - \partial(\mathsf{Q}^\omega) \\ &= 1 - \prod_{n\geq 1} \left(1 - 2^{-n}\right) \approx 0.71. \end{aligned}$$

These computations were mentioned in §8.3.5. Showing that they hold for $\Delta(\mathsf{B})$ will have to wait until Sárközy's Density Theorem 11.15 is proved.

Then one can use known asymptotics for $a(n)$ to assert that $\Delta(\mathsf{B})$ exists, and thus must equal $\partial(\mathsf{B})$.

9.8. Discrete multiplicative number systems

In discrete systems $\mathcal{A}$ the values of $\|\mathsf{a}\|$, for $\mathsf{a} \in \mathsf{A}$, are powers of some fixed integer λ. One can view any additive number system as a discrete multiplicative number system, so that the study of multiplicative number systems is a natural generalization of the study of the additive case. Conversely, the discrete multiplicative number systems can be viewed as additive number systems. Although the global asymptotic density results of Part 2 can be applied to discrete systems, one obtains superior local asymptotic density results by treating discrete systems as additive systems and applying the results of Part 1.

Definition 9.39. A multiplicative number system $\mathcal{A}$ is *discrete* if there is a positive integer λ such that, for any $\mathsf{a} \in \mathsf{A}$, $\|\mathsf{a}\|$ is an integer power of λ.

9.8.1. An additive number system $\mathcal{A} = (\mathsf{A}, \mathsf{P}, +, 0, \|\ \|)$ can be viewed as a discrete multiplicative number system. Define an associated discrete multiplicative number system $\widetilde{\mathcal{A}}$ as follows, after selecting a positive integer $\lambda > 1$. Let $\widetilde{\mathcal{A}} = (\widetilde{\mathsf{A}}, \widetilde{\mathsf{P}}, \cdot, 1, \widetilde{\|\ \|})$ where $\widetilde{\mathsf{A}} = \mathsf{A}$, $\widetilde{\mathsf{P}} = \mathsf{P}$, $\cdot = +$, $1 = 0$, and let

$$\widetilde{\|\mathsf{a}\|} \;=\; \lambda^{\|\mathsf{a}\|}.$$

$\widetilde{\mathcal{A}}$ is readily seen to be a discrete multiplicative number system. $\mathcal{A}$ and $\widetilde{\mathcal{A}}$ have the same underlying free monoid. Only the norm has been changed. The global counting functions $\widetilde{A}(x)$ and $A(x)$ are related by

$$\widetilde{A}(x) \;=\; A(\log_\lambda x).$$

The set of additives of b in $\mathcal{A}$ is the same as the set of multiples of b in $\widetilde{\mathcal{A}}$, that is,

$$\mathsf{b} + \mathsf{A} \;=\; \mathsf{b}\widetilde{\mathsf{A}}.$$

Corresponding partition sets are also the same, that is, if $\mathsf{P}_1, \ldots, \mathsf{P}_k$ is a partition of P (and hence of $\widetilde{\mathsf{P}}$) then, for any choice of γ_i,

$$\gamma_1 \mathsf{P}_1 + \cdots + \gamma_k \mathsf{P}_k \;=\; \mathsf{P}_1^{\gamma_1} \cdots \mathsf{P}_k^{\gamma_k}.$$

Furthermore, the local counting functions are closely related, for if $\mathsf{B} \subseteq \mathsf{A}$ then

$$\begin{aligned} b(n) \;&=\; \widetilde{b}(\lambda^n) \\ \widetilde{b}(n) \;&=\; \begin{cases} b(\log_\lambda n) & \text{if } \log_\lambda n \in \mathbb{N} \\ 0 & \text{otherwise,} \end{cases} \end{aligned}$$

and from this it follows that the asymptotic densities are the same:

$$\delta = \widetilde{\delta}$$
$$\Delta = \widetilde{\Delta}.$$

Let $\mathbf{B}(x)$ be the generating series of B in $\mathcal{A}$, and $\widetilde{\mathbf{B}}(x)$ the corresponding series of B in $\widetilde{\mathcal{A}}$. Then,

$$\mathbf{B}(x) = \sum_{n\geq 0} b(n)x^n$$
$$\widetilde{\mathbf{B}}(x) = \sum_{n\geq 1} \widetilde{b}(n)n^{-x},$$

with the translations

$$\mathbf{B}(x) = \widetilde{\mathbf{B}}(-\log_\lambda x)$$
$$\widetilde{\mathbf{B}}(x) = \mathbf{B}(\lambda^{-x}).$$

The radius of convergence ρ of $\mathcal{A}$ and the abscissa of convergence $\widetilde{\alpha}$ of $\widetilde{\mathcal{A}}$ are related by

$$\widetilde{\alpha} = -\log_\lambda \rho$$
$$\rho = \lambda^{-\widetilde{\alpha}}.$$

From this one can conclude that the Dirichlet densities agree, that is,

$$\partial = \widetilde{\partial}.$$

Thus one can study additive number systems as discrete multiplicative number systems insofar as density is concerned. Note that, for any λ chosen as specified above, $\mathcal{A}$ is reduced iff λ is the largest integer such that the norms of elements of $\widetilde{\mathcal{A}}$ are powers of λ.

9.8.2. A discrete multiplicative number system $\mathcal{A} = \big(\mathsf{A}, \mathsf{P}, \cdot, 1, \|\ \|\big)$ can be viewed as an additive number system. Define an associated additive number system $\widetilde{\mathcal{A}}$ after choosing a particular integer λ such that every element of $\mathcal{A}$ has size a power of λ. Let $\widetilde{\mathcal{A}} = \big(\widetilde{\mathsf{A}}, \widetilde{\mathsf{P}}, +, 0, \widetilde{\|\ \|}\big)$ where $\widetilde{\mathsf{A}} = \mathsf{A}$, $\widetilde{\mathsf{P}} = \mathsf{P}$, $+ = \cdot$, $0 = 1$, and let

$$\widetilde{\|\mathsf{a}\|} = \log_\lambda \|\mathsf{a}\|.$$

It is easy to check that $\widetilde{\mathcal{A}}$ is indeed an additive number system. Again $\mathcal{A}$ and $\widetilde{\mathcal{A}}$ have the same underlying free monoid. Only the norm has been changed. The global counting functions $\widetilde{A}(x)$ and $A(x)$ are related by

$$\widetilde{A}(x) = A(\lambda^x).$$

The set of multiples of b in $\mathcal{A}$ is the same as the set of additives of b in $\widetilde{\mathcal{A}}$, that is,

$$\mathsf{bA} = \mathsf{b} + \widetilde{\mathsf{A}}.$$

Corresponding partition sets are also the same, that is, if $\mathsf{P}_1, \ldots, \mathsf{P}_k$ is a partition of P (and hence of $\widetilde{\mathsf{P}}$) then, for any choice of γ_i,

$$\mathsf{P}_1^{\gamma_1} \cdots \mathsf{P}_k^{\gamma_k} = \gamma_1 \mathsf{P}_1 + \cdots + \gamma_k \mathsf{P}_k.$$

Furthermore, the counting functions are closely related, for if $\mathsf{B} \subseteq \mathsf{A}$, then

$$\begin{aligned} \widetilde{b}(n) &= b(\lambda^n) \\ b(n) &= \begin{cases} \widetilde{b}(\log_\lambda n) & \text{if } \log_\lambda n \in N \\ 0 & \text{otherwise,} \end{cases} \end{aligned}$$

and from this it follows that the asymptotic densities are the same:

$$\begin{aligned} \delta &= \widetilde{\delta} \\ \Delta &= \widetilde{\Delta}. \end{aligned}$$

Let $\mathbf{B}(x)$ be the generating series of B in $\mathcal{A}$, and $\widetilde{\mathbf{B}}(x)$ the corresponding series of B in $\widetilde{\mathcal{A}}$. Then

$$\begin{aligned} \mathbf{B}(x) &= \sum_{n \geq 1} b(n) n^{-x} \\ \widetilde{\mathbf{B}}(x) &= \sum_{n \geq 0} \widetilde{b}(n) x^n, \end{aligned}$$

with the translations as before

$$\begin{aligned} \mathbf{B}(x) &= \widetilde{\mathbf{B}}(\lambda^{-x}) \\ \widetilde{\mathbf{B}}(x) &= \mathbf{B}(-\log_\lambda x). \end{aligned}$$

The radius of convergence $\widetilde{\rho}$ of $\widetilde{\mathcal{A}}$ and the abscissa of convergence α of $\mathcal{A}$ are related by

$$\begin{aligned} \widetilde{\rho} &= \lambda^{-\alpha} \\ \alpha &= -\log_\lambda \widetilde{\rho}. \end{aligned}$$

From this one can conclude that the Dirichlet densities agree, that is,

$$\partial = \widetilde{\partial}.$$

Thus one can study discrete multiplicative number systems as additive number systems insofar as density is concerned. Note that $\widetilde{\mathcal{A}}$ is reduced iff λ is the largest integer such that the norms of elements of $\mathcal{A}$ are powers of λ.

Now the local asymptotic density theorems are translated from the additive setting in Part 1 into the discrete multiplicative setting. In the following two theorems it is assumed that

> *$\mathcal{A}$ is a discrete multiplicative number system, and λ is the largest positive integer such that every member of A has norm a power of λ.*

Theorem 9.40 (See Theorem 4.3). *If $\alpha = 0$ then the condition*

$$\lim_{n\to\infty} \frac{a(\lambda^{n-1})}{a(\lambda^n)} = 1$$

is necessary and sufficient in order for all partition sets of $\mathcal{A}$ to have local asymptotic density. If so, then, for a partition set $\prod \mathsf{P}_i{}^{\gamma_i}$,

$$\delta\Big(\prod \mathsf{P}_i{}^{\gamma_i}\Big) = \begin{cases} 0 & \text{if some } \gamma_i \text{ is small} \\ 1 & \text{otherwise.} \end{cases}$$

Theorem 9.41 (See Theorem 5.13). *Suppose*

$$\frac{a(\lambda^{n-1})}{a(\lambda^n)} \to \lambda^{-\alpha} \qquad \text{as } n \to \infty.$$

If there exist M and C such that

$$\frac{a(\lambda^{n-k})}{a(\lambda^n)} \leq C\lambda^{-\alpha k}, \quad \text{for } M \leq k \leq n,$$

then all partition sets B of $\mathcal{A}$ have local asymptotic density, and

$$\delta(\mathsf{B}) = \partial(\mathsf{B}).$$

9.9. When sets bA have global asymptotic density

It is important to understand when a set bA of multiples has global asymptotic density. For this purpose a closer look at the structure of the positive reals, and solutions of Cauchy's functional equation, are needed.

9.9.1. Subgroups of the reals. Two of the natural groups that occur when considering the set $\mathbb{R}$ of real numbers are the additive group $(\mathbb{R}, +)$ of all reals and the multiplicative group $((0,\infty), \cdot)$ of positive reals. The exponential function

$$e^x : \mathbb{R} \to (0, \infty)$$

is an isomorphism between these two groups, and furthermore it is a homeomorphism between the corresponding topological spaces, that is, it and its inverse, $\log x$, are both continuous.

A classification of the subgroups of the multiplicative group of positive reals is needed. First this is done in the additive group of reals.

Lemma 9.42. *Let D be a subgroup of the additive group of reals. Then either D is cyclic or D is dense in the reals.*

PROOF. Certainly $0 \in D$, and if $D = \{0\}$ then it is cyclic. So assume $D \neq \{0\}$. Let $\lambda = \inf\{d \in D : d > 0\}$.

Suppose $\lambda > 0$. Then $\lambda \in D$, for otherwise one can choose $r, s \in D$ with $\lambda < r < s$ and $s - r < \lambda$, contradicting the definition of λ since $s - r \in D$. Consequently $\{n\lambda : n \in \mathbb{Z}\} \subseteq D$. But then there can be no element $r \in D$ satisfying $n\lambda < r < (n+1)\lambda$, for otherwise $0 < r - n\lambda < \lambda$ and $r - n\lambda \in D$, again contradicting the definition of λ. Thus $D = \{n\lambda : n \in \mathbb{Z}\}$, the cyclic subgroup of the additive reals generated by λ.

Next suppose that $\lambda = 0$. For reals r, s with $0 < r < s$, choose some $d \in D$ with $0 < d < s - r$. (This is possible since $D \neq \{0\}$.) Then there is a positive integer n such that $r \leq nd \leq s$. This suffices to show that D is dense in the reals. □

Corollary 9.43. *Let D be a subgroup of the multiplicative group of the positive reals $(0, \infty)$. Then either D is cyclic or D is dense in $(0, \infty)$.*

PROOF. The log function from $(0, \infty)$ to $\mathbb{R}$ is such that, for $D \subseteq (0, \infty)$,

- D is a subgroup of $(0, \infty)$ iff $\log(D) = \{\log(x) : x \in D\}$ is a subgroup of $\mathbb{R}$,
- D is a cyclic subgroup of $(0, \infty)$ iff $\log(D)$ is a cyclic subgroup of $\mathbb{R}$, and
- D is a dense subset of $(0, \infty)$ iff $\log(D)$ is a dense subset of $\mathbb{R}$.

Now apply Lemma 9.42. □

9.9.2. Solutions to the Cauchy functional equation.

Definition 9.44. The *Cauchy functional equation* is

$$f(x + y) = f(x) + f(y).$$

The solutions $f(x) = \lambda x$, for $\lambda \in \mathbb{R}$, are the obvious ones on the reals, but there are others.[2]

Proposition 9.45. *Suppose D is a dense subgroup of the additive reals and $f : D \to \mathbb{R}$ is a nondecreasing function that solves Cauchy's functional equation*

$$f(x + y) = f(x) + f(y)$$

on D. Then, for some $\lambda \geq 0$, $f(x) = \lambda x$ on D.

PROOF. Suppose f is a solution of Cauchy's equation on D. By substituting $x - y$ for x, and transposing, one has $f(x - y) = f(x) - f(y)$. Thus $f(0) = 0$ and $f(-x) = -f(x)$. Also $f(nx) = nf(x)$, for $n \in \mathbb{Z}$.

Choose any $d > 0$ in D. Then, for any $n \in \mathbb{N}$, there is a d_1 in D with $0 < nd_1 < d$. This means $f(d_1) \leq f(d)/n$. So $d_1 > 0$ can be found such that $f(d_1)$ is as small as desired. By the monotonicity of f this means f is continuous at 0.

[2] There are a number of conditions guaranteeing that the only solutions are the λx just mentioned—for example, Cauchy knew that the continuity of f suffices.

If, for some $d \in D$ with $d \neq 0$, one has $f(d) = 0$ then, for all $n \in \mathbb{N}$, $f(nd) = 0$. Since f is assumed to be monotonic, it follows that $f(x) = 0$, for all $x \in D$.

So now assume, for $d \in D$, $f(d) = 0$ implies $d = 0$. For $0 < r < s$, with $r, s \in D$, choose m such that

$$mr \le s < (m+1)r.$$

Then $mf(r) \le f(s) \le (m+1)f(r)$ leads to

$$\frac{mf(r)}{(m+1)r} \le \frac{f(s)}{s} \le \frac{(m+1)f(r)}{mr},$$

so

$$\frac{m}{m+1} \le \frac{f(s)/s}{f(r)/r} \le \frac{m+1}{m}.$$

Thus

$$\lim_{\substack{r\to 0\\ r\in D}} \frac{f(s)/s}{f(r)/r} = 1.$$

Thus $f(s)/s = \lambda$ where

$$\lambda = \lim_{\substack{r\to 0\\ r\in D}} \frac{f(r)}{r}.$$

□

Corollary 9.46. *Suppose D is a dense subgroup of the multiplicative group of positive reals and $f : D \to (0, \infty)$ is a nondecreasing function that solves the functional equation*

$$f(xy) = f(x)f(y)$$

on D. Then, for some $\lambda \ge 0$, $f(x) = x^\lambda$ on D.

Proof. Let $E = \log(D)$, the image of D under the log function. Then E is a dense subgroup of $\mathbb{R}$. Define g on E by

$$g(x) = \log\big(f(e^x)\big).$$

Then $g(x+y) = g(x) + g(y)$, so g is a solution to Cauchy's equation on a dense subgroup of the reals. Thus $g(x) = \lambda x$, for some λ. But then

$$\log\big(f(e^x)\big) = \lambda x,$$

so $f(x) = x^\lambda$ on D. □

9.9.3. Applications to RV_α. Now comes the main result tying density to RV_α.

Proposition 9.47 ([**17**], 1997). *Suppose every set of multiples* bA *has global asymptotic density. Then either* $\mathcal{A}$ *is discrete or* $\mathcal{A} \in \mathsf{RV}_\alpha$.

PROOF. Let D be the set of x in $(0, \infty)$ such that

$$f(x) = \lim_{t\to\infty} \frac{A(tx)}{A(t)}$$

exists and is positive. Note that $f(x)$ is nondecreasing on D. It will be proved that D is either

- all of $(0, \infty)$, or
- a cyclic subgroup of the multiplicative group of positive reals.

First it is proved that D is a subgroup of the multiplicative group of positive reals and $f : D \to (0, \infty)$ is a homomorphism.

Clearly $1 \in D$ and $f(1) = 1$. So suppose $x, y \in D$. Then

$$\begin{aligned} f(x)/f(y) &= \lim_{t\to\infty} \frac{A(tx)}{A(t)} \Big/ \lim_{t\to\infty} \frac{A(ty)}{A(t)} \\ &= \lim_{t\to\infty} \frac{A(tx)}{A(ty)} \\ &= \lim_{t\to\infty} \frac{A(tx/y)}{A(t)}. \end{aligned}$$

Thus $x/y \in D$ and $f(x/y) = f(x)/f(y)$, establishing the fact that D is a subgroup and f is a homomorphism.

Let $\mathsf{b} \in \mathsf{A}$. From Lemma 9.15 and Proposition 9.18,

$$\lim_{t\to\infty} \frac{A(t/\|\mathsf{b}\|)}{A(t)} = \Delta(\mathsf{bA}) > 0,$$

so $1/\|\mathsf{b}\| \in D$, and thus $\|\mathsf{b}\| \in D$.

D is either cyclic or dense in $(0, \infty)$ by Corollary 9.43. If D is cyclic then, since it is not the trivial subgroup $\{1\}$, choose a generator $c > 1$ of D. For $\mathsf{b} \in \mathsf{A}$, since $\|\mathsf{b}\| \in D$ it follows that $\|\mathsf{b}\|$ is a power of c, so c is an integer. This means $\mathcal{A}$ is discrete.

If D is not cyclic then it must be dense in $(0, \infty)$. Since f is nondecreasing and satisfies $f(xy) = f(x)f(y)$ on the dense subgroup D of the positive reals, from Corollary 9.46 it follows that $f(x) = x^\lambda$ on D, for some $\lambda \geq 0$. Now, for any $u \in (0, \infty)$, and $x, y \in D$ with $x < u < y$,

$$\frac{A(tx)}{A(t)} \leq \frac{A(tu)}{A(t)} \leq \frac{A(ty)}{A(t)}$$

and thus

$$x^\lambda \leq \liminf_{t\to\infty} \frac{A(tu)}{A(t)} \leq \limsup_{t\to\infty} \frac{A(tu)}{A(t)} \leq y^\lambda.$$

Since x, y can be chosen arbitrarily close together it follows that

$$\lim_{t\to\infty} \frac{A(tu)}{A(t)} = u^\lambda,$$

so indeed $D = (0, \infty)$ and $f(x) = x^\lambda$ on D. This says $\mathcal{A} \in \mathsf{RV}_\lambda$. By Corollary 7.24, $\lambda = \alpha$. □

9.10. The strictly multiplicative case and RV_α

Now the connection between RV_α and the existence of global asymptotic density of partition sets can be made. The discrete systems are disguised additive systems, and have a quite different theory from that of the other multiplicative systems.

Definition 9.48. A multiplicative number system $\mathcal{A}$ is *strictly multiplicative* if it is not discrete.

Proposition 9.49. *Let $\mathcal{A}$ be a strictly multiplicative number system. Then all sets of multiples* bA *have global asymptotic density iff $\mathcal{A} \in \mathsf{RV}_\alpha$. If so then one has*

$$\Delta(\mathsf{bA}) = \|\mathsf{b}\|^{-\alpha} = \partial(\mathsf{bA}).$$

PROOF. This follows from Proposition 9.17 and Proposition 9.47. □

Corollary 9.50. *If all partition sets of a strictly multiplicative number system $\mathcal{A}$ have asymptotic density then $\mathcal{A} \in \mathsf{RV}_\alpha$.*

PROOF. This follows from Proposition 9.49 and Lemma 9.26. □

9.11. The discrete case and RV_α

The condition $\mathcal{A} \in \mathsf{RV}_\alpha$ played a central role in the last section on strictly multiplicative systems. However, if a discrete system $\mathcal{A}$ is in RV_α then the next theorem shows that α must be 0.

Proposition 9.51 ([**17**], 1997). *Let $\mathcal{A}$ be a discrete multiplicative number system with $\mathcal{A} \in \mathsf{RV}_\alpha$. Then $\alpha = 0$, and the corresponding reduced additive system $\widetilde{\mathcal{A}}$ has $\widetilde{A}(n) \in \mathsf{RT}_1$.*

PROOF. Let λ be the largest positive integer such that, for every $\mathsf{a} \in \mathsf{A}$, $\|\mathsf{a}\|$ is an integer power of λ. The only possible value of $\lim_{t\to\infty} A(\frac{3}{2}t)/A(t) = \left(\frac{3}{2}\right)^\alpha$ is 1 since $A(\frac{3}{2}\lambda^n) = A(\lambda^n)$, for all nonnegative integers n, so α must be 0. Then, from $A(\lambda^{n-1})/A(\lambda^n) \to 1$ as $n \to \infty$, follows $\widetilde{A}(n-1)/\widetilde{A}(n) \to 1$ as $n \to \infty$. □

Remark 9.52. The condition $\mathcal{A} \in \mathsf{RV}_\alpha$ is not necessary for the partition sets of a discrete $\mathcal{A}$ to have global asymptotic density. Consider, for example, the case when $\mathcal{A}$ has exactly 2^n indecomposables of size 2^n, and no indecomposables of other sizes. In this case partition sets have local asymptotic density, but $A(x)$ does not have regular variation at infinity, by Corollary 7.24 and Proposition 9.51, since $\alpha = 1$.

9.12. Analog of Schur's Tauberian Theorem

Results that can be used to obtain global asymptotic density from Dirichlet density are given in the Tauberian theorems of this section and §11.1. The Tauberian theorem in this section is a Dirichlet series analog of Schur's Tauberian Theorem for power series—applications are given in this chapter as well as in a generalization of Oppenheim's asymptotics in §11.5.

Theorem 9.53. *Given $\alpha \in \mathbb{R}$, suppose $\mathbf{S}(x), \mathbf{T}(x)$ are two Dirichlet series such that*

(a) *$\mathbf{T}(x)$ has nonnegative coefficients,*
(b) *$T(x) \in \mathsf{RV}_\alpha$, and*
(c) *the abscissa of absolute convergence of $\mathbf{S}(x)$ is less than α.*

Let $\mathbf{R}(x) = \mathbf{S}(x) * \mathbf{T}(x)$. Then,

$$\lim_{x\to\infty} \frac{R(x)}{T(x)} = \mathbf{S}(\alpha).$$

If $\mathbf{S}(\alpha) > 0$ then this can be expressed by the asymptotic relation

$$R(x) \sim \mathbf{S}(\alpha) \cdot T(x). \tag{9.53.1}$$

Proof. Implicit in hypothesis (b) is the fact that $T(x)$ is eventually positive. Let α_s be the abscissa of absolute convergence of $\mathbf{S}(x)$. Then $\alpha_s < \alpha$ by assumption. Choose β such that $\alpha_s < \beta < \alpha$. Since $2^{-\alpha} < 2^{-\beta}$, by (b) choose $M_0 \geq 1$ such that, for $x \geq M_0$, $T(x) > 0$ holds as well as

$$\frac{T(x)}{T(2x)} < 2^{-\beta}. \tag{9.53.2}$$

Note that $T(x)$ is positive on $[M_0, \infty)$ and, by (a), nondecreasing on $(0, \infty)$.
Claim: For some constant C,

$$\frac{T(x)}{T(y)} \leq C(x/y)^\beta, \quad \text{for} \quad y \geq M_0 \quad \text{and} \quad 1 \leq x \leq y.$$

Suppose $y \geq M_0$ and $1 \leq x \leq y$.

If $\beta \leq 0$ then, since $T(x)$ is nondecreasing, $C = 1$ satisfies the claim. So now assume $\beta > 0$. The case $T(x) = 0$ is obvious.

Assume $T(x) > 0$. From the monotonicity of $T(x)$,

$$\frac{T(x)}{T(y)} \leq \frac{T(x)}{T(2x)} \cdots \frac{T(2^{m-1}x)}{T(2^m x)},$$

where $2^m x \leq y < 2^{m+1}x$.

If $x \geq M_0$ then, by (9.53.2),

$$\begin{aligned} \frac{T(x)}{T(y)} &\leq (2^{-\beta})^m \\ &\leq (2^{-\beta})^{\left(\log_2(y/x)\right)-1} \\ &= 2^\beta (x/y)^\beta. \end{aligned} \tag{9.53.3}$$

If $x < M_0$ then, since $M_0 \leq y$,

$$\begin{aligned} \frac{T(x)}{T(y)} &\leq \frac{T(M_0)}{T(y)} \\ &\leq 2^{\beta}(M_0/y)^{\beta} \qquad \text{by (9.53.3)} \\ &\leq (2M_0)^{\beta}(x/y)^{\beta}. \end{aligned}$$

So $C = \max\left(1, (2M_0)^{\beta}\right)$ works in all cases. This finishes the proof of the claim. Note that it yields

$$\frac{T(x/n)}{T(x)} \leq Cn^{-\beta} \quad \text{when} \quad x \geq M_0 \quad \text{and} \quad 1 \leq n \leq x.$$

From the triangle inequality, with $x \geq M_0$,

$$\left|\mathbf{S}(\alpha) - \frac{R(x)}{T(x)}\right| \leq \underbrace{\left|\mathbf{S}(\alpha) - \sum_{n\leq x} s(n)n^{-\alpha}\right|}_{\text{I}} + \underbrace{\left|\sum_{n\leq x} s(n)n^{-\alpha} - \frac{R(x)}{T(x)}\right|}_{\text{II}}.$$

Clearly term I goes to 0 as $x \to \infty$. Thus it is sufficient to show that term II vanishes as $x \to \infty$. For any $M \geq M_0$ and any $x \geq M$,

$$\begin{aligned} \left|\sum_{n\leq x} s(n)n^{-\alpha} - \frac{R(x)}{T(x)}\right| &= \left|\sum_{n\leq x} s(n)n^{-\alpha} - \frac{1}{T(x)} \cdot \sum_{n\leq x} r(n)\right| \\ &= \left|\sum_{n\leq x} s(n)n^{-\alpha} - \frac{1}{T(x)} \cdot \sum_{m\cdot n\leq x} t(m)s(n)\right| \\ &= \left|\sum_{n\leq x} s(n)n^{-\alpha} - \frac{1}{T(x)} \cdot \sum_{n\leq x} T(x/n)s(n)\right| \\ &= \left|\sum_{n\leq x} s(n)\left(n^{-\alpha} - \frac{T(x/n)}{T(x)}\right)\right| \\ &\leq \underbrace{\left|\sum_{n\leq M} s(n)\left(n^{-\alpha} - \frac{T(x/n)}{T(x)}\right)\right|}_{\text{III}} \\ &\qquad + \underbrace{\left|\sum_{M<n\leq x} s(n)\left(n^{-\alpha} - \frac{T(x/n)}{T(x)}\right)\right|}_{\text{IV}}. \end{aligned}$$

Term III goes to 0 as $x \to \infty$ since (b) guarantees that, for any fixed n,

$$\lim_{x\to\infty} \frac{T(x/n)}{T(x)} = n^{-\alpha}.$$

Thus it is sufficient to show that Term IV goes to 0 as $M \to \infty$. For term IV, by the Claim,

$$\begin{aligned}\left|\sum_{M<n\leq x} s(n)\Big(n^{-\alpha} - \frac{T(x/n)}{T(x)}\Big)\right| &\leq \sum_{n>M} |s(n)|n^{-\alpha} + \sum_{M<n\leq x} |s(n)|\frac{T(x/n)}{T(x)} \\ &\leq \sum_{n>M} |s(n)|n^{-\alpha} + C\sum_{n>M} |s(n)|n^{-\beta}.\end{aligned}$$

The sums on the right side go to 0 as $M \to \infty$ since they are the tail ends of convergent series. This finishes the proof. □

An immediate consequence of the analog of Schur's Theorem is that the Dirichlet product of two Dirichlet series is 'in RV_α' if one of them has nonnegative coefficients and is 'in RV_α', and the other, positive at α, has abscissa of convergence less than α.

Corollary 9.54. *Given $\alpha \in \mathbb{R}$, suppose $T(x) \in \mathsf{RV}_\alpha$ and the abscissa of convergence of $\mathbf{S}(x)$ is less than α. If $\mathbf{T}(x)$ has nonnegative coefficients and $\mathbf{S}(\alpha) > 0$ then, for $\mathbf{R}(x) = \mathbf{S}(x) * \mathbf{T}(x)$,*

$$R(x) \in \mathsf{RV}_\alpha.$$

PROOF. From Theorem 9.53,

$$R(x) \sim \mathbf{S}(\alpha) \cdot T(x),$$

so, by Proposition 7.21, $R(x) \in \mathsf{RV}_\alpha$. □

9.12.1. Adding and deleting indecomposables. Membership in RV_α is preserved when removing or adding sets of indecomposables that are not too large, provided $\alpha > 0$. In particular, adding or removing finitely many indecomposables preserves membership in RV_α when $\alpha > 0$.

Proposition 9.55. *Suppose $\mathsf{Q} \subseteq \mathsf{P}$ is such that $\mathbf{A}_\mathsf{Q}(x)$ has abscissa of convergence $\alpha_q < \alpha$. Then*

$$\mathcal{A} \in \mathsf{RV}_\alpha \quad \textit{iff} \quad \mathcal{A}_{\mathsf{P}\setminus\mathsf{Q}} \in \mathsf{RV}_\alpha,$$

where $\mathcal{A}_{\mathsf{P}\setminus\mathsf{Q}}$ is the subsystem of $\mathcal{A}$ generated by $\mathsf{P} \setminus \mathsf{Q}$.

PROOF. By Corollary 8.28, the abscissa of convergence of $[\mathbf{1}/\mathbf{A}_\mathsf{Q}](x)$ is at most α_q, and $[\mathbf{1}/\mathbf{A}_\mathsf{Q}](\alpha) > 0$. Then, using

$$\begin{aligned}\mathbf{A}(x) &= \mathbf{A}_\mathsf{Q}(x) * \mathbf{A}_{\mathsf{P}\setminus\mathsf{Q}}(x) \\ \mathbf{A}_{\mathsf{P}\setminus\mathsf{Q}}(x) &= [\mathbf{1}/\mathbf{A}_\mathsf{Q}](x) * \mathbf{A}(x),\end{aligned}$$

it follows from Corollary 9.54 that $\mathbf{A}(x) \in \mathsf{RV}_\alpha$ iff $\mathbf{A}_{\mathsf{P}\setminus\mathsf{Q}}(x) \in \mathsf{RV}_\alpha$. □

9.12.2. Comparing $\mathbf{A}(x)$ with $e^{\mathbf{P}(x)}$. The structure of the Dirichlet series for $e^{\mathbf{P}(x)}$ can be more transparent than that of $\mathbf{A}(x)$, so it is useful to know how information about the former transfers to the latter.

Lemma 9.56. *The function*

$$\sum_{m\geq 2} \mathbf{P}(mx)/m$$

has a Dirichlet series expansion with abscissa of convergence $\alpha/2$.

PROOF. For $x > \alpha/2$ and $m \geq 2$, $\mathbf{P}(mx)$ converges, and

$$\begin{aligned} \mathbf{P}(mx) &= \sum_{n\geq 2} p(n) n^{-mx} \\ &= \sum_{n\geq 2} p(n) n^{-(m-2)x} n^{-2x} \\ &\leq 2^{-(m-2)x} \sum_{n\geq 2} p(n) n^{-2x} \\ &= 2^{-(m-2)x} \mathbf{P}(2x). \end{aligned}$$

Thus, for $x > \alpha/2$,

$$\begin{aligned} \sum_{m\geq 2} \mathbf{P}(mx)/m &\leq \sum_{m\geq 2} 2^{-(m-2)x} \mathbf{P}(2x) \\ &= \Big(\sum_{m\geq 0} 2^{-mx}\Big) \mathbf{P}(2x) \\ &= \frac{\mathbf{P}(2x)}{1-2^{-x}}, \end{aligned}$$

so $\sum_{m\geq 2} \mathbf{P}(mx)/m$ converges, for $x > \alpha/2$. Consequently Lemma 7.16 shows that $\sum_{m\geq 2} \mathbf{P}(mx)/m$ has a Dirichlet series expansion with abscissa of (absolute) convergence $\alpha/2$. $\square$

The notation $\big[n^{-u}\big] f(u)$ refers to the coefficient of n^{-u} in the Dirichlet series expansion of $f(u)$. In the following,

$$\sum_{n\leq x} \big[n^{-u}\big] e^{\mathbf{P}(u)} \in \mathsf{RV}_\alpha$$

expresses the fact that the cumulative coefficient sum function of the Dirichlet series expansion of $e^{\mathbf{P}(u)}$ is in RV_α.

Proposition 9.57. *If $\alpha > 0$ then*

$$\mathcal{A} \in \mathsf{RV}_\alpha \quad \textit{iff} \quad \sum_{n \le x} \left[n^{-u}\right] e^{\mathbf{P}(u)} \in \mathsf{RV}_\alpha .$$

If $\mathcal{A} \in \mathsf{RV}_\alpha$ then

$$A(x) \ \sim\ C \cdot \sum_{n \le x} \left[n^{-u}\right] e^{\mathbf{P}(u)},$$

where $C = \exp\Big(\sum_{m \ge 2} \mathbf{P}(m\alpha)/m\Big)$.

Proof. From

$$\begin{aligned} \mathbf{A}(u) \ &=\ \exp\Big(\sum_{m \ge 1} \mathbf{P}(mu)/m\Big) \\ &=\ e^{\mathbf{P}(u)} \cdot \exp\Big(\sum_{m \ge 2} \mathbf{P}(mu)/m\Big) \end{aligned} \tag{9.57.1}$$

follows

$$e^{\mathbf{P}(u)} \ =\ \mathbf{A}(u) \cdot \exp\Big(-\sum_{m \ge 2} \mathbf{P}(mu)/m\Big). \tag{9.57.2}$$

Since $\alpha > 0$, $e^{\mathbf{P}(u)}$ has a Dirichlet series expansion with abscissa of (absolute) convergence α by Lemma 7.18 and Corollary 8.18 (a).

Now suppose $\sum_{n \le x} \left[n^{-u}\right] e^{\mathbf{P}(u)} \in \mathsf{RV}_\alpha$. Then apply the analog of Schur's Tauberian Theorem to (9.57.1), in view of Lemma 9.56, to obtain

$$A(x) \ \sim\ C \cdot \sum_{n \le x} \left[n^{-u}\right] e^{\mathbf{P}(u)},$$

where $C = \exp\Big(\sum_{m \ge 2} \mathbf{P}(m\alpha)/m\Big)$. Then $\mathcal{A} \in \mathsf{RV}_\alpha$.

A similar argument holds if $\mathcal{A} \in \mathsf{RV}_\alpha$ by applying the same Tauberian theorem to (9.57.2). □

9.12.3. An application to density.

The analog of Schur's Tauberian Theorem easily leads to a density theorem for multiplicative number systems, the analog of Compton's Theorem 3.49.

Definition 9.58. For B a subset of A, let $[\mathbf{B}/\mathbf{A}](x)$ be the Dirichlet series given by

$$[\mathbf{B}/\mathbf{A}](x) \ =\ \mathbf{B}(x) * [\mathbf{1}/\mathbf{A}(x)].$$

Lemma 9.59. *For $x \ge 0$ such that $\mathbf{A}(x)$ converges,*

$$[\mathbf{B}/\mathbf{A}](x) \ =\ \mathbf{B}(x)/\mathbf{A}(x)$$

and $[\mathbf{B}/\mathbf{A}](x)$ converges absolutely.

PROOF. For $x \geq 0$ with $\mathbf{A}(x) < \infty$, one has $\mathbf{B}(x) < \infty$. For such x, $[\mathbf{1}/\mathbf{A}](x)$ converges absolutely by Corollary 8.28, so

$$\begin{aligned}
[\mathbf{B}/\mathbf{A}](x) &= \mathbf{B}(x) * [\mathbf{1}/\mathbf{A}](x) && \text{definition} \\
&= \mathbf{B}(x) \cdot [\mathbf{1}/\mathbf{A}](x) && \text{by Proposition 7.3 (f)} \\
&= \frac{\mathbf{B}(x)}{\mathbf{A}(x)} && \text{by Corollary 8.28.}
\end{aligned}$$

From the absolute convergence of $\mathbf{B}(x)$ and $[\mathbf{1}/\mathbf{A}](x)$ it follows that $[\mathbf{B}/\mathbf{A}](x)$ converges absolutely. □

Theorem 9.60. *Let $\mathcal{A}$ be a multiplicative number system and let $\mathsf{B} \subseteq \mathsf{A}$. If $\mathcal{A} \in \mathsf{RV}_\alpha$, and the abscissa of absolute convergence of $[\mathbf{B}/\mathbf{A}](x)$ is less than α, then*

$$\Delta(\mathsf{B}) = \partial(\mathsf{B}) = [\mathbf{B}/\mathbf{A}](\alpha).$$

PROOF. Let $\mathbf{S}(x) = [\mathbf{B}/\mathbf{A}](x)$ and $\mathbf{T}(x) = \mathbf{A}(x)$. Then the hypotheses of the analog of Schur's Theorem are satisfied, and

$$\begin{aligned}
\partial(\mathsf{B}) &= \lim_{x \to \alpha} \frac{\mathbf{B}(x)}{\mathbf{A}(x)} && \text{definition} \\
&= \lim_{x \to \alpha} [\mathbf{B}/\mathbf{A}](x) && \text{Lemma 9.59} \\
&= [\mathbf{B}/\mathbf{A}](\alpha) && \text{as } \alpha_s < \alpha \\
&= \lim_{n \to \infty} \frac{B(n)}{A(n)} && \text{Theorem 9.53} \\
&= \Delta(\mathsf{B}) && \text{definition.}
\end{aligned}$$

□

9.12.4. The number of Abelian groups. Erdös and Szekeres [**24**] found asymptotics for the number of Abelian groups of size at most n.

Theorem 9.61 (Erdös and Szekeres, 1934)**.** *Let $\mathcal{A}$ be the multiplicative number system for finite Abelian groups. Then*

$$A(x) \sim \left(\prod_{n=2}^{\infty} \zeta(n)\right) \cdot x.$$

PROOF. From §8.3.5,

$$\mathbf{A}(x) = \prod_{\substack{p \text{ prime} \\ n \geq 1}} \left(1 - (p^n)^{-x}\right)^{-1}.$$

This can be written as $\mathbf{A}(x) = \mathbf{S}(x) \cdot \zeta(x)$, where

$$\mathbf{S}(x) = \prod_{\substack{p \text{ prime} \\ n \geq 2}} \left(1 - (p^n)^{-x}\right)^{-1}.$$

Clearly the global counting function for $\zeta(x)$ is in RV_1 since it is just $\lfloor x \rfloor$, and using Lemma 7.11 it is easy to check that $\mathbf{S}(x)$ has abscissa of convergence $\alpha_s = 1/2$. Thus, by Theorem 3.42,

$$A(x) \;\sim\; \mathbf{S}(1) \cdot \lfloor x \rfloor \;\sim\; \mathbf{S}(1) \cdot x.$$

By rearranging the terms of the infinite product expression for $\mathbf{S}(x)$ one has

$$\mathbf{S}(x) \;=\; \prod_{n\geq 2} \prod_{p \text{ prime}} \left(1 - (p^n)^{-x}\right)^{-1} \;=\; \prod_{n\geq 2} \zeta(nx).$$

Thus $\mathbf{S}(1) = \prod_{n=2}^{\infty} \zeta(n)$. □

The constant $\zeta(2)\zeta(3)\cdots$ is approximately 2.295.

9.13. Simple partition sets

In this section it is proved that some of the simplest of the partition sets of $\mathcal{A}$, namely those obtained by specifying the exact number of times that each member of a given finite set of indecomposables can occur in a decomposition, and otherwise leaving the number of occurrences of all other indecomposables unrestricted, have Dirichlet density that can be described by a simple formula. The global asymptotic density exists and agrees with the Dirichlet density of simple partition sets provided $\mathcal{A} \in \mathsf{RV}_\alpha$.

Definition 9.62. Let $\mathsf{p}_1, \ldots, \mathsf{p}_k \in \mathsf{P}$ and let $m_1, \ldots, m_k$ be nonnegative integers. Then the partition set

$$\mathsf{B} \;=\; \mathsf{p}_1^{m_1} \cdots \mathsf{p}_k^{m_k} \left(\mathsf{P} \smallsetminus \{\mathsf{p}_1, \ldots, \mathsf{p}_k\}\right)^{\omega}$$

is called a *simple* partition set.

Lemma 9.63. *Every simple partition set*

$$\mathsf{B} \;=\; \mathsf{p}_1^{m_1} \cdots \mathsf{p}_k^{m_k} \left(\mathsf{P} \smallsetminus \{\mathsf{p}_1, \ldots, \mathsf{p}_k\}\right)^{\omega}$$

has Dirichlet density, and

$$\partial(\mathsf{B}) \;=\; [\mathbf{B}/\mathbf{A}](\alpha) \;=\; \prod_{i=1}^{k} \|\mathsf{p}_i\|^{-m_i\alpha}(1 - \|\mathsf{p}_i\|^{-\alpha}).$$

PROOF. Since

$$\mathsf{A} \;=\; \mathsf{p}_1^{\omega} \cdots \mathsf{p}_k^{\omega} \cdot \left(\mathsf{P} \smallsetminus \{\mathsf{p}_1, \ldots, \mathsf{p}_k\}\right)^{\omega},$$

from Corollary 9.31 and Lemma 9.32, for $x > 0$,

$$\begin{aligned}
\mathbf{A}(x) &= \mathbf{p}_1^{\omega}(x) \cdots \mathbf{p}_k^{\omega}(x) \cdot \left(\mathbf{P} \smallsetminus \{\mathbf{p}_1, \ldots, \mathbf{p}_k\}\right)^{\omega}(x)\\
&= \left(1 - \|\mathsf{p}_1\|^{-x}\right)^{-1} \cdots \left(1 - \|\mathsf{p}_k\|^{-x}\right)^{-1} \cdot \left(\mathbf{P} \smallsetminus \{\mathbf{p}_1, \ldots, \mathbf{p}_k\}\right)^{\omega}(x),\\
\mathbf{B}(x) &= \mathbf{p}_1^{m_1}(x) \cdots \mathbf{p}_k^{m_k}(x) \cdot \left(\mathbf{P} \smallsetminus \{\mathbf{p}_1, \ldots, \mathbf{p}_k\}\right)^{\omega}(x)\\
&= \|\mathsf{p}_1\|^{-m_1 x} \cdots \|\mathsf{p}_k\|^{-m_k x} \cdot \left(\mathbf{P} \smallsetminus \{\mathbf{p}_1, \ldots, \mathbf{p}_k\}\right)^{\omega}(x),
\end{aligned}$$

and thus, for $x > \alpha$,

$$[\mathbf{B}/\mathbf{A}](x) \;=\; \prod_{i=1}^{k} \|\mathsf{p}_i\|^{-m_i x}(1 - \|\mathsf{p}_i\|^{-x}).$$

This is a continuous function, so

$$\partial(\mathsf{B}) \;=\; [\mathbf{B}/\mathbf{A}](\alpha) \;=\; \prod_{i=1}^{k} \|\mathsf{p}_i\|^{-m_i \alpha}(1 - \|\mathsf{p}_i\|^{-\alpha}).$$

□

Proposition 9.64. *Suppose $\mathcal{A}$ is in RV_α. Then every simple partition set*

$$\mathsf{B} \;=\; \mathsf{p}_1^{m_1}\cdots\mathsf{p}_k^{m_k}\big(\mathsf{P}\smallsetminus\{\mathsf{p}_1,\ldots,\mathsf{p}_k\}\big)^{\omega}$$

has global asymptotic density equal to its Dirichlet density, that is,

$$\Delta(\mathsf{B}) \;=\; \partial(\mathsf{B}) \;=\; [\mathbf{B}/\mathbf{A}](\alpha) \;=\; \prod_{i=1}^{k} \|\mathsf{p}_i\|^{-m_i \alpha}\big(1 - \|\mathsf{p}_i\|^{-\alpha}\big). \tag{9.64.1}$$

PROOF. From Lemma 9.63 it is known that the last two equalities in (9.64.1) hold. Now Theorem 9.60 applies since $[\mathbf{B}/\mathbf{A}](x)$ has $-\infty$ as its abscissa of absolute convergence. □

9.14. The asymptotic density of P^γ

In §9.7.1 the Dirichlet density of P^γ was determined. Now the global asymptotic density is determined, given that $\mathcal{A} \in \mathsf{RV}_\alpha$ and $\mathbf{A}(\alpha) = \infty$.

Lemma 9.65. *Suppose* $\mathsf{B} \subseteq \mathsf{C} \subseteq \mathsf{A}$ *and* $\Delta(\mathsf{C}) = 0$. *Then* $\Delta(\mathsf{B}) = 0$.

PROOF. This follows from $B(n) \le C(n)$ and $\displaystyle\lim_{n\to\infty}\frac{C(n)}{A(n)} \;=\; 0.$ □

Lemma 9.66. *Suppose* $\mathsf{B} \subseteq \mathsf{B}_k \subseteq \mathsf{A}$. *Then*

$$\lim_{k\to\infty}\Delta(\mathsf{B}_k) \;=\; 0 \quad \textit{implies} \quad \Delta(\mathsf{B}) \;=\; 0.$$

PROOF. This follows from $B(n) \le B_k(n)$ and $\displaystyle\lim_{k\to\infty}\lim_{n\to\infty}\frac{B_k(n)}{A(n)} = 0.$ □

Lemma 9.67. *If $\mathcal{A} \in \mathsf{RV}_\alpha$ and $\mathbf{A}(\alpha) = \infty$ then, for γ an exponent,*

$$\Delta(\mathsf{P}^\gamma) \;=\; \begin{cases} 0 & \textit{if } \gamma \textit{ is small} \\ 1 & \textit{if } \gamma \textit{ is large.} \end{cases}$$

PROOF. First the case that $\alpha = 0$ is considered. Given $m \in \mathbb{N}$, choose $\mathsf{p} \in \mathsf{P}$ and note that

$$\Delta\big(\mathsf{p}^m\mathsf{A}\big) \;=\; 1,$$

by the formula in Proposition 9.17. As

$$\mathsf{p}^m\mathsf{A} \subseteq \mathsf{P}^{\geq m},$$

it follows that the density of $\mathsf{P}^{\geq m}$ is also 1. The other cases of $\Delta(\mathsf{P}^\gamma)$ (when $\alpha = 0$) follow from this.

Now suppose that $\alpha > 0$. Define P_n to be $\mathsf{P}_{(\leq n)}$, the (finite) set of indecomposables in P of norm at most n. For $m, n \in \mathbb{N}$, let

$$\mathsf{B}_{m,n} = \mathsf{P}_n^m \cdot (\mathsf{P} \smallsetminus \mathsf{P}_n)^\omega.$$

From

$$\begin{aligned}\mathbf{B}_{m,n}(x) &= \mathbf{P}_n^m(x) \cdot (\mathbf{P} \smallsetminus \mathbf{P}_n)^\omega(x)\\ \mathbf{A}(x) &= \mathbf{P}_n^\omega(x) \cdot (\mathbf{P} \smallsetminus \mathbf{P}_n)^\omega(x)\end{aligned}$$

follows, for $x > \alpha$,

$$\frac{\mathbf{B}_{m,n}(x)}{\mathbf{A}(x)} = \frac{\mathbf{P}_n^m(x)}{\mathbf{P}_n^\omega(x)} = \Big(\sum_{\mathsf{a}\in\mathsf{P}_n^m} \|\mathsf{a}\|^{-x}\Big) \cdot \prod_{\mathsf{p}\in\mathsf{P}_n} \big(1 - \|\mathsf{p}\|^{-x}\big).$$

Since the expansion of the last line is a Dirichlet polynomial, by Theorem 9.60 the asymptotic density of $\mathsf{B}_{m,n}$ exists and

$$\Delta(\mathsf{B}_{m,n}) = \partial(\mathsf{B}_{m,n}) = [\mathbf{B}_{m,n}/\mathbf{A}](\alpha).$$

Thus,

$$\begin{aligned}\Delta(\mathsf{B}_{m,n}) &= \frac{\mathbf{P}_n^m(\alpha)}{\mathbf{P}_n^\omega(\alpha)} && \text{from above}\\ &\leq \frac{\mathbf{P}_n(\alpha)^m}{e^{\mathbf{P}_n(\alpha)}} && \text{by Corollaries 8.23 and 9.33}\\ &\to 0 \quad \text{as } n \to \infty,\end{aligned}$$

the last line following from the fact that $\mathbf{P}(\alpha) = \infty$, and

$$\mathbf{P}_n(\alpha) = p(2)2^{-\alpha} + \cdots + p(n)n^{-\alpha} \to \mathbf{P}(\alpha) \quad \text{as } n \to \infty.$$

Then $\Delta(\mathsf{P}^m) = 0$ follows from $\mathsf{P}^m \subseteq \mathsf{B}_{0,n} \cup \cdots \cup \mathsf{B}_{m,n}$.

From this it is easy to show that $\Delta(\mathsf{P}^\gamma)$ is as described. □

Corollary 9.68. *Suppose $\mathcal{A} \in \mathsf{RV}_\alpha$ and $\mathbf{A}(\alpha) = \infty$. Then, for any partition set $\mathsf{B} = \mathsf{P}_1^{\gamma_1} \cdots \mathsf{P}_k^{\gamma_k}$ with all exponents γ_i small, $\Delta(\mathsf{B}) = 0$.*

PROOF. Let each γ_i be either m_i or $\leq m_i$, and let $m = \sum m_i$. Then $\mathsf{B} \subseteq \mathsf{P}^{\leq m}$, and since $\Delta(\mathsf{P}^{\leq m}) = 0$, the result follows. □

9.15. Adding an indecomposable

Given a multiplicative number system $\mathcal{A} = \big(\mathsf{A}, \mathsf{P}, \cdot, 1, \|\ \|\big)$ with counting functions $a(n), p(n)$ and abscissa of convergence α, augment $\mathcal{A}$ by a new indecomposable q to form the multiplicative number system $\mathcal{A}_{\mathsf{q}}$ by simply adding q to P to form $\mathsf{P}_{\mathsf{q}} = \mathsf{P} \cup \{\mathsf{q}\}$, by letting A_{q} be the set of formal products $\mathsf{q}^m\mathsf{a}$, for m a nonnegative integer, and $\mathsf{a} \in \mathsf{A}$, and extending the norm by the definition

$$\|\mathsf{q}^m\mathsf{a}\| \ = \ \|\mathsf{q}\|^m\|\mathsf{a}\|.$$

Let $a_{\mathsf{q}}, p_{\mathsf{q}}$ be the counting functions of $\mathcal{A}_{\mathsf{q}}$. Then

$$p_{\mathsf{q}}(n) \ = \ \begin{cases} p(n) + 1 & \text{if} \quad n = \|\mathsf{q}\| \\ p(n) & \text{if} \quad n \neq \|\mathsf{q}\|. \end{cases}$$

From the fundamental identity of $\mathcal{A}_{\mathsf{q}}$ follows

$$\begin{aligned} \sum_{n\geq 1} a_{\mathsf{q}}(n)n^{-x} \ &= \ \prod_{n\geq 2}(1-n^{-x})^{-p_{\mathsf{q}}(n)} \\ &= \ \big(1 - \|\mathsf{q}\|^{-x}\big)^{-1} \cdot \prod_{n\geq 2}(1-n^{-x})^{-p(n)} \end{aligned}$$

and thus

$$\mathbf{A}_{\mathsf{q}}(x) \ = \ \big(1 - \|\mathsf{q}\|^{-x}\big)^{-1} \cdot \mathbf{A}(x). \tag{9.68.1}$$

One can easily see from Corollary 8.18 (b) that $\mathcal{A}_{\mathsf{q}}$ has the same abscissa of convergence α as $\mathcal{A}$ since $\mathbf{P}(x)$ and $\mathbf{P}_{\mathsf{q}}(x)$ have the same abscissa of convergence.

Theorem 9.69. *Suppose all partition sets of $\mathcal{A}$ have global asymptotic density. Let $\mathcal{A}_{\mathsf{q}}$ be an extension of $\mathcal{A}$ by a new indecomposable q. If $\mathcal{A}$ is strictly multiplicative then all partition sets of $\mathcal{A}_{\mathsf{q}}$ have global asymptotic density.*

PROOF. Without loss of generality assume that an arbitrary partition of $\mathsf{P}_{\mathsf{q}} = \mathsf{P} \cup \{\mathsf{q}\}$ is written in the form

$$\mathsf{P}_1 \cup \{\mathsf{q}\}, \mathsf{P}_2, \ldots, \mathsf{P}_k$$

where $\mathsf{P}_1, \ldots, \mathsf{P}_k$ is a partition of P, that is, the new element q is in the first partition class. The fact that any partition set

$$\mathsf{B} \ = \ (\mathsf{P}_1 \cup \{\mathsf{q}\})^{\gamma_1} \cdots {\mathsf{P}_k}^{\gamma_k}$$

has global asymptotic density in $\mathcal{A}_{\mathsf{q}}$ is to be proved. From (9.68.1)

$$A_{\mathsf{q}}(x) \ = \ A(x) \ + \ A\Big(\frac{x}{\|\mathsf{q}\|}\Big) \ + \ A\Big(\frac{x}{\|\mathsf{q}\|^2}\Big) \ + \ \cdots$$

Claim 1:

$$A_{\mathsf{q}}(x) \;\in\; \mathsf{RV}_\alpha.$$

By Corollary 9.50, $A(x) \in \mathsf{RV}_\alpha$. For $x > 0$ and $\varepsilon > 0$, choose M such that, for $t \geq M$,

$$\left|A(tx) - x^\alpha A(t)\right| \;<\; \varepsilon A(t).$$

Then

$$\begin{aligned}
\left|A_{\mathsf{q}}(tx) - x^\alpha A_{\mathsf{q}}(t)\right| \;&=\; \left|\sum_{k\geq 0}\left[A\Big(\frac{tx}{\|\mathsf{q}\|^k}\Big) - x^\alpha A\Big(\frac{t}{\|\mathsf{q}\|^k}\Big)\right]\right| \\
&\leq\; \sum_{t/\|\mathsf{q}\|^k \geq M} \left|A\Big(\frac{tx}{\|\mathsf{q}\|^k}\Big) - x^\alpha A\Big(\frac{t}{\|\mathsf{q}\|^k}\Big)\right| \\
&\qquad +\; \sum_{t/\|\mathsf{q}\|^k < M} \left|A\Big(\frac{tx}{\|\mathsf{q}\|^k}\Big) - x^\alpha A\Big(\frac{t}{\|\mathsf{q}\|^k}\Big)\right| \\
&\leq\; \sum_{t/\|\mathsf{q}\|^k \geq M} \varepsilon A\Big(\frac{t}{\|\mathsf{q}\|^k}\Big) \\
&\qquad +\; \sum_{t/\|\mathsf{q}\|^k < M} \left|A\Big(\frac{tx}{\|\mathsf{q}\|^k}\Big) - x^\alpha A\Big(\frac{t}{\|\mathsf{q}\|^k}\Big)\right| \\
&\leq\; \varepsilon A_{\mathsf{q}}(t) \;+\; \sum_{t/\|\mathsf{q}\|^k < M} \left|A\Big(\frac{tx}{\|\mathsf{q}\|^k}\Big) - x^\alpha A\Big(\frac{t}{\|\mathsf{q}\|^k}\Big)\right|.
\end{aligned}$$

Since $\lim\limits_{t\to\infty} A_{\mathsf{q}}(t) = \infty$, the claim follows.

Claim 2:
Let Δ_{q} be the global asymptotic density function of $\mathcal{A}_{\mathsf{q}}$. Then A has global asymptotic density in $\mathcal{A}_{\mathsf{q}}$ given by

$$\Delta_{\mathsf{q}}(\mathsf{A}) \;=\; 1 - \|\mathsf{q}\|^{-\alpha}.$$

This follows from Proposition 9.64 since A is a simple partition set of $\mathcal{A}_{\mathsf{q}}$ (note that $\mathsf{A} = \mathsf{q}^0\mathsf{P}^\omega$) and since $\mathcal{A}_{\mathsf{q}} \in \mathsf{RV}_\alpha$ by Claim 1.

Claim 3:
For B a partition set of $\mathcal{A}$, and, for any nonnegative integer m,

$$\Delta_{\mathsf{q}}(\mathsf{q}^m\mathsf{B}) \;=\; \Delta(\mathsf{B}) \cdot \|\mathsf{q}\|^{-m\alpha}\big(1 - \|\mathsf{q}\|^{-\alpha}\big).$$

This follows from

$$\begin{aligned}
\frac{|\mathsf{q}^m\mathsf{B}|_{\leq n}}{A_{\mathsf{q}}(n)} \;&=\; \frac{B(n/\|\mathsf{q}\|^m)}{A_{\mathsf{q}}(n)} \\
&=\; \frac{B(n/\|\mathsf{q}\|^m)}{A(n/\|\mathsf{q}\|^m)} \cdot \frac{A(n/\|\mathsf{q}\|^m)}{A(n)} \cdot \frac{A(n)}{A_{\mathsf{q}}(n)}
\end{aligned}$$

since

$$\begin{aligned}
\lim_{n\to\infty} \frac{B(n/\|\mathsf{q}\|^m)}{A(n/\|\mathsf{q}\|^m)} &= \Delta(\mathsf{B}) && \text{as } \Delta(\mathsf{B}) \text{ exists}\\
\lim_{n\to\infty} \frac{A(n/\|\mathsf{q}\|^m)}{A(n)} &= \|\mathsf{q}\|^{-m\alpha} && \text{as } \mathcal{A} \in \mathsf{RV}_\alpha\\
\lim_{n\to\infty} \frac{A(n)}{A_{\mathsf{q}}(n)} &= 1 - \|\mathsf{q}\|^{-\alpha} && \text{by Claim 2.}
\end{aligned}$$

Claim 4:
With the same assumptions as in the previous claim,

$$\Delta_{\mathsf{q}}(\mathsf{q}^{\leq m}\mathsf{B}) \;=\; \Delta(\mathsf{B}) \cdot \big(1 - \|\mathsf{q}\|^{-(m+1)\alpha}\big).$$

This follows from noting that $\mathsf{q}^{\leq m}\mathsf{B}$ is the disjoint union of the $\mathsf{q}^j\mathsf{B}$, for $0 \leq j \leq m$, and thus

$$\begin{aligned}
\Delta_{\mathsf{q}}(\mathsf{q}^{\leq m}\mathsf{B}) &= \sum_{0\leq j\leq m} \Delta_{\mathsf{q}}(\mathsf{q}^j\mathsf{B})\\
&= \sum_{0\leq j\leq m} \Delta(\mathsf{B}) \cdot \|\mathsf{q}\|^{-j\alpha}\big(1 - \|\mathsf{q}\|^{-\alpha}\big) && \text{by Claim 3}\\
&= \Delta(\mathsf{B}) \cdot \big(1 - \|\mathsf{q}\|^{-(m+1)\alpha}\big).
\end{aligned}$$

Claim 5:
For any partition set B of $\mathcal{A}$,

$$\Delta_{\mathsf{q}}(\mathsf{q}^{\omega}\mathsf{B}) \;=\; \Delta(\mathsf{B}).$$

Given $\varepsilon > 0$ choose M such that, for $n \geq M$,

$$\Big|B(n) - \Delta(\mathsf{B}) \cdot A(n)\Big| \;<\; \varepsilon A(n).$$

Then

$$\begin{aligned}
&\Big||\mathsf{q}^{\omega}\mathsf{B}|_{\leq n} - \Delta(\mathsf{B}) \cdot A_{\mathsf{q}}(n)\Big|\\
&\quad= \Big|\sum_{k\geq 0}\Big(B(n/\|\mathsf{q}\|^k) - \Delta(\mathsf{B}) \cdot A(n/\|\mathsf{q}\|^k)\Big)\Big|\\
&\quad\leq \sum_{n/\|\mathsf{q}\|^k\geq M} \Big|B(n/\|\mathsf{q}\|^k) - \Delta(\mathsf{B}) \cdot A(n/\|\mathsf{q}\|^k)\Big|\\
&\qquad\quad + \sum_{n/\|\mathsf{q}\|^k< M} \Big|B(n/\|\mathsf{q}\|^k) - \Delta(\mathsf{B}) \cdot A(n/\|\mathsf{q}\|^k)\Big|\\
&\quad\leq \sum_{n/\|\mathsf{q}\|^k\geq M} \varepsilon A(n/\|\mathsf{q}\|^k) \;+\; \sum_{n/\|\mathsf{q}\|^k< M} \Big|B(n/\|\mathsf{q}\|^k) - \Delta(\mathsf{B}) \cdot A(n/\|\mathsf{q}\|^k)\Big|\\
&\quad\leq \varepsilon A_{\mathsf{q}}(n) \;+\; \sum_{n/\|\mathsf{q}\|^k< M} \Big|B(n/\|\mathsf{q}\|^k) - \Delta(\mathsf{B}) \cdot A(n/\|\mathsf{q}\|^k)\Big|,
\end{aligned}$$

which establishes the claim since $A_{\mathsf{q}}(n) \to \infty$.

Claim 6:
Partition sets of $\mathcal{A}_{\mathsf{q}}$ have global asymptotic density.

Let B be a partition set of $\mathcal{A}_{\mathsf{q}}$, say

$$\mathsf{B} = (\mathsf{P}_1 \cup \{\mathsf{q}\})^{\gamma_1} \cdots \mathsf{P}_k^{\gamma_k}.$$

Then B is in the closure under disjoint union and set difference (between comparable sets) of partition sets of the form

$$\mathsf{q}^{\gamma}\mathsf{P}_1^{\gamma_1} \cdots \mathsf{P}_k^{\gamma_k}$$

since, letting $\mathsf{S} = \mathsf{P}_2^{\gamma_2} \cdots \mathsf{P}_k^{\gamma_k}$,

$$\begin{aligned}
\big(\mathsf{P}_1 \cup \{\mathsf{q}\}\big)^m \mathsf{S} &= \bigcup_{0 \le j \le m} \mathsf{q}^j {\mathsf{P}_1}^{m-j} \mathsf{S} \\
\big(\mathsf{P}_1 \cup \{\mathsf{q}\}\big)^{\le m} \mathsf{S} &= \bigcup_{0 \le j \le m} \big(\mathsf{P}_1 \cup \{\mathsf{q}\}\big)^j \mathsf{S} \\
\big(\mathsf{P}_1 \cup \{\mathsf{q}\}\big)^{\omega} \mathsf{S} &= \mathsf{q}^{\omega} {\mathsf{P}_1}^{\omega} \mathsf{S} \\
\big(\mathsf{P}_1 \cup \{\mathsf{q}\}\big)^{\ge m} \mathsf{S} &= \big(\mathsf{P}_1 \cup \{\mathsf{q}\}\big)^{\omega} \mathsf{S} \setminus \big(\mathsf{P}_1 \cup \{\mathsf{q}\}\big)^{\le m-1} \mathsf{S}.
\end{aligned}$$

This finishes the proof of the theorem. □

9.16. First conjecture

From [**7**] one has, for $\mathcal{A}$ an additive number system, $\delta(\mathsf{P}) = 0$ implies $\mathbf{A}(\rho) = \infty$ (see Appendix C). The corresponding assertion for multiplicative number systems is the following. (Note: Warlimont [**55**] has recently proved that $\Delta(\mathsf{P}) = 0$ implies $\alpha < \infty$.)

Conjecture 9.70. $\Delta(\mathsf{P}) = 0$ implies $\mathbf{A}(\alpha) = \infty$.

CHAPTER 10

The Case $\alpha = 0$

If $\alpha = 0$ then the Dirichlet density of each partition set is either 0 or 1, and there is a simple criterion to determine which is the case. Furthermore, the fortunate situation exists that

$$\mathcal{A} \in \mathsf{RV}_0$$

is necessary and sufficient in order that all partition sets have global asymptotic density. For $\mathcal{A} \in \mathsf{RV}_0$, the global asymptotic density agrees with the Dirichlet density on partition sets. Thus much of the interesting work on density when $\alpha = 0$ focuses on criteria to determine when $\mathcal{A} \in \mathsf{RV}_0$.

10.1. The fundamental results when $\alpha = 0$

First Dirichlet density is considered.

Theorem 10.1. *If $\alpha = 0$ then, for any partition set $\prod \mathsf{P}_i^{\gamma_i}$,*

$$\partial\Big(\prod \mathsf{P}_i^{\gamma_i}\Big) = \begin{cases} 0 & \text{if some } \gamma_i \text{ is small} \\ 1 & \text{otherwise.} \end{cases}$$

PROOF. First apply Proposition 9.14 to show that, for any $\mathsf{b} \in \mathsf{A}$,

$$\partial(\mathsf{bA}) = 1. \tag{10.1.1}$$

Now suppose $\mathsf{B} = \prod \mathsf{P}_i^{\gamma_i}$ is a partition set with all γ_i large. Choose any $\mathsf{b} \in \mathsf{B}$. Then $\mathsf{bA} \subseteq \mathsf{B}$ ensures, by (10.1.1), that $\partial(\mathsf{B}) = 1$.

Finally, suppose some γ_i is small, say γ_j. If γ_j is either m or $\leq m$ note that

$$\mathsf{B} \subseteq \mathsf{P}_j^{\leq m}(\mathsf{P} \smallsetminus \mathsf{P}_j)^{\omega}. \tag{10.1.2}$$

The complement of the right side is

$$\mathsf{P}_j^{>m}(\mathsf{P} \smallsetminus \mathsf{P}_j)^{\omega},$$

which has Dirichlet density 1 since it is a partition set with all exponents large. Thus $\mathsf{P}_j^{\leq m}(\mathsf{P} \smallsetminus \mathsf{P}_j)^{\omega}$ has Dirichlet density 0, and then, by (10.1.2), B has Dirichlet density 0. □

Theorem 10.2. *If $\alpha = 0$, the condition $\mathcal{A} \in \mathsf{RV}_0$ is necessary and sufficient for all partition sets of $\mathcal{A}$ to have global asymptotic density. If $\mathcal{A} \in \mathsf{RV}_0$ then, for any partition set $\prod \mathsf{P}_i^{\gamma_i}$,*

$$\Delta\Big(\prod \mathsf{P}_i^{\gamma_i}\Big) = \begin{cases} 0 & \text{if some } \gamma_i \text{ is small} \\ 1 & \text{otherwise,} \end{cases}$$

so global asymptotic density agrees with Dirichlet density on partition sets.

PROOF. Suppose that $\mathcal{A} \in \mathsf{RV}_0$. Then, for $\mathsf{b} \in \mathsf{A}$,

$$\lim_{t\to\infty} \frac{A(t/\|\mathsf{b}\|)}{A(t)} = \|\mathsf{b}\|^0 = 1,$$

so, by Lemma 9.15, $\Delta(\mathsf{bA}) = 1$. Now proceed parallel to the proof of Theorem 10.1 to derive the claimed values of $\Delta\big(\prod \mathsf{P}_i^{\gamma_i}\big)$. This proves that $\mathcal{A} \in \mathsf{RV}_0$ is sufficient.

For the necessity of the condition $\mathcal{A} \in \mathsf{RV}_0$, assume all partition sets have global asymptotic density and split the argument into two cases. If $\mathcal{A}$ is strictly multiplicative then Proposition 9.49 says that $\mathcal{A} \in \mathsf{RV}_0$. So assume $\mathcal{A}$ is discrete. Choose λ to be the largest positive integer such that members of $\mathcal{A}$ have norms that are powers of λ. Then the associated additive system $\widetilde{\mathcal{A}}$ is reduced, its radius of convergence is $\rho = 1$, and all of its partition sets have global asymptotic density. By Theorem 4.3, $\widetilde{A}(n) \in \mathsf{RT}_1$, so

$$\lim_{n\to\infty} \frac{\widetilde{A}(n-1)}{\widetilde{A}(n)} = 1$$

and thus $\lim_{n\to\infty} \dfrac{A(\lambda^{n-1})}{A(\lambda^n)} = 1$. Now, for $t > 0$, let $n_t = \lfloor \log_\lambda t \rfloor$. Then

$$\frac{A(\lambda^{n_t-1})}{A(\lambda^{n_t+1})} \le \frac{A(t/\lambda)}{A(t)} \le 1,$$

so $\lim_{t\to\infty} \dfrac{A(t/\lambda)}{A(t)} = 1$. By Lemma 7.25, $\mathcal{A} \in \mathsf{RV}_0$.

Now Theorem 10.1 and the first part of this proof show that ∂ and Δ agree on partition sets. □

The next to the last step of the proof of Theorem 10.2 involves a method to show $\mathcal{A} \in \mathsf{RV}_0$ that is worth repeating.

Lemma 10.3. *If $\Delta(\mathsf{bA}) = 1$, for some $\mathsf{b} \in \mathsf{A}$ with $\mathsf{b} \neq 1$, then $\mathcal{A} \in \mathsf{RV}_0$.*

PROOF. This is immediate from Lemma 7.25 in view of Lemma 9.15. □

10.2. Odlyzko's Product Theorem

Definition 10.4. Given two multiplicative number systems $\mathcal{A}_1$ and $\mathcal{A}_2$, define the *product* $\mathcal{A}_1 \times \mathcal{A}_2$ as the multiplicative number system whose underlying monoid is the direct product of the two monoids, and the norm is the product of the coordinate norms, that is,

$$\|(\mathsf{a}_1, \mathsf{a}_2)\| \;=\; \|\mathsf{a}_1\|_1 \cdot \|\mathsf{a}_2\|_2 \,.$$

The set of indecomposables of $\mathcal{A}_1 \times \mathcal{A}_2$ is

$$\{(\mathsf{p}_1, 1) : \mathsf{p}_1 \in \mathsf{P}_1\} \;\cup\; \{(1, \mathsf{p}_2) : \mathsf{p}_2 \in \mathsf{P}_2\}.$$

Let $a_i(n), p_i(n)$ be the counting functions of $\mathcal{A}_i$, and let $a(n), p(n)$ be the counting functions of $\mathcal{A}_1 \times \mathcal{A}_2$. Then

$$\begin{aligned} p(n) &= p_1(n) \;+\; p_2(n) \\ a(n) &= \sum_{i \cdot j = n} a_1(i) \cdot a_2(j). \end{aligned}$$

In case $\mathcal{A}_1$ and $\mathcal{A}_2$ are subsystems of a given multiplicative number system $\mathcal{A}$ with $\mathsf{P}_1 \cap \mathsf{P}_2 = \varnothing$ then $\mathcal{A}_1 \times \mathcal{A}_2$ is isomorphic to the subsystem of $\mathcal{A}$ generated by the indecomposables $\mathsf{P}_1 \cup \mathsf{P}_2$. The following theorem is from [**14**].

Theorem 10.5 (Odlyzko, 1997). *Let $\mathcal{A}_1, \ldots, \mathcal{A}_n$ be multiplicative number systems in RV_0. Then the product $\mathcal{A}_1 \times \cdots \times \mathcal{A}_n$ is also in RV_0.*

PROOF. It suffices to consider $n = 2$. So suppose $\mathcal{A}_1$ and $\mathcal{A}_2$ are in RV_0, and let $\mathcal{A} = \mathcal{A}_1 \times \mathcal{A}_2$. The claim is that $\mathcal{A}$ is in RV_0. From $\mathbf{A}(x) = \mathbf{A}_1(x) * \mathbf{A}_2(x)$ follows

$$A(x) \;=\; \sum_{k \le x} a_1(k) \cdot A_2(x/k).$$

Let $\varepsilon > 0$ be given. Since $A_2(x) \in \mathsf{RV}_0$ there is an M such that

$$A_2(2x) \;\le\; (1 + \varepsilon/2) \cdot A_2(x) \quad \text{for } x \ge M.$$

Then

$$\begin{aligned} A(2x) &= \sum_{k \le 2x} a_1(k) \cdot A_2(2x/k) \\ &= \Big(\sum_{k \le x/M} a_1(k) \cdot A_2(2x/k) \Big) \;+\; \Big(\sum_{x/M < k \le 2x} a_1(k) \cdot A_2(2x/k) \Big) \\ &\le (1 + \varepsilon/2) \cdot \Big(\sum_{k \le x} a_1(k) \cdot A_2(x/k) \Big) \\ &\qquad + \; A_2(2M) \cdot \big(A_1(2x) - A_1(x/M)\big) \\ &= (1 + \varepsilon/2) \cdot A(x) \;+\; \mathrm{o}\big(A_1(x)\big), \end{aligned}$$

since $A_1(x) \in \mathsf{RV}_0$. Thus

$$\lim_{x \to \infty} \frac{A(2x)}{A(x)} \;=\; 1,$$

so, by Lemma 7.25, $A(x) \in \mathsf{RV}_0$. ∎

10.3. Further results on RV_0

If P is finite then the asymptotics of §8.5 can be used to show $\mathcal{A} \in \mathsf{RV}_0$. If $\log\|\mathsf{p}_n\| \sim cn$, where $\mathsf{p}_1, \mathsf{p}_2, \ldots$ is an enumeration of P, then one can apply some deep results of Richmond [**47**], based on the work of Roth and Szekeres [**51**], to conclude that $\mathcal{A} \in \mathsf{RV}_0$. Along with the recent polynomial bound result for the additive case, Bell [**6**] also announced a polylog bound result for the multiplicative case, namely, if $P(n) = \mathrm{O}\big((\log n)^c\big)$ then $A(x) \in \mathsf{RV}_0$. This covers the two previous results. The polylog bound result is the best possible in the sense that if $f(n)$ is any *nondecreasing* function such that $P(n) = \mathrm{O}\big(f(n)\big)$ implies $\mathcal{A} \in \mathsf{RV}_0$ then $f(n)$ must be polylog bounded.

Recalling the examples of Boolean algebras, Abelian p-groups, $x^m = x$ rings and $\mathcal{H}_3$, the count function $P(n)$ is polylog bounded in each case, so they are in RV_0, and all partition sets in these systems have density as described in Theorems 10.1 and 10.2.

In particular, consider the following question, for $\mathcal{K}$ being one of the just mentioned classes, given $k \geq 0$ and a directly indecomposable $\mathbf{D} \in \mathcal{K}$: *What is the probability that a finite algebra chosen at random from $\mathcal{K}$ has at least k copies of* $\mathbf{D}$ *in its (unique) factorization into indecomposables?* To answer this, let p denote the indecomposable element that corresponds to $\mathbf{D}$ in the multiplicative number system derived from $\mathcal{K}$. Then the asymptotic density of the set

$$\mathsf{B} \;=\; \mathsf{p}^{\geq k} \cdot (\mathsf{P} \setminus \{\mathsf{p}\})^{\omega}$$

gives the probability. By Theorem 10.2 the answer is $\Delta(\mathsf{B}) = 1$ since all exponents are large. So a randomly chosen finite member of $\mathcal{K}$ will almost surely have at least k copies of $\mathbf{D}$ in its factorization.

For a slightly more involved example, consider the class $\mathcal{K}$ of *pointed* Abelian p-groups $(G, \cdot, 1, c)$. This simply means that there is a distinguished element c. Then there are (up to isomorphism) n indecomposable pointed Abelian p-groups of size p^n. In this case the global counting function satisfies $P(n) \leq \frac{(\log_p n)(1+\log_p n)}{2} + 1$, so a randomly selected finite pointed Abelian group almost certainly has at least k copies of any given indecomposable, for any choice of $k \geq 0$.

The properties that have been considered in this section (with respect to their probability of occurring) can be expressed in first-order logic. As will be seen in Chapter 12, when the multiplicative number system of a class of algebraic structures is in RV_0, any first-order sentence has a probability of either 0 or 1 of holding in a randomly chosen finite member of the class.

Problem 10.6. Extend the class of multiplicative number systems $\mathcal{A}$ for which one can show that $\mathcal{A} \in \mathsf{RV}_0$.

10.4. Second conjecture

From [**8**] one has, for an additive number system $\mathcal{A}$, $p(n) \in \mathsf{RT}_1$ implies $a(n) \in \mathsf{RT}_1$. The corresponding assertion for multiplicative systems is the following.

Conjecture 10.7. If $P(x) \in \mathsf{RV}_0$ then $A(x) \in \mathsf{RV}_0$.

CHAPTER 11

The Case $0 < \alpha < \infty$

The search continues for conditions guaranteeing that all partition sets of $\mathcal{A}$ have asymptotic density, now assuming that α is positive.[1] If $\mathcal{A}$ is discrete then one can apply the results of Part 1 (see §9.8), so this chapter is primarily aimed at strictly multiplicative systems.

For $\mathcal{A}$ strictly multiplicative, if all partition sets of $\mathcal{A}$ have global asymptotic density then, from Corollary 9.50, $\mathcal{A}$ must be in RV_α. However, $\mathcal{A} \in \mathsf{RV}_\alpha$ is not sufficient to show that all partition sets have global asymptotic density unless $\alpha = 0$. For $0 < \alpha < \infty$, the goal is to find solutions to

$$\mathcal{A} \in \mathsf{RV}_\alpha + \Big(\quad\Big) \qquad \text{implies} \qquad \Delta(\mathsf{B}) \text{ exists, for partition sets } \mathsf{B}.$$

11.1. Analog of Compton's Tauberian Theorem

The Tauberian theorem in this section parallels the one for additive systems proved by Compton—it will be used to prove Sárközy's Density Theorem.

Theorem 11.1. *Let* $\mathbf{S}(x), \mathbf{T}(x)$ *be two Dirichlet series such that, for some* $\alpha \in \mathbb{R}$,

(a) $T(x) \in \mathsf{RV}_\alpha$,
(b) $\mathbf{S}(x)$ *converges absolutely at* $x = \alpha$, *and,*
(c) *for some positive* K *and* C,

$$0 \le \frac{|T(x/n)|}{T(x)} \le Cn^{-\alpha}, \qquad \text{for } K \le n \le x.$$

Let $\mathbf{R}(x) = \mathbf{S}(x) * \mathbf{T}(x)$. *Then,*

$$\lim_{x\to\infty} \frac{R(x)}{T(x)} = \mathbf{S}(\alpha).$$

If $\mathbf{S}(\alpha) > 0$ *then the conclusion can be expressed by the asymptotic relation*

$$R(x) \sim \mathbf{S}(\alpha) \cdot T(x).$$

[1] This chapter is a reworking of the results of Burris and Sárközy [**17**] so that they give a close analog to the work of Compton from Chapter 5, plus a detailed treatment of a generalization of Oppenheim's asymptotics from [**40**], [**41**].

PROOF. The proof is, for the most part, identical to the first part of the proof of Theorem 9.53. Implicit in condition (a) is the fact that $T(x)$ is eventually positive.

First,

$$\left|\mathbf{S}(\alpha) - \frac{R(x)}{T(x)}\right| \le \underbrace{\left|\mathbf{S}(\alpha) - \sum_{n\le x} s(n)n^{-\alpha}\right|}_{\text{I}} + \underbrace{\left|\sum_{n\le x} s(n)n^{-\alpha} - \frac{R(x)}{T(x)}\right|}_{\text{II}}.$$

Clearly term I goes to 0 as $x \to \infty$. Thus it suffices to show that term II vanishes as $x \to \infty$. For any M, and any $x \ge M$,

$$\begin{aligned}
&\left|\sum_{n\le x} s(n)n^{-\alpha} - \frac{R(x)}{T(x)}\right| \\
&\quad = \left|\sum_{n\le x} s(n)n^{-\alpha} - \sum_{mn\le x} s(n)t(m)\Big/T(x)\right| \\
&\quad = \left|\sum_{n\le x} s(n)n^{-\alpha} - \sum_{n\le x} s(n)T(x/n)\Big/T(x)\right| \\
&\quad = \left|\sum_{n\le x} s(n)\left(n^{-\alpha} - \frac{T(x/n)}{T(x)}\right)\right| \\
&\quad \le \underbrace{\left|\sum_{n\le M} s(n)\left(n^{-\alpha} - \frac{T(x/n)}{T(x)}\right)\right|}_{\text{III}} + \underbrace{\left|\sum_{M<n\le x} s(n)\left(n^{-\alpha} - \frac{T(x/n)}{T(x)}\right)\right|}_{\text{IV}}.
\end{aligned}$$

For any fixed M, term III goes to 0 as $x \to \infty$ since condition (a) guarantees that, for any fixed n,

$$\lim_{x\to\infty} \frac{T(x/n)}{T(x)} = n^{-\alpha}.$$

Thus it suffices to show that term IV goes to 0 as $M \to \infty$. Now consider

$$\left|\sum_{M<n\le x} s(n)\left(n^{-\alpha} - \frac{T(x/n)}{T(x)}\right)\right| \le \underbrace{\left|\sum_{n>M} s(n)n^{-\alpha}\right|}_{\text{V}} + \underbrace{\sum_{M<n\le x} |s(n)|\left|\frac{T(x/n)}{T(x)}\right|}_{\text{VI}}.$$

Term V goes to 0 as $M \to \infty$ since it is the tail end of the convergent series $\sum s(n)n^{-\alpha}$. Thus it suffices to show that term VI goes to 0 as $M \to \infty$.

Assume $M \ge K$, so that hypothesis (c) can be used to show

$$\sum_{M<n\le x} |s(n)|\frac{|T(x/n)|}{T(x)} \le C \cdot \sum_{M<n\le x} |s(n)|n^{-\alpha} \le C \cdot \sum_{n>M} |s(n)|n^{-\alpha}.$$

As $\sum s(n)n^{-\alpha}$ is assumed to be absolutely convergent, the expression on the right side goes to 0 as $M \to \infty$, so term VI goes to 0 as $M \to \infty$. This is all that is needed to prove the theorem. □

The following lemmas give practical tools to verify that a given $\mathbf{T}(x)$ satisfies condition (c) of Theorem 11.1.

Lemma 11.2. *Suppose f and f_0 are functions on $(0, \infty)$ such that, for some $\alpha \in \mathbb{R}$,*

(a) $f(x) \sim x^\alpha f_0(x)$,
(b) *$f(x)$ is bounded on finite intervals $(0, r)$,*
(c) *$f_0(x) \in \mathsf{RV}_0$, and*
(d) *$f_0(x)$ is eventually nondecreasing.*

Then $f(x) \in \mathsf{RV}_\alpha$, and there are positive constants C and K such that

$$0 \ \le\ \frac{|f(x/n)|}{f(x)} \ \le\ Cn^{-\alpha}, \qquad \text{for } K \le n \le x.$$

Proof. $f(x) \in \mathsf{RV}_\alpha$ by Proposition 7.21. Choose $K_0 \ge 1$ such that $f_0(x)$ is nondecreasing and positive on $[K_0, \infty)$, and redefine $f_0(x) = f_0(K_0)$, for $0 < x \le K_0$. This does not affect the above hypotheses, and now $f_0(x)$ is positive and nondecreasing on $(0, \infty)$. Choose $K \ge K_0$ such that $x \ge K$ implies

$$\big|f(x) - x^\alpha f_0(x)\big| \ \le\ \tfrac{1}{2}x^\alpha f_0(x) \quad \text{and } f(x) > 0.$$

Then, for $x \ge K$,

$$\tfrac{1}{2}x^\alpha f_0(x) \ \le\ f(x) \ \le\ \tfrac{3}{2}x^\alpha f_0(x). \tag{11.2.1}$$

Choose D such that, for $1 \le x \le K$,

$$|f(x)| \ \le\ Dx^\alpha f_0(x). \tag{11.2.2}$$

Let $C = \max(2D, 3)$ and suppose $K \le n \le x$.
Case 1: $x/n \ge K$. Then

$$\begin{aligned} \frac{|f(x/n)|}{f(x)} &\le \Big(\tfrac{3}{2}(x/n)^\alpha f_0(x/n)\Big)\Big/\Big(\tfrac{1}{2}x^\alpha f_0(x)\Big) && (11.2.1)\\ &\le 3n^{-\alpha} && f_0 \text{ is nondecreasing}\\ &\le Cn^{-\alpha} && \text{definition of } C. \end{aligned}$$

Case 2: $x/n < K$. Then

$$\begin{aligned} \frac{|f(x/n)|}{f(x)} &\le \Big(D(x/n)^\alpha f_0(x/n)\Big)\Big/\Big(\tfrac{1}{2}x^\alpha f_0(x)\Big) && (11.2.1) \text{ and } (11.2.2)\\ &\le 2Dn^{-\alpha} && f_0 \text{ is nondecreasing}\\ &\le Cn^{-\alpha} && \text{definition of } C. \end{aligned}$$

This concludes the proof. □

Lemma 11.3. *Suppose $s(n) = cn^{\mu}$, with $c > 0$ and $\mu \geq 0$. Then $S(x) \in \mathsf{RV}_{\mu+1}$, and there are positive constants C and K such that*

$$\frac{S(x/n)}{S(x)} \leq Cn^{-(\mu+1)}, \qquad \text{for } K \leq n \leq x.$$

PROOF. $S(x) = \sum_{n \leq x} cn^{\mu} \sim c\frac{x^{\mu+1}}{\mu+1}$, so Lemma 11.2 applies. □

11.2. When is $\Delta(\mathsf{B}) = 0$?

In this section it will be proved that if $\mathcal{A} \in \mathsf{RV}_{\alpha}$ then any partition set with Dirichlet density 0 also has global asymptotic density 0.

Lemma 11.4. *Suppose $\mathsf{B} \subseteq \mathsf{B}_k \subseteq \mathsf{A}$. Then*

$$\lim_{k\to\infty} \partial(\mathsf{B}_k) = 0 \quad \textit{implies} \quad \partial(\mathsf{B}) = 0.$$

PROOF. This follows from $\mathbf{B}(x) \leq \mathbf{B}_k(x)$ and $\lim_{k\to\infty} \lim_{x\to\alpha} \dfrac{\mathbf{B}_k(x)}{\mathbf{A}(x)} = 0.$ □

Lemma 11.5. *Suppose $\mathcal{A} \in \mathsf{RV}_{\alpha}$ and $\mathsf{Q} \subseteq \mathsf{P}$. If $\mathbf{A}_{\mathsf{Q}}(\alpha) = \infty$ then every partition set of the form*

$$\mathsf{B} = \mathsf{Q}^{\gamma} \cdot (\mathsf{P} \setminus \mathsf{Q})^{\omega}$$

has global asymptotic density, and

$$\Delta(\mathsf{B}) = \partial(\mathsf{B}) = \begin{cases} 0 & \text{if } \gamma \text{ is small} \\ 1 & \text{if } \gamma \text{ is large.} \end{cases}$$

PROOF. Just repeat the proof of Lemma 9.67 for the case $\alpha > 0$, replacing P_n by Q_n, and noting that $\Delta(\mathsf{B}_n) = \partial(\mathsf{B}_n)$ leads to $\partial(\mathsf{B}_n) \to 0$ as well as $\Delta(\mathsf{B}_n) \to 0$. □

Remark 11.6. In this chapter the condition $\mathbf{A}_{\mathsf{Q}}(\alpha) = \infty$ is equivalent to $\mathbf{Q}(\alpha) = \infty$ since $\alpha > 0$ (see Corollary 8.18).

Lemma 11.7. *Let Q be a nonempty subset of P. Then*

$$\partial_{\mathsf{Q}}(\mathsf{Q}^{\gamma}) = 0 \quad \textit{iff} \quad \gamma \textit{ is small and } \mathbf{A}_{\mathsf{Q}}(\alpha) = \infty.$$

PROOF. If $\mathbf{A}_{\mathsf{Q}}(\alpha) < \infty$ then, by Lemma 9.36, $\partial_{\mathsf{Q}}(\mathsf{Q}^{\gamma}) > 0$. Thus $\partial_{\mathsf{Q}}(\mathsf{Q}^{\gamma}) = 0$ implies $\mathbf{A}_{\mathsf{Q}}(\alpha) = \infty$. So, proving the lemma reduces to assuming $\mathbf{A}_{\mathsf{Q}}(\alpha) = \infty$ and proving $\partial_{\mathsf{Q}}(\mathsf{Q}^{\gamma}) = 0$ iff γ is small, which is clear from Lemma 9.36. □

Lemma 11.8. *For a partition set $\mathsf{B} = \mathsf{P}_1^{\gamma_1} \cdots \mathsf{P}_k^{\gamma_k}$,*

$$\partial(\mathsf{B}) = 0 \quad \textit{iff} \quad \textit{for some } i,\ \gamma_i \textit{ is small and } \mathbf{A}_{\mathsf{P}_i}(\alpha) = \infty.$$

PROOF. From Theorem 9.37, $\partial(\mathsf{B}) = 0$ iff $\partial_{\mathsf{P}_i}(\mathsf{P}_i^{\gamma_i}) = 0$, for some i. Now apply Lemma 11.7. □

Proposition 11.9. *If $\mathcal{A} \in \mathsf{RT}_\alpha$ then, for any partition set* B,

$$\partial(\mathsf{B}) \;=\; 0 \quad \textit{implies} \quad \Delta(\mathsf{B}) \;=\; 0.$$

PROOF. Let $\mathsf{B} = \mathsf{P}_1^{\gamma_1} \cdots \mathsf{P}_k^{\gamma_k}$ be a partition set such that $\partial(\mathsf{B}) = 0$. By Lemma 11.8, for some i, γ_i is small and $\mathbf{P}_i^{\gamma_i}(\alpha) = \infty$. Then, by Lemma 11.5,

$$\Delta\big(\mathsf{P}_i^{\gamma_i} \cdot (\mathsf{P} \smallsetminus \mathsf{P}_i)^\omega\big) \;=\; 0,$$

and, as $\mathsf{B} \;\subseteq\; \mathsf{P}_i^{\gamma_i} \cdot (\mathsf{P} \smallsetminus \mathsf{P}_i)^\omega$, by Lemma 9.65 one has $\Delta(\mathsf{B}) \;=\; 0.$ □

11.3. The density of $\overline{\mathsf{B}}$

Conditions that guarantee that the global asymptotic density of any partition set B exists, and equals the Dirichlet density, are sought. For partition sets with zero Dirichlet density no extra conditions are needed since $\partial(\mathsf{B}) = 0$ implies $\Delta(\mathsf{B}) = 0$.

The more difficult case occurs when $\partial(\mathsf{B}) > 0$. The Tauberian theorem used to show that $\Delta(\mathsf{B})$ exists requires that $[\mathbf{B}/\mathbf{A}](x) \in \mathcal{DS}_\alpha$. One way to fulfill this requirement is to consider a possibly larger partition set $\overline{\mathsf{B}}$ that has the same Dirichlet density as B, and $[\overline{\mathbf{B}}/\mathbf{A}](x) \in \mathcal{DS}_\alpha$. The Tauberian theorem gives, under suitable hypotheses, $\Delta(\overline{\mathsf{B}}) = \partial(\overline{\mathsf{B}}) = [\overline{\mathbf{B}}/\mathbf{A}](\alpha)$. Then, using the zero density of $\overline{\mathsf{B}} \smallsetminus \mathsf{B}$, one has $\Delta(\mathsf{B}) = \partial(\mathsf{B}) = [\overline{\mathbf{B}}/\mathbf{A}](\alpha)$.

Lemma 11.10. *Let Q be a nonempty subset of P. If $\mathbf{A}_\mathsf{Q}(\alpha) < \infty$ then, for any exponent γ,*

$$[\mathbf{Q}^\gamma/\mathbf{A}_\mathsf{Q}](x) \in \mathcal{DS}_\alpha \quad \textit{and} \quad \partial_\mathsf{Q}(\mathsf{Q}^\gamma) \;=\; [\mathbf{Q}^\gamma/\mathbf{A}_\mathsf{Q}](\alpha).$$

PROOF. Lemma 9.36 shows that $\partial_\mathsf{Q}(\mathsf{Q}^\gamma)$ is $[\mathbf{Q}^\gamma](\alpha)/\mathbf{A}_\mathsf{Q}(\alpha)$. Then, applying Lemma 9.59 to $\mathcal{A}_\mathsf{Q}$ gives the desired conclusions. □

Lemma 11.11. *Suppose $\mathsf{B} \;=\; \mathsf{P}_1^{\gamma_1} \cdots \mathsf{P}_k^{\gamma_k}$ is a partition set such that $\partial(\mathsf{B}) > 0$. If $\gamma_i = \omega$ whenever $\mathsf{A}_{\mathsf{P}_i}(\alpha) = \infty$ then*

$$[\mathbf{B}/\mathbf{A}](x) \in \mathcal{DS}_\alpha \quad \textit{and} \quad \partial(\mathsf{B}) \;=\; [\mathbf{B}/\mathbf{A}](\alpha).$$

PROOF. From Theorem 9.37

$$\partial(\mathsf{B}) \;=\; \partial_{\mathsf{P}_1}(\mathsf{P}_1^{\gamma_1}) \cdots \partial_{\mathsf{P}_k}(\mathsf{P}_k^{\gamma_k}),$$

and, since $\partial(\mathsf{B}) > 0$, it follows that each $\partial_{\mathsf{P}_i}(\mathsf{P}_i^{\gamma_i}) > 0$. Thus, for $x > \alpha$,

$$\begin{aligned}
[\mathbf{B}/\mathbf{A}](x) &= \frac{\mathbf{B}(x)}{\mathbf{A}(x)} && \text{by Lemma 9.59}\\
&= \prod_{1 \le j \le k} \frac{\mathbf{P}_j^{\gamma_j}(x)}{\mathbf{A}_{\mathsf{P}_j}(x)} && \text{by Proposition 9.30}\\
&= [\mathbf{P}_1^{\gamma_1}/\mathbf{A}_{\mathsf{P}_1}](x) \cdots [\mathbf{P}_k^{\gamma_k}/\mathbf{A}_{\mathsf{P}_k}](x) && \text{by Lemma 9.59.}
\end{aligned}$$

Each $[\mathbf{P}_j^{\gamma_j}/\mathbf{A}_{\mathsf{P}_j}](x)$ is in $\mathcal{DS}_\alpha$. To see this first note that the large exponents γ_j with $\mathsf{A}_{\mathsf{P}_j}(\alpha) = \infty$ are ω, and clearly $[\mathbf{P}_i^\omega/\mathbf{A}_{\mathsf{P}_i}](x) = 1$ is in $\mathcal{DS}_\alpha$, and secondly, Lemma 11.10 applies in the other cases. Then Lemma 7.14 gives $[\mathbf{B}/\mathbf{A}](x) \in \mathcal{DS}_\alpha$, so $\partial(\mathsf{B}) \;=\; [\mathbf{B}/\mathbf{A}](\alpha)$. □

Definition 11.12. For Q a nonempty subset of P and γ an exponent, let

$$\overline{\mathsf{Q}^\gamma} = \begin{cases} \mathsf{Q}^\omega & \text{if } \gamma \text{ is large and } \mathbf{A}_{\mathsf{Q}}(\alpha) = \infty \\ \mathsf{Q}^\gamma & \text{otherwise.} \end{cases}$$

For any partition set $\mathsf{B} = \mathsf{P}_1^{\gamma_1} \cdots \mathsf{P}_k^{\gamma_k}$, let

$$\overline{\mathsf{B}} = \overline{\mathsf{P}_1^{\gamma_1}} \cdots \overline{\mathsf{P}_k^{\gamma_k}}.$$

$\overline{\mathsf{B}}$ is also a partition set and $\mathsf{B} \subseteq \overline{\mathsf{B}}$. Any large exponent γ_i of B is promoted to ω in $\overline{\mathsf{B}}$, provided $\mathsf{A}_{\mathsf{P}_i}(\alpha) = \infty$.

Lemma 11.13. *Suppose $\mathcal{A} \in \mathsf{RV}_\alpha$. For any partition set* B,

(a) $\Delta(\overline{\mathsf{B}} \setminus \mathsf{B}) = \partial(\overline{\mathsf{B}} \setminus \mathsf{B}) = 0$, *and*
(b) $\partial(\mathsf{B}) > 0$ *implies*

$$[\overline{\mathbf{B}}/\mathbf{A}](x) \in \mathcal{DS}_\alpha \quad \textit{and} \quad \partial(\overline{\mathsf{B}}) = [\overline{\mathbf{B}}/\mathbf{A}](\alpha).$$

Proof. Let $\mathsf{B} = \mathsf{P}_1^{\gamma_1} \cdots \mathsf{P}_k^{\gamma_k}$.

(a) For γ a large exponent that is not ω, say γ is $\geq m$ with $m > 0$, let $\widehat{\gamma}$ be $< m$, a small exponent. Then

$$\overline{\mathsf{B}} \setminus \mathsf{B} \subseteq \bigcup_{\substack{\gamma_i \text{ large}, \neq \omega \\ \mathbf{P}_i(\alpha) = \infty}} \left(\mathsf{P}_i^{\widehat{\gamma_i}} \cdot (\mathsf{P} \setminus \mathsf{P}_i)^\omega\right),$$

and, for γ_i large, $\neq \omega$ with $\mathbf{P}_i(\alpha) = \infty$, by Lemma 11.5 one has

$$\Delta\left(\mathsf{P}_i^{\widehat{\gamma_i}} \cdot (\mathsf{P} \setminus \mathsf{P}_i)^\omega\right) = \partial\left(\mathsf{P}_i^{\widehat{\gamma_i}} \cdot (\mathsf{P} \setminus \mathsf{P}_i)^\omega\right) = 0,$$

as $\widehat{\gamma_i}$ is small. Thus $\Delta(\overline{\mathsf{B}} \setminus \mathsf{B}) = \partial(\overline{\mathsf{B}} \setminus \mathsf{B}) = 0$.

(b) This follows from Lemma 11.11 as $\partial(\overline{\mathsf{B}}) \geq \partial(\mathsf{B}) > 0$. □

Lemma 11.14. *Suppose $\mathcal{A} \in \mathsf{RV}_\alpha$. For any partition set* B,

(a) $\partial(\overline{\mathsf{B}}) = \partial(\mathsf{B})$, *and*
(b) $\Delta(\overline{\mathsf{B}}) = \Delta(\mathsf{B})$ (*if either side exists*).

Proof. $\overline{\mathsf{B}}$ is the disjoint union of B and $\overline{\mathsf{B}} \setminus \mathsf{B}$, and, by Lemma 11.13 (a), $\Delta(\overline{\mathsf{B}} \setminus \mathsf{B}) = \partial(\overline{\mathsf{B}} \setminus \mathsf{B}) = 0$. □

11.4. Sárközy's Density Theorem

The following theorem and its corollaries were proved by András Sárközy in [**17**]. The theorem is a multiplicative global analog of Compton's Density Theorem 5.13 for additive number systems.

Theorem 11.15 (Sárközy, 1997)**.** *Suppose $\mathcal{A} \in \mathsf{RV}_\alpha$. If there exist positive constants C and K such that*

$$\frac{A(x/n)}{A(x)} \leq Cn^{-\alpha}, \qquad \textit{for } K \leq n \leq x,$$

then all partition sets B of $\mathcal{A}$ have global asymptotic density, and one has $\Delta(\mathsf{B}) = \partial(\mathsf{B})$.

PROOF. Let B be a partition set. By Theorem 9.37, $\partial(\mathsf{B})$ exists.

Case 1: $\partial(\mathsf{B}) = 0$.
Then, from Proposition 11.9, $\Delta(\mathsf{B}) = 0$, so in this case the theorem holds.

Case 2: $\partial(\mathsf{B}) > 0$.
By Lemma 11.13 (b), $[\overline{\mathbf{B}}/\mathbf{A}](\alpha)$ is absolutely convergent. Then $\mathbf{A}(x)$ and $[\overline{\mathbf{B}}/\mathbf{A}](x)$ satisfy the hypotheses, for $\mathbf{T}(x)$ and $\mathbf{S}(x)$ respectively, of Theorem 11.1, and $\overline{\mathbf{B}}(x)$ plays the role of $\mathbf{R}(x)$. Thus

$$\lim_{n\to\infty} \frac{\overline{B}(n)}{A(n)} \;=\; [\overline{\mathbf{B}}/\mathbf{A}](\alpha),$$

so $\Delta(\overline{\mathsf{B}}) = \partial(\overline{\mathsf{B}}) = [\overline{\mathbf{B}}/\mathbf{A}](\alpha)$. Now apply Lemma 11.14. □

Remark 11.16. The condition $\dfrac{A(x/n)}{A(x)} \le Cn^{-\alpha}$, for $K \le n \le x$, implies (by setting $x = n$) that $A(n) \ge \dfrac{1}{C}n^{\alpha}$, for $n \ge K$. Thus one has $A(n)/n^{\alpha+1} \ge (Cn)^{-1}$, so $\mathbf{A}^{+}(\alpha+1) = \infty$.

Now a corollary of Theorem 11.15 is derived that is easy to apply to numerous situations.

Corollary 11.17. *Suppose*
(a) $A(x) \sim x^{\alpha}A_0(x)$,
(b) $A_0(x) \in \mathsf{RV}_0$, *and*
(c) $A_0(x)$ *is eventually nondecreasing.*

Then all partition sets of $\mathcal{A}$ have global asymptotic density which equals the Dirichlet density.

PROOF. This follows from Lemma 11.2 and Theorem 11.15. □

Remark 11.18. Conditions (b) and (c) have interesting connections with the theory of regular variation at infinity. Every function in RV_{α} is asymptotically equal to a strictly increasing member of RV_{α}, *provided* $\alpha > 0$. This result breaks down, for $\alpha = 0$, as Karamata constructed a counterexample—see [**25**].

Corollary 11.19. *Suppose the global count function $A(x)$ satisfies*

$$A(x) \;\sim\; cx^{\alpha},$$

for some $c > 0$ and $\alpha > 0$. Then all partition sets of $\mathcal{A}$ have global asymptotic density which equals the Dirichlet density.

PROOF. This follows from Corollary 11.17. □

Thus all partition sets of the usual multiplicative number system of positive integers have global asymptotic density. Indeed, Corollary 11.19 shows that two of the popular axioms used in the study of prime number theorems in abstract number systems are conditions guaranteeing that partition sets B satisfy $\Delta(\mathsf{B}) = \partial(\mathsf{B})$.

- **Knopfmacher's Axiom A** from [**29**] is
$$A(x) \;=\; ax^{\mu} \;+ \mathrm{O}\big(x^{\nu}\big) \qquad \text{where } a > 0,\ 0 \le \nu < \mu.$$
- **Beurling's Condition** from [**10**] is
$$A(x) \;=\; ax \;+ \mathrm{O}\Big(\frac{x}{\log^{\nu} x}\Big) \qquad \text{where } a > 0,\ \nu > \tfrac{3}{2}.$$

From the work of Weber at the end of the 1800s, it is known that the multiplicative system of *integral ideals of any algebraic number field* satisfy both of these axioms.

11.4.1. Abelian groups. From the Erdös and Szekeres result given in Theorem 9.61 one has $A(x) \sim cx$, where $c = \prod_{n\ge 2}\zeta(n)$. Thus all partition sets of the multiplicative number system of finite Abelian groups have global asymptotic density that is equal to their Dirichlet density. The computations in §8.3.5, regarding the probability that there is an element of order 2 in a randomly chosen finite Abelian group, have been justified.

11.5. Generalizing Oppenheim's asymptotics

Theorem 11.20. *Let $\mathcal{A}$ be a multiplicative number system with*
$$p(n) \;=\; an^{\mu} + \mathrm{O}(n^{\eta}),$$
where $a > 0$, $\mu > -1$, and $\eta < \mu$. Then, for some positive C,
$$A(x) \;\sim\; Cx^{\mu+1}\frac{e^{2\sqrt{a\log x}}}{(\log x)^{3/4}}.$$
Thus all partition sets of $\mathcal{A}$ have a global asymptotic density which equals the Dirichlet density.

This result gives us a rich collection of examples of multiplicative number systems with partition sets having global asymptotic density. It is in many ways quite analogous to the result of Theorem 5.17 concerning additive systems, and indeed when applied to discrete multiplicative systems it can be modified in a straightforward manner to give a global (hence weaker) version of Theorem 5.17.

The method of proof used is based primarily on the two Oppenheim papers [**40**], [**41**] of 1926 and 1927 that are concerned with the number of ways an integer can be factored. If $\mu = 0$ then eventually $p(n) = a$ under the above hypotheses. So this theorem generalizes the case $p(n) = 1$ studied by Oppenheim. Since only the dominant term of the asymptotics is of interest, some shortcuts are employed, including the use of the notation $\stackrel{*}{=}$ and $\stackrel{*}{\sim}$ from Chapter 5. At the point where Oppenheim has reduced the problem to integrating around a small semicircle about the pole $z = 1$, there is a switch to the asymptotics developed in Lemma 5.18, for the additive case. (The direction followed by Oppenheim at this point was considerably more complicated, involving a deft application of Bessel functions.)

The proof of the asymptotic estimate in Theorem 11.20 is based on following Oppenheim's method of analyzing the integral of the function being pursued, and then pulling the estimates for the integral back to the original function. For this purpose, the following straightforward generalization of Lemma 2 on p. 280 of Apostol [**1**] is used.

Lemma 11.21. *Let $f(x)$ be a nondecreasing nonnegative function that is eventually positive on $(0,\infty)$, and let $F(x) = \int_1^x f(u)du$. If*

$$F(x) \sim x^c L(x),$$

where $c > 0$ and $L(x) \in \mathsf{RV}_0$, then

$$f(x) \sim c x^{c-1} L(x).$$

PROOF. From the monotonicity of $f(x)$, for $a > 1$,

$$F(ax) - F(x) = \int_x^{ax} f(u)du \geq (a-1)xf(x),$$

and thus

$$\begin{aligned}\frac{f(x)}{x^{c-1}L(x)} &\leq \frac{1}{a-1}\left[\frac{F(ax)}{x^c L(x)} - \frac{F(x)}{x^c L(x)}\right] \\ &= \frac{1}{a-1}\left[\frac{F(ax)}{(ax)^c L(ax)} \cdot \frac{L(ax)}{L(x)} \cdot a^c - \frac{F(x)}{x^c L(x)}\right],\end{aligned}$$

so,

$$\limsup_{x\to\infty} \frac{f(x)}{x^{c-1}L(x)} \leq \frac{1}{a-1}\big(a^c - 1\big).$$

Taking the limit as a goes to 1 from the right gives

$$\limsup_{x\to\infty} \frac{f(x)}{x^{c-1}L(x)} \leq c,$$

and then

$$\limsup_{x\to\infty} \frac{f(x)}{cx^{c-1}L(x)} \leq 1.$$

Now let $0 < b < 1$. Then

$$F(x) - F(bx) = \int_{bx}^{x} f(u)du \leq (1-b)xf(x),$$

and thus

$$\begin{aligned}\frac{f(x)}{x^{c-1}L(x)} &\geq \frac{1}{1-b}\left[\frac{F(x)}{x^c L(x)} - \frac{F(bx)}{x^c L(x)}\right] \\ &= \frac{1}{1-b}\left[\frac{F(x)}{x^c L(x)} - \frac{F(bx)}{(bx)^c L(bx)} \cdot \frac{L(bx)}{L(x)} \cdot b^c\right].\end{aligned}$$

From this

$$\liminf_{x\to\infty} \frac{f(x)}{x^{c-1}L(x)} \geq \frac{1}{1-b}(1-b^c),$$

and taking the limit as b goes to 1 from the left gives

$$\liminf_{x\to\infty} \frac{f(x)}{x^{c-1}L(x)} \geq c.$$

Thus

$$\liminf_{x\to\infty} \frac{f(x)}{cx^{c-1}L(x)} \geq 1.$$

Combining the above one has

$$\lim_{x\to\infty} \frac{f(x)}{cx^{c-1}L(x)} = 1.$$

□

Some information about being able to change the order of integration is also needed. Suppose a Dirichlet series $\mathbf{S}(x) = \sum_{n\geq 1} s(n)n^{-x}$ with nonnegative $s(n)$ and with abscissa of convergence $0 \leq \alpha < \infty$ is given.

Let $S^*(x)$ be defined by

$$S^*(x) = \begin{cases} S(x) & \text{if } x \text{ is not an integer} \\ S(x) - \frac{1}{2}s(x) & \text{if } x \text{ is an integer.} \end{cases}$$

Perron's integral formula for $S^*(x)$ is

$$S^*(x) = \frac{1}{2\pi i}\int_{c-i\infty}^{c+i\infty} e^{\mathbf{S}(z)}\frac{x^z}{z}dz, \tag{11.21.1}$$

and it is valid for any $c > \alpha$.

Lemma 11.22. *For* $\mathbf{S}(x)$ *and* $S^*(x)$ *as above, let*

$$\mathbb{S}(x) = \int_1^x S^*(u)du.$$

Then one can interchange integral signs, giving

$$\mathbb{S}(x) = \frac{1}{2\pi i}\int_{c-i\infty}^{c+i\infty} e^{\mathbf{S}(z)}\frac{x^{z+1}}{z(z+1)}dz.$$

PROOF. Choose any real $b > 1$ and consider the strip R in the x, t plane given by $1 \leq x \leq b$ and $-\infty < t < \infty$. Then, with $z = c + it$,

(a) the integrand $e^{\mathbf{S}(z)}\frac{x^z}{z}$ of (11.21.1) is continuous in R, and

(b) the integral $\frac{1}{2\pi i}\int_{c-i\infty}^{c+i\infty} e^{\mathbf{S}(z)}\frac{x^z}{z}dz$ converges uniformly to $S^*(x)$, for $1 \le x \le b$.

Item (b) follows from using the estimates given in Apostol [**1**], pp. 243–246, to show that

$$S^*(x) - \frac{1}{2\pi i}\int_{c-iT}^{c+iT} e^{\mathbf{S}(z)}\frac{x^z}{z}dz \;=\; \mathrm{O}(1/T)$$

uniformly, for $1 \le x \le b$.

Using (a) and (b) above it follows that

$$\int_0^x \left(\int_{c-i\infty}^{c+i\infty} e^{\mathbf{S}(z)}\frac{u^z}{z}dz\right)du \;=\; \int_{c-i\infty}^{c+i\infty} e^{\mathbf{S}(z)}\frac{x^{z+1}}{z(z+1)}dz.$$

□

Now the preparations have been made to prove Theorem 11.20.

Proof of the Theorem. From the hypotheses on $p(n)$,

$$p(n) \;=\; an^\mu + f(n)$$

where, for some constant c,

$$|f(n)| \;\le\; cn^\eta.$$

With $\varepsilon(x)$ defined by

$$\varepsilon(x) \;=\; \sum_{n\ge 2} f(n)n^{-x},$$

$\varepsilon(x)$ is a Dirichlet series with abscissa of convergence at most $\eta+1$, which is less than $\mu+1$. Now

$$\begin{aligned}\mathbf{P}(z) \;&=\; \sum_{n\ge 2} an^\mu n^{-z} \;+\; \varepsilon(z)\\ &=\; -a \;+\; a\zeta(z-\mu) \;+\; \varepsilon(z),\end{aligned}$$

so

$$e^{\mathbf{P}(z)} \;\stackrel{*}{=}\; e^{a\zeta(z-\mu)}\cdot e^{\varepsilon(z)}. \tag{11.22.1}$$

The strategy to prove the theorem is to do a detailed analysis of the asymptotics of

$$\sum_{n\le x}\left[n^{-z}\right]e^{a\zeta(z-\mu)},$$

and then to use this to find the asymptotics of

$$\sum_{n\le x}\left[n^{-z}\right]e^{\mathbf{P}(z)}.$$

This in turn gives the desired asymptotics of $A(x)$.

To determine the asymptotics of

$$J(x) = \sum_{n \le x} [n^{-z}] e^{a\zeta(z-\mu)},$$

the Perron integral formula

$$J^*(x) = \frac{1}{2\pi i} \int_{c-i\infty}^{c+i\infty} e^{a\zeta(z-\mu)} \frac{x^z}{z} dz, \qquad \text{for } c > \mu + 1$$

is used. Letting

$$\mathbb{J}(x) = \int_1^x J^*(u) du,$$

from Lemma 11.22 it follows that

$$\begin{aligned} \mathbb{J}(x) &= \frac{1}{2\pi i} \int_{c-i\infty}^{c+i\infty} e^{a\zeta(z-\mu)} \frac{x^{z+1}}{z(z+1)} dz, && \text{for } c > \mu + 1 \\ &\stackrel{*}{=} \int_{c-i\infty}^{c+i\infty} \frac{e^{a\zeta(z)} x^{z+\mu+1}}{(z+\mu)(z+\mu+1)} dz, && \text{for } c > 1. \end{aligned}$$

The goal now is to find asymptotics for $\mathbb{J}(x)$.

For any $\varepsilon > 0$, Hardy and Littlewood, as well as Weyl, proved that, for t sufficiently large, and, for $\Im(z) = t$, $\Re(z) \ge 1$, (see Oppenheim [**41**], p. 125)

$$e^{\zeta(z)} = \mathrm{O}(|t|^{\varepsilon}). \tag{11.22.2}$$

From this one sees that, for any contour of the form in Figure 1, the integral

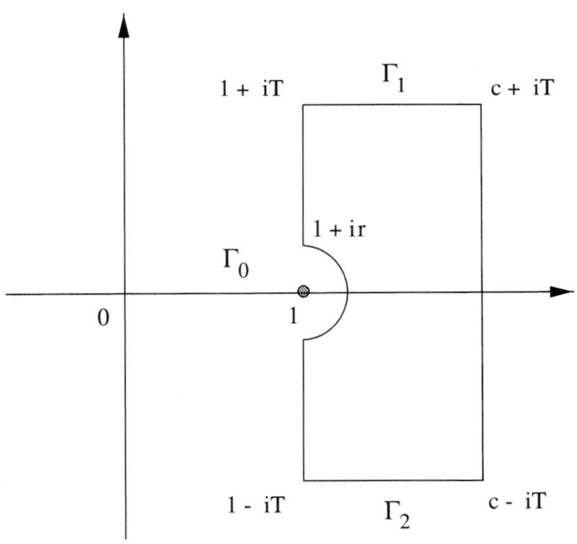

FIGURE 1

of $\dfrac{e^{a\zeta(z)} x^{z+\mu+1}}{(z+\mu)(z+\mu+1)}$, calculated along either the top Γ_1 or the bottom Γ_2, goes to zero as $T \to \infty$ since, for $t > T_0$, and for a suitable T_0, by (11.22.2) one has

$$\left|\frac{e^{a\zeta(z)}x^{z+\mu+1}}{(z+\mu)(z+\mu+1)}\right| \le x^{c+\mu+1}\frac{|t|^{1/2}}{|t|^2}. \tag{11.22.3}$$

Thus the contour of the integral for $\mathbb{J}(x)$ can be changed to the vertical line $z = 1 + it$ with a small semicircle of radius r around the pole $z = 1$. This contour is subdivided into five portions as indicated in Figure 2.

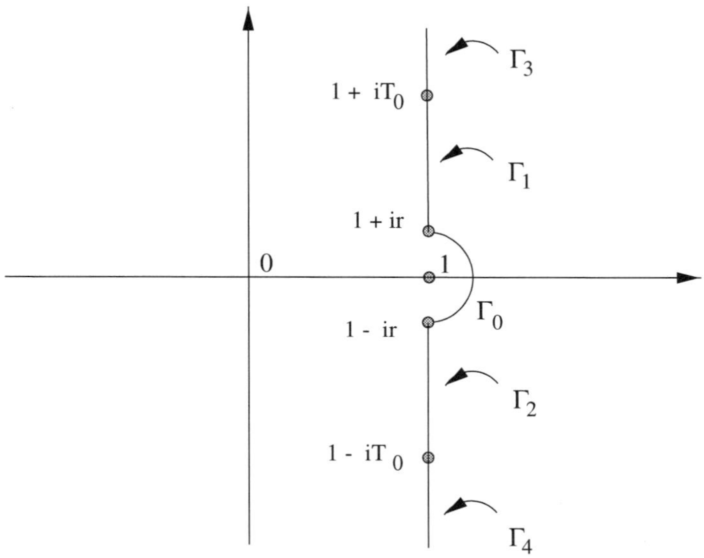

FIGURE 2

For $0 \le k \le 4$, with the Γ_k as in Figure 2, let

$$\mathbb{J}_k(x) = \int_{\Gamma_k} \frac{e^{a\zeta(z)}x^{z+\mu+1}}{(z+\mu)(z+\mu+1)}dz.$$

Then $\mathbb{J}_3(x)$ and $\mathbb{J}_4(x)$ are $\mathrm{O}(x^{\mu+2})$ by the estimate in (11.22.3). Since, in a strip about the line $z = 1$,

$$\zeta(z) = \frac{1}{z-1} + \gamma + \varepsilon(z),$$

where $\varepsilon(z)$ is holomorphic in this strip, it follows that the integrand is $O\big(x^{\mu+2}\big)$ on Γ_1 and Γ_2. So $\mathbb{J}_1(x)$ and $\mathbb{J}_2(x)$ are $O\big(x^{\mu+2}\big)$. Thus

$$\mathbb{J}(x) \overset{*}{=} \int_{\Gamma_0} \frac{e^{a\zeta(z)}x^{z+\mu+1}}{(z+\mu)(z+\mu+1)}dz + \mathrm{O}(x^{\mu+2}).$$

The following will be used, for z on Γ_0:

$$\begin{aligned} e^{a\zeta(z)} &\overset{*}{=} e^{a/(z-1)} \cdot \big(1 + \mathrm{O}(r)\big) \\ x^{z+\mu+1} &= x^{\mu+2}e^{(z-1)\log x} \end{aligned}$$

$$\frac{1}{(z+\mu)(z+\mu+1)} \quad \stackrel{*}{=} \quad 1 + \mathrm{O}(r).$$

Thus

$$\begin{aligned}
\mathbb{J}(x) &\stackrel{*}{=} x^{\mu+2} \int_{\Gamma_0} e^{a/(z-1)+(z-1)\log x} \big(1 + \mathrm{O}(r)\big) dz \; + \mathrm{O}\big(x^{\mu+2}\big) \\
&= x^{\mu+2} \Bigg(\underbrace{\int_{\Gamma_0} e^{a/(z-1)+(z-1)\log x} dz}_{\mathbb{K}_1(x)} \\
&\qquad + \underbrace{\int_{\Gamma_0} e^{a/(z-1)+(z-1)\log x} \cdot \mathrm{O}(r) dz}_{\mathbb{K}_2(x)} \Bigg) + \mathrm{O}\big(x^{\mu+2}\big).
\end{aligned}$$

Let Γ_0 be parametrized by $z = 1 + re^{i\theta}$, for $-\pi/2 \leq \theta \leq \pi/2$. Then

$$\mathbb{K}_1(x) \;=\; ir \int_{-\pi/2}^{\pi/2} e^{\frac{a}{r} e^{-i\theta} + (r \log x) e^{i\theta}} \cdot e^{i\theta} d\theta.$$

Now choose $\dfrac{a}{r} \;=\; r \log x$ and thus

$$r \;=\; \sqrt{\frac{a}{\log x}} \qquad \text{and} \qquad \frac{a}{r} \;=\; \sqrt{a \log x}.$$

Using (5.18.1) from Lemma 5.18

$$\begin{aligned}
\mathbb{K}_1(x) &\stackrel{*}{=} r \int_{-\pi/2}^{\pi/2} e^{\sqrt{a \log x}\left(e^{-i\theta} + e^{i\theta}\right)} \cdot e^{i\theta} d\theta \\
&= r \int_{-\pi/2}^{\pi/2} e^{2\sqrt{a \log x}(\cos\theta)} \cdot \cos\theta d\theta \qquad (11.22.4) \\
&\stackrel{*}{=} \sqrt{\frac{a}{\log x}} \cdot \frac{e^{2\sqrt{a \log x}}}{\sqrt{2\sqrt{a \log x}}} \cdot \left(1 + \mathrm{O}\Big(\frac{1}{\sqrt{2\sqrt{a \log x}}}\Big)\right) \\
&\stackrel{*}{=} \frac{e^{2\sqrt{a \log x}}}{(\log x)^{3/4}} \cdot \left(1 + \mathrm{O}\Big(\frac{1}{(\log x)^{1/4}}\Big)\right). \qquad (11.22.5)
\end{aligned}$$

Since the integrand of $\mathbb{K}_1(x)$ as given in (11.22.4) is real and nonnegative it follows that

$$\mathbb{K}_2(x) \;=\; \mathrm{O}(r)\mathbb{K}_1(x) \;=\; \mathrm{O}\Big(\frac{1}{\sqrt{\log x}}\Big)\mathbb{K}_1(x).$$

From this follows

$$\begin{aligned}
\mathbb{J}(x) &\stackrel{*}{=} x^{\mu+2} \cdot \mathbb{K}_1(x) \cdot \left(1 + \mathrm{O}\Big(\frac{1}{\sqrt{\log x}}\Big)\right) \;+\; \mathrm{O}\big(x^{\mu+2}\big) \\
&\stackrel{*}{\sim} x^{\mu+2} \cdot \frac{e^{2\sqrt{a \log x}}}{(\log x)^{3/4}}.
\end{aligned}$$

Then, by Lemma 11.21,

$$J^*(x) = \frac{1}{2\pi i}\int_{c-i\infty}^{c+i\infty} \frac{e^{a\zeta(z)}x^{z+\mu}}{z+\mu}dz \stackrel{*}{\sim} x^{\mu+1}\frac{e^{2\sqrt{a\log x}}}{(\log x)^{3/4}}.$$

Now

$$J^*(x) \le J(x) \le J^*(x+1),$$

and the above asymptotics, for $J^*(x)$, guarantee that

$$\lim_{x\to\infty} J^*(x)/J^*(x+1) = 1.$$

This suffices to show that $J(x) \sim J^*(x)$, and thus one has $J(x) \in \mathsf{RV}_{\mu+1}$ as $J^*(x) \in \mathsf{RV}_{\mu+1}$.

Since

$$\sum_{n\le x}\left[n^{-z}\right]e^{a\zeta(z-\mu)} = J(x) \in \mathsf{RV}_{\mu+1}$$

$$\alpha_\varepsilon \le \eta+1 < \mu+1,$$

the analog of Schur's Tauberian Theorem (9.53) can be applied to (11.22.1) to show

$$\sum_{n\le x}\left[n^{-z}\right]e^{\mathbf{P}(z)} \stackrel{*}{\sim} \sum_{n\le x}\left[n^{-z}\right]e^{a\zeta(z-\mu)},$$

and hence that $\sum_{n\le x}\left[n^{-z}\right]e^{\mathbf{P}(z)} \in \mathsf{RV}_{\mu+1}$. Applying Proposition 9.57,

$$A(x) \stackrel{*}{\sim} \sum_{n\le x}\left[n^{-z}\right]e^{\mathbf{P}(z)}.$$

Thus

$$A(x) \stackrel{*}{\sim} \sum_{n\le x}\left[n^{-z}\right]e^{\mathbf{P}(z)} \stackrel{*}{\sim} J(x) \sim J^*(x) \stackrel{*}{\sim} x^{\mu+1}\frac{e^{2\sqrt{a\log x}}}{(\log x)^{3/4}}.$$

From this follows $A(x) \stackrel{*}{\sim} x^{\mu+1}A_0(x)$, where $A_0(x) = \dfrac{e^{2\sqrt{a\log x}}}{(\log x)^{3/4}}$. Now apply Corollary 11.17 of Sárközy's Density Theorem to finish the proof of the theorem. □

Having the abscissa of convergence α of $\mathbf{A}(z)$ strictly between 0 and ∞ offers simpler possibilities to analyze $A(n)$ than does the case that $\alpha = 0$, the reason being that if $\alpha = 0$ then all the terms $\mathbf{P}(mz)/m$ also have abscissa of convergence 0. Then, when $\mathbf{P}(z)$ contributes a pole at 0, these other terms will contribute to this pole. The analysis of this section succeeded because $e^{\mathbf{P}(z)}$ contributed the dominant term to the asymptotics, and the asymptotics of the global count function $\sum_{n\le x}\left[n^{-z}\right]e^{\mathbf{P}(z)}$ were found because $\mathbf{P}(z)$ has a simple pole at α.

Problem 11.23. If $0 < \alpha < \infty$ and $\mathbf{P}(z)$ has only one pole, does it follow that $A(x)$ is covered by Corollary 11.17?

Problem 11.24. Extend the class of multiplicative number systems $\mathcal{A}$ with $\alpha > 0$ for which one can show that all partition sets of $\mathcal{A}$ have global asymptotic density.

Problem 11.25. Find necessary and sufficient conditions on two multiplicative number systems $\mathcal{A}_1$ and $\mathcal{A}_2$ for all partition sets of $\mathcal{A}_1 \times \mathcal{A}_2$ to have global asymptotic density.

11.6. Third conjecture

The conditions $p(n) \in \mathsf{RT}_\rho$ and $\liminf_n np(n)\rho^n > 1$ suffice, for additive number systems, to show that all partition sets have local asymptotic density (see [**8**]).

The corresponding assertion for multiplicative number systems is the following.

Conjecture 11.26. If

(a) $P(x) \in \mathsf{RV}_\alpha$, where $\alpha > 0$, and

(b) $\liminf_{x\to\infty} \dfrac{P(x)\log x}{x^\alpha} > 1$

then all partition sets have global asymptotic density.

A slightly stronger conjecture would conclude that $A(x) \sim A_0(x)x^\alpha$ with $A_0(x) \in \mathsf{RV}_0$ and eventually nondecreasing. Also, one can make the conjecture with x replaced by n in (b).

CHAPTER 12

First-Order Limit Laws

This chapter is concerned with applying the previous results to obtain logical limit laws for classes $\mathcal{K}$ of L-structures, where L is a first-order language. Idziak and Tyszkiewicz [**28**] show that one can rarely hope to have limit laws for monadic second-order logic in the multiplicative setting, so only first-order logic is considered in this chapter.

12.1. Asymptotic density of subsets of $\mathcal{K}$

Definition 12.1. Let L be a first-order language. For $\mathcal{H}$ a class of finite L-structures, and $n \geq 1$, let

$$\mathcal{H}_{(n)} \;=\; \{\mathbf{S} \in \mathcal{H} : |S| = n\},$$

the subclass of structures of size n in $\mathcal{H}$; and let

$$\mathcal{H}_{(\leq n)} \;=\; \{\mathbf{S} \in \mathcal{H} : |S| \leq n\},$$

the subclass of structures of size at most n in $\mathcal{H}$.

Definition 12.2. For $\mathcal{H}$ a class of structures, let $\widetilde{\mathcal{H}}$ be the set of isomorphism types $\widetilde{\mathbf{S}}$ of members $\mathbf{S}$ of $\mathcal{K}$.

Definition 12.3. For $\mathcal{H} \subseteq \mathcal{K}$, the *local asymptotic density* and the *global asymptotic density* of $\mathcal{H}$ in $\mathcal{K}$ (if they exist) are given by

$$\delta(\mathcal{H}) \;=\; \lim_{\substack{n\to\infty \\ \mathcal{K}_{(n)}\neq\varnothing}} \frac{|\widetilde{\mathcal{H}_{(n)}}|}{|\widetilde{\mathcal{K}_{(n)}}|}$$

$$\Delta(\mathcal{H}) \;=\; \lim_{n\to\infty} \frac{|\widetilde{\mathcal{H}_{(\leq n)}}|}{|\widetilde{\mathcal{K}_{(\leq n)}}|}.$$

The goal is to find interesting classes of structures $\mathcal{K}$ such that, for any first-order sentence φ, the subclass of $\mathcal{K}$ defined by φ, namely

$$\mathcal{K}_{\varphi} \;=\; \{\mathbf{S} \in \mathcal{K} : \mathbf{S} \models \varphi\},$$

has asymptotic density. In this situation $\mathcal{K}$ has a *first-order limit law.* If the densities of the $\mathcal{K}_{\varphi}$ exist, and are always 0 or 1, then $\mathcal{K}$ has a *first-order 0–1 law.*

To apply the results of the previous section one needs a class $\mathcal{K}$ of finite structures that naturally leads to a multiplicative number system with *direct product* playing the role of multiplication.

Definition 12.4. The *direct product* $\mathbf{S}$ of L-structures $\mathbf{S}_1, \ldots, \mathbf{S}_k$ is defined by letting $S = S_1 \times \cdots \times S_k$, the Cartesian product of the sets $S_1, \ldots, S_k$, and then interpreting the function, constant and relation symbols of L coordinatewise, that is:

- $g(s_1, \ldots, s_n)(i) = g(s_1(i), \ldots, s_n(i))$, for g an n-ary function symbol of L, and $s_1, \ldots, s_n \in S$,
- $c(i) = c_i$, where c_i is the interpretation of a constant symbol c in $\mathbf{S}_i$,
- $r(s_1, \ldots, s_n)$ holds in $\mathbf{S}$ iff for $1 \le i \le k$, $r(s_1(i), \ldots, s_n(i))$ holds in $\mathbf{S}_i$, where r is an n-ary relation symbol in L.

The importance of classes of finite structures closed under direct product was noted by Compton.

Definition 12.5. A finite structure $\mathbf{S}$ with $|S| > 1$ in a class $\mathcal{K}$ of structures is said to be *$\mathcal{K}$-indecomposable* if it is not isomorphic to a direct product of two or more smaller members of $\mathcal{K}$.

Let L be a first-order language. A class $\mathcal{K}$ of finite L-structures that is closed under isomorphism is said to be *adequate* if:

(a) $\mathcal{K}$ is closed under direct product,
(b) members of $\mathcal{K}$ with at least two elements can be uniquely decomposed into (a direct product of) $\mathcal{K}$-indecomposable members of $\mathcal{K}$, and
(c) up to isomorphism there is a unique one-element structure $\mathbf{1}$ in $\mathcal{K}$, and $\mathbf{1} \times \mathbf{T} \cong \mathbf{T}$, for $\mathbf{T} \in \mathcal{K}$.

Clearly, adequate classes $\mathcal{K}$ form a multiplicative number system $\mathcal{A}_{\mathcal{K}}$ on $\widetilde{\mathcal{K}}$ by defining:

$$\begin{aligned} \|\widetilde{\mathbf{S}}\| &= |S| \\ \widetilde{\mathbf{S}} \times \widetilde{\mathbf{T}} &= \widetilde{\mathbf{S} \times \mathbf{T}}. \end{aligned}$$

When working with direct products, the key tools to establish the connection between the models of a first-order sentence and partition sets are (1) the Feferman-Vaught Theorem, and (2) Skolem's analysis of the expressive power of first-order sentences in Boolean algebra. These topics are developed in the next two sections.

12.2. The Feferman-Vaught Theorem

Definition 12.6. Given a set I, $Su(I)$ is the *set of all subsets of* I, and $\mathbf{Su}(I) = (Su(I), \cup, \cap, ', 0, 1)$ is the *Boolean algebra of all subsets of* I.

Informally, the Feferman-Vaught theorem says that if a first-order sentence φ is given then one can find finitely many sentences φ_j such that the truth of φ in a direct product $\prod_i \mathbf{A}_i$ can be determined by knowing in which of the $\mathbf{A}_i$ each φ_j holds. Here is the basic result.[1]

[1] This famous result is written-up in several places, in varying degrees of generalization, for example, in Chang and Keisler [**19**], p. 342, with historical comments on p. 526. They prove the theorem in the more general setting of *reduced* products.

Theorem 12.7 (Feferman-Vaught). *Given a first-order language* L *and a first-order* L*-sentence* φ *one can (effectively) find a formula* $\Phi(X_1,\ldots,X_m)$ *in the language of Boolean algebra, and* L*-sentences* φ_j, $1 \le j \le m$, *such that, for any family* $\mathbf{S}_i$, $i \in I$, *of* L*-structures, one has*

$$\prod_{i\in I}\mathbf{S}_i \models \varphi \quad \textit{iff} \quad \mathbf{Su}(I) \models \Phi([\![\varphi_1]\!],\ldots,[\![\varphi_m]\!]),$$

where

$$[\![\varphi_j]\!] = \{i \in I : \mathbf{S}_i \models \varphi_j\}.$$

The sequence of formulas

$$\langle \Phi; \varphi_1, \ldots, \varphi_m \rangle$$

is called a Feferman-Vaught sequence *of* φ.

PROOF. The proof gives a slightly stronger result, namely for $\varphi(\vec{x})$ a first-order formula there is a Feferman-Vaught sequence $\langle \Phi; \varphi_1, \ldots, \varphi_m \rangle$ such that, for any sequence $\vec{f}$ of elements of a direct product $\prod_{i\in I}\mathbf{S}_i$,

$$\prod_{i\in I}\mathbf{S}_i \models \varphi(\vec{f}) \quad \text{iff} \quad \mathbf{Su}(I) \models \Phi([\![\varphi_1(\vec{f})]\!],\ldots,[\![\varphi_m(\vec{f})]\!]).$$

The following steps recursively define Feferman-Vaught sequences $\mathsf{FV}[\varphi]$ for first-order formulas φ.

(a) If $\varphi(\vec{x})$ is atomic then let

$$\mathsf{FV}[\varphi] = \langle X \approx 1; \varphi \rangle.$$

(b) If $\varphi(\vec{x}) = \neg\,\psi(\vec{x})$ and

$$\mathsf{FV}[\psi] = \langle \Psi; \psi_1, \ldots, \psi_m \rangle$$

then let

$$\mathsf{FV}[\varphi] = \langle \neg\,\Psi; \psi_1, \ldots, \psi_m \rangle.$$

(c) If $\varphi(\vec{x}) = \psi(\vec{x}) \,\Box\, \xi(\vec{x})$, where $\Box \in \{\wedge, \vee, \rightarrow, \leftrightarrow\}$, and

$$\mathsf{FV}[\psi] = \langle \Psi(\vec{X}); \psi_1, \ldots, \psi_m \rangle$$
$$\mathsf{FV}[\xi] = \langle \Xi(\vec{Y}); \xi_1, \ldots, \xi_n \rangle$$

then let

$$\mathsf{FV}[\psi \,\Box\, \xi] = \langle \Psi(\vec{X}) \,\Box\, \Xi(\vec{Y}^\star); \psi_1, \ldots, \psi_m, \xi_1, \ldots, \xi_n \rangle,$$

where $\vec{Y}^\star = \langle X_{m+1}, \ldots, X_{m+n} \rangle$.

(d) If $\varphi(\vec{x}) = \exists x\, \psi(x, \vec{x})$ and

$$\mathsf{FV}[\psi] = \langle \Psi(\vec{X}); \psi_1, \ldots, \psi_m \rangle$$

then let $M_1, \ldots, M_{2^m}$ be a listing of the subsets of $\{1, \ldots, m\}$, and define

$$\mathsf{FV}[\varphi] = \langle \Phi(X_1, \ldots, X_{2^m}); \varphi_1, \ldots, \varphi_{2^m} \rangle,$$

where

$$\begin{aligned}
\Phi &= \exists Y_1 \cdots \exists Y_{2^m}\Big[\mathsf{Part}(\vec{Y}) \wedge \bigwedge_{1\le j\le 2^m} (Y_j \subseteq X_j) \wedge \\
&\qquad\qquad \Psi\Big(\bigcup_{1\in M_j} Y_j, \dots, \bigcup_{m\in M_j} Y_j\Big)\Big] \\
\mathsf{Part}(\vec{Y}) &= \Big(\bigwedge_{1\le j<k\le 2^m} Y_j \cap Y_k \approx 0\Big) \wedge \Big(\bigcup_{1\le j\le 2^m} Y_j \approx 1\Big) \\
\varphi_j(\vec{x}\,) &= \exists x\Big(\bigwedge_{k\in M_j} \psi_k(x,\vec{x}\,) \wedge \bigwedge_{k\notin M_j} \neg\,\psi_k(x,\vec{x}\,)\Big).
\end{aligned}$$

(e) If $\varphi(\vec{x}\,) = \forall x\, \psi(x,\vec{x}\,)$ then let

$$\mathsf{FV}[\varphi] = \mathsf{FV}[\neg\,\exists x\,\neg\,\psi].$$

Now, to verify that this actually defines Feferman-Vaught sequences, that is, sequences of formulas with the desired property, one uses proof by induction on formulas.

(a) Suppose $\varphi(\vec{x}\,)$ is atomic. Then the assertion

$$\prod_{i\in I} \mathbf{S}_i \models \varphi(\vec{f}\,) \quad \text{iff} \quad \mathbf{Su}(I) \models [\![\varphi(\vec{f}\,)]\!] = 1$$

is just the definition of atomic formulas in direct products—they hold iff they hold in each coordinate.

(b) Suppose $\varphi(\vec{x}\,) = \neg\,\psi(\vec{x}\,)$ and

$$\mathsf{FV}[\psi] = \big\langle \Psi; \psi_1, \dots, \psi_m \big\rangle$$

is such that

$$\prod_{i\in I} \mathbf{S}_i \models \psi(\vec{f}\,) \quad \text{iff} \quad \mathbf{Su}(I) \models \Psi\big([\![\psi_1(\vec{f}\,)]\!], \dots, [\![\psi_m(\vec{f}\,)]\!]\big).$$

Then

$$\begin{aligned}
\prod_{i\in I} \mathbf{S}_i \models \varphi(\vec{f}\,) \quad &\text{iff} \quad \prod_{i\in I} \mathbf{S}_i \not\models \psi(\vec{f}\,) \\
&\text{iff} \quad \mathbf{Su}(I) \not\models \Psi\big([\![\psi_1(\vec{f}\,)]\!], \dots, [\![\psi_m(\vec{f}\,)]\!]\big) \\
&\text{iff} \quad \mathbf{Su}(I) \models \neg\,\Psi\big([\![\psi_1(\vec{f}\,)]\!], \dots, [\![\psi_m(\vec{f}\,)]\!]\big),
\end{aligned}$$

and thus $\mathsf{FV}[\varphi]$ is a Feferman-Vaught sequence for φ.

(c) Suppose $\varphi(\vec{x}\,) = \psi(\vec{x}\,) \,\Box\, \xi(\vec{x}\,)$ and

$$\begin{aligned}
\mathsf{FV}[\psi] &= \big\langle \Psi(\vec{X}); \psi_1, \dots, \psi_m \big\rangle \\
\mathsf{FV}[\xi] &= \big\langle \Xi(\vec{Y}); \xi_1, \dots, \xi_n \big\rangle
\end{aligned}$$

are such that

$$\prod_{i\in I}\mathbf{S}_i \models \psi(\vec{f}) \quad \text{iff} \quad \mathbf{Su}(I) \models \Psi\big([\![\psi_1(\vec{f})]\!],\dots,[\![\psi_m(\vec{f})]\!]\big)$$
$$\prod_{i\in I}\mathbf{S}_i \models \xi(\vec{f}) \quad \text{iff} \quad \mathbf{Su}(I) \models \Xi\big([\![\xi_1(\vec{f})]\!],\dots,[\![\xi_n(\vec{f})]\!]\big).$$

Then, with '$\Box$' being the translation from symbol to English, for example, '$\wedge$' is the word and,

$$\begin{aligned}
\prod_{i\in I}\mathbf{S}_i \models \varphi(\vec{f}) \quad &\text{iff} \quad \prod_{i\in I}\mathbf{S}_i \models \psi(\vec{f})\,\Box\,\xi(\vec{f})\\
&\text{iff} \quad \prod_{i\in I}\mathbf{S}_i \models \psi(\vec{f}) \quad \text{`}\Box\text{'} \quad \prod_{i\in I}\mathbf{S}_i \models \xi(\vec{f})\\
&\text{iff} \quad \mathbf{Su}(I) \models \Psi\big([\![\psi_1(\vec{f})]\!],\dots,[\![\psi_m(\vec{f})]\!]\big)\\
&\qquad \text{`}\Box\text{'} \quad \mathbf{Su}(I) \models \Xi\big([\![\xi_1(\vec{f})]\!],\dots,[\![\xi_n(\vec{f})]\!]\big)\\
&\text{iff} \quad \mathbf{Su}(I) \models \Psi\big([\![\psi_1(\vec{f})]\!],\dots,[\![\psi_m(\vec{f})]\!]\big)\\
&\qquad\qquad \Box\,\Xi\big([\![\xi_1(\vec{f})]\!],\dots,[\![\xi_n(\vec{f})]\!]\big),
\end{aligned}$$

and thus $\mathsf{FV}[\varphi]$ is a Feferman-Vaught sequence for φ.

(d) Suppose $\varphi(\vec{x}\,) = \exists x\,\psi(x,\vec{x}\,)$ and

$$\mathsf{FV}[\psi] = \big\langle \Psi(\vec{X});\psi_1,\dots,\psi_m\big\rangle$$

is such that

$$\prod_{i\in I}\mathbf{S}_i \models \psi(f,\vec{f}) \quad \text{iff} \quad \mathbf{Su}(I) \models \Psi\big([\![\psi_1(f,\vec{f})]\!],\dots,[\![\psi_m(f,\vec{f})]\!]\big).$$

Assume

$$\prod_{i\in I}\mathbf{S}_i \models \varphi(\vec{f}),$$

that is, $\prod_{i\in I}\mathbf{S}_i \models \exists x\,\psi(x,\vec{f})$. Choose $f\in\prod_{i\in I}S_i$ such that

$$\prod_{i\in I}\mathbf{S}_i \models \psi(f,\vec{f}).$$

By the induction hypothesis,

$$\mathbf{Su}(I) \models \Psi\big([\![\psi_1(f,\vec{f})]\!],\dots,[\![\psi_m(f,\vec{f})]\!]\big).$$

For $1\le j\le 2^m$, let

$$I_j = \Big[\!\!\Big[\bigwedge_{k\in M_j}\psi_k(f,\vec{f}) \wedge \bigwedge_{k\notin M_j}\neg\,\psi_k(f,\vec{f})\Big]\!\!\Big].$$

The I_j partition I, and $I_j \subseteq [\![\varphi_j(\vec{f})]\!]$, for $1 \le j \le 2^m$. Furthermore, $\psi_k(f, \vec{f})$ is logically equivalent to

$$\bigwedge_{k \in M_j} \psi_j(f, \vec{f}) \wedge \bigwedge_{k \notin M_j} \neg \psi_j(f, \vec{f}).$$

Thus the assignment of I_j to Y_j shows that

$$\mathbf{Su}(I) \models \Phi([\![\varphi_1]\!], \ldots, [\![\varphi_{2^m}]\!]).$$

Conversely, if

$$\prod_{i \in I} \mathbf{Su}(I) \models \Phi([\![\varphi_1(\vec{f})]\!], \ldots, [\![\varphi_{2^m}(\vec{f})]\!])$$

then choose subsets I_j of I to witness the Y_j in Φ. The sets I_j partition I, because of the condition $\mathsf{Part}(\vec{Y})$, and each $I_j \subseteq [\![\varphi_j(\vec{f})]\!]$, since $Y_j \subseteq X_j$ is a condition in Φ. Since

$$\varphi_j(\vec{f}) = \exists x \Big(\bigwedge_{k \in M_j} \psi_k(x, \vec{f}) \wedge \bigwedge_{k \notin M_j} \neg \psi_k(x, \vec{f}) \Big),$$

it follows that

$$I_j \subseteq \Big[\!\!\Big[\exists x \Big(\bigwedge_{k \in M_j} \psi_k(x, \vec{f}) \wedge \bigwedge_{k \notin M_j} \neg \psi_k(x, \vec{f}) \Big) \Big]\!\!\Big],$$

so one can choose $f \in \prod_{i \in I} S_i$ such that, for $1 \le j \le 2^m$,

$$I_j \subseteq \Big[\!\!\Big[\Big(\bigwedge_{k \in M_j} \psi_k(f, \vec{f}) \wedge \bigwedge_{k \notin M_j} \neg \psi_k(f, \vec{f}) \Big) \Big]\!\!\Big].$$

Then

$$I_j = \Big[\!\!\Big[\Big(\bigwedge_{k \in M_j} \psi_k(f, \vec{f}) \wedge \bigwedge_{k \notin M_j} \neg \psi_k(f, \vec{f}) \Big) \Big]\!\!\Big]$$

as the formulas inside the $[\![\ ]\!]$ symbols defining the I_j are mutually contradictory. Since the formula $\psi_k(x, \vec{x})$ is logically equivalent to

$$\bigwedge_{k \in M_j} \psi_j(x, \vec{x}) \wedge \bigwedge_{k \notin M_j} \neg \psi_j(x, \vec{x}),$$

from

$$\mathbf{Su}(I) \models \Psi\Big(\bigcup_{1 \in M_j} I_j, \ldots, \bigcup_{m \in M_j} I_j \Big)$$

one has

$$\mathbf{Su}(I) \models \Psi([\![\psi_1(f, \vec{f})]\!], \ldots, [\![\psi_m(f, \vec{f})]\!]),$$

and thus, by the induction hypothesis,

$$\prod_{i\in I} \mathbf{S}_i \models \psi(f, \vec{f}).$$

Then one has $\prod_{i\in I} \mathbf{S}_i \models \exists x\, \psi(x, \vec{f})$, and thus

$$\prod_{i\in I} \mathbf{S}_i \models \varphi(\vec{f}).$$

(e) This follows from the previous cases of negation and existential quantifier since $\forall x\, \psi$ is logically equivalent to $\neg \exists x \neg \psi$.

Finally, notice that if φ is a sentence then

$$\mathsf{FV}[\varphi] = \langle \Phi; \varphi_1, \ldots, \varphi_m \rangle$$

is such that all φ_j are also sentences. □

12.3. Skolem's analysis of the first-order calculus of classes

Boolean algebra originated in the study of all subsets of a given set.

Definition 12.8. The class $\mathcal{BA}_0$ is the collection of Boolean algebras of the form $\mathbf{Su}(X)$.

In 1919 Skolem published a paper [**52**] to clarify and complete the work of Schröder (from the 1890s) on the *elimination problem* of the Calculus of Classes. In modern terms the Calculus of Classes is concerned with the class $\mathcal{BA}_0$. In essence Schröder wanted to show that any first-order formula $\varphi(\vec{x})$ in the language of Boolean algebra is equivalent modulo $\mathcal{BA}_0$ to a quantifier-free formula $\psi(\vec{x})$. (See the definition below.) He realized that there were complications due to having to refer to the number of elements in the set X, but he failed to get a clear grasp of the possibilities.

Definition 12.9. Let $\varphi(\vec{x})$ and $\psi(\vec{x})$ be two L-formulas, and let $\mathcal{K}$ be a class of L-structures. Then $\varphi(\vec{x})$ *is equivalent to* $\psi(\vec{x})$ *modulo* $\mathcal{K}$, written

$$\varphi(\vec{x}) \equiv \psi(\vec{x}) \quad \text{modulo } \mathcal{K},$$

if, for every member $\mathbf{S}$ of $\mathcal{K}$ and every tuple of elements $\vec{s}$ from S, $\varphi(\vec{s})$ holds in $\mathbf{S}$ iff $\psi(\vec{s})$ holds in $\mathbf{S}$.

Thus formulas that are equivalent modulo $\mathcal{K}$ are satisfied by the same tuples in any member of $\mathcal{K}$.

Skolem gave an elegant analysis of the situation by introducing new unary predicates $|\ |_{\geq n}$, where $|Y|_{\geq n}$ asserts that 'Y has at least n elements in it'. With these predicates he was able to fully carry out quantifier elimination modulo $\mathcal{BA}_0$. The details of Skolem's proof are provided here. For the remainder of this section the symbol $\equiv$ means *equivalent modulo* $\mathcal{BA}_0$.

Given a first-order formula $\varphi(\vec{y})$ in the language $\cup, \cap, ', 0, 1$ of $\mathcal{BA}_0$ augmented by the unary predicates $|\ |_{\geq n}$:

- First put φ in *prenex form*, say
$$\mathsf{Q}_1 x_1 \cdots \mathsf{Q}_n x_n \psi(\vec{x}, \vec{y}),$$
where the Q_i are quantifiers, and ψ is quantifier-free;
- If the last quantifier Q_n is $\forall$, change it to $\neg\exists\neg$. This gives an equivalent (perhaps the same) formula
$$\mathsf{Q}_1 x_1 \cdots \mathsf{Q}_{n-1} x_{n-1} \pm \exists x_n \psi^\star,$$
where the notation $\pm\phi$ means either ϕ or $\neg\phi$, whichever is appropriate.

  Now attention is focused on $\exists x_n \psi^\star$, with the intention of eliminating the existential quantifier.[2]
- The next step is to put the quantifier-free formula $\psi^\star$ into *disjunctive form*:
$$\psi^\star \equiv \bigvee \bigwedge \pm\text{atomic}.$$
- Then one can move the existential quantifier inside the disjunction:
$$\begin{aligned} \exists x_n \varphi^\star &\equiv \exists x_n \bigvee \bigwedge \pm\text{atomic} \\ &\equiv \bigvee \exists x_n \bigwedge \pm\text{atomic}. \end{aligned}$$

  Thus one can now concentrate on eliminating $\exists x_n$ from each of the disjuncts $\exists x_n \bigwedge \pm$atomic.
- Fixing attention on one of the disjuncts, x_n may not appear in some of the atomic formulas, so pull those atomic formulas outside the scope of the quantifier:
$$\begin{aligned} &\exists x_n \bigwedge \pm \text{ atomic } \equiv \\ &\qquad \Big[\exists x_n \bigwedge \pm\text{atomic}(x_1, \cdots, x_n, \vec{y})\Big] \wedge \bigwedge \pm\text{atomic}(x_1, \cdots, x_{n-1}, \vec{y}). \end{aligned}$$
- Thus, to carry out the elimination of quantifiers, it suffices to eliminate quantifiers in the cases of the form
$$\exists x_n \bigwedge \pm\text{atomic}(x_1, \cdots, x_n, \vec{y}),$$
where x_n actually appears in each of the atomic formulas.

Definition 12.10. For $\vec{y} = (y_1, \cdots, y_k)$, a $\vec{y}$*-constituent* is a term of the form
$$\pm y_1 \cap \cdots \cap \pm y_k,$$
where $\pm y_i$ can be either y_i or y_i'.

---

[2]When this has been accomplished, return to the above formula, replacing $\exists x_n \psi^\star$ with a quantifier-free equivalent ξ to obtain
$$\mathsf{Q}_1 x_1 \cdots \mathsf{Q}_{n-1} x_{n-1} \pm \xi.$$
Now repeat the process until there are no quantifiers left.

Definition 12.11. A *Boolean combination* of formulas $\varphi_1, \cdots, \varphi_k$ is a formula built up from the φ_i using $\wedge$, $\vee$, and $\neg$.

Theorem 12.12 (Skolem, 1919). *The first-order theory of $\mathcal{BA}_0$ with the numerical predicates $|\ |_{\geq n}$ admits elimination of quantifiers. Indeed,*

(a) *every formula $\varphi(\vec{y})$ is equivalent modulo $\mathcal{BA}_0$ to a formula which is a Boolean combination of numerical assertions about the $\vec{y}$-constituents, and*

(b) *every sentence is equivalent modulo $\mathcal{BA}_0$ to a Boolean combination of assertions about the size of the domain.*

Proof. All subformulas of a given φ which involve $\approx$ can be put in the form $t \approx 0$, and hence in the form $\neg|t|_{\geq 1}$. Thus $\approx$ *can be eliminated.* So it suffices to eliminate the quantifier in a formula of the form

$$\exists x \bigwedge \pm \big|t_i(x, \vec{y})\big|_{\geq i}.$$

To do this put the term t_i in disjunctive form

$$\bigcup \pm x \cap \pm y_1 \cdots \cap \pm y_k.$$

Then by repeated use of

$$|s \cup t|_{\geq k} \ \equiv\ \bigvee_{i=0}^{k} |s|_{\geq i} \wedge |t|_{\geq k-i},$$

for s and t *disjoint*, it suffices to show how to eliminate the quantifier from formulas of the form

$$\exists x \bigwedge \pm \big| \pm x \cap C_j(\vec{y})\big|_{\geq i},$$

where each $C_j(\vec{y})$ denotes one of the $\vec{y}$-constituents.

Now collect the atomic formulas involving each particular constituent together to obtain

$$\exists x \bigwedge_j \Big(\bigwedge_i \pm\big| \pm x \cap C_j(\vec{y})\big|_{\geq i}\Big).$$

Then an easy argument, based on the fact that the $C_j(\vec{y})$ are disjoint, shows that this is equivalent modulo $\mathcal{BA}_0$ to

$$\bigwedge_j \exists x \Big(\bigwedge_i \pm\big| \pm x \cap C_j(\vec{y})\big|_{\geq i}\Big).$$

Consequently what is left is the problem of showing how to eliminate a quantifier from a formula of the form

$$\exists x \Big(\bigwedge_i \pm\big| \pm x \cap C(\vec{y})\big|_{\geq i}\Big),$$

for a single constituent $C(\vec{y})$. Since $|x|_{\geq i+1}$ implies $|x|_{\geq i}$ in the Calculus of Classes, such a formula can be reduced to one involving at most 4 conjuncts:

$$\exists x \Big(\big|x \cap C(\vec{y})\big|_{\geq i_1} \wedge \neg\big|x \cap C(\vec{y})\big|_{\geq i_2} \wedge \big|x' \cap C(\vec{y})\big|_{\geq i_3} \wedge \neg\big|x' \cap C(\vec{y})\big|_{\geq i_4}\Big).$$

It can also be the case that some of these ±atomics are omitted. This formula says that, for some x, one has

$$i_1 \leq \left|x \cap C(\vec{y})\right| < i_2 \quad \text{and} \quad i_3 \leq \left|x' \cap C(\vec{y})\right| < i_4.$$

Whether or not such an x exists depends only on the size of $C(\vec{y})$, so this can be transformed into a Boolean combination of statements of the form

$$\left|C(\vec{y})\right|_{\geq i},$$

in case there are some y's; and otherwise one has a Boolean combination of statements of the form

$$|1|_{\geq i},$$

that is, assertions about the size of the universe 1.

This establishes the elimination of quantifiers, and the claims about the expressive power of formulas and sentences. □

Definition 12.13. Let Ψ be a first-order sentence in the language of Boolean algebra. The smallest integer n such that Ψ is equivalent modulo $\mathcal{BA}_0$ to a Boolean combination of the assertions $|1|_{\geq 1}, \ldots, |1|_{\geq n}$ will be denoted by n_Ψ.

Definition 12.14. $\mathbf{B}_n$ is the Boolean algebra of subsets of $\{1, \ldots, n\}$.

Corollary 12.15. *For any Boolean algebra sentence* Ψ,

$$n \geq n_\Psi \quad \Rightarrow \quad \mathbf{B}_n \models \Psi, \text{ or}$$
$$n \geq n_\Psi \quad \Rightarrow \quad \mathbf{B}_n \models \neg\Psi.$$

PROOF. From Skolem's theorem, Ψ is equivalent modulo $\mathcal{BA}_0$ to a Boolean combination of the sentences $|1|_{\geq 1}, \ldots, |1|_{\geq n_\Psi}$. Also, for any $n \geq n_\Psi$, all of the statements $|1|_{\geq 1}, \ldots, |1|_{\geq n_\Psi}$ are true in $\mathbf{B}_n$. □

12.4. $\mathcal{K}_\varphi$ is a disjoint union of partition classes

This section is based on the 1996 paper [**15**] of Burris and Idziak.

Definition 12.16. A partition $\mathcal{P}_1, \ldots, \mathcal{P}_k$ of $\mathcal{P}$ is assumed to have the property that each $\mathcal{P}_i$ is closed under isomorphism. A *partition class* of $\mathcal{K}$ is a class of the form $\prod \mathcal{P}_i^{\gamma_i}$, where the $\mathcal{P}_i$ partition $\mathcal{P}$.

Proposition 12.17. *Let* $\mathcal{K}$ *be an adequate class of* L*-structures. For* φ *a first-order* L*-sentence, there is a number* n *and a partition* $\mathcal{P}_1, \ldots, \mathcal{P}_k$ *of the class* $\mathcal{P}$ *of indecomposables of* $\mathcal{K}$ *such that* $\mathcal{K}_\varphi$ *is a disjoint union of partition classes* $\mathcal{P}_1^{\gamma_1} \cdots \mathcal{P}_k^{\gamma_k}$, *where each* γ_i *is one of* $0, 1, \ldots, n-1$, *or* $(\geq n)$.

PROOF. Let $\langle \Phi, \varphi_1, \ldots, \varphi_m \rangle$ be a Feferman-Vaught sequence, for φ. Then choose a Feferman-Vaught sequence $\langle \Phi_j, \varphi_{j1}, \ldots, \varphi_{jm_j} \rangle$, for each φ_j. Let $\mathcal{P}_1, \ldots, \mathcal{P}_k$ be the partition of $\mathcal{P}$ defined by putting two $\mathcal{K}$-indecomposables in the same partition class iff they satisfy the same φ_{ij}. Let n be the maximum of the n_Ψ where Ψ ranges over the sentences $\Phi_k(\varepsilon_1, \ldots, \varepsilon_{m_k})$, with each ε_j being either 0 or 1.

Given γ_i as in the theorem, first one can use the Feferman-Vaught theorem to show that any two members of $\mathcal{P}_i^{\gamma_i}$ satisfy the same φ_j. To see this, choose $\mathbf{P}_1, \ldots, \mathbf{P}_p \in \mathcal{P}_i$ and $\mathbf{Q}_1, \ldots, \mathbf{Q}_q \in \mathcal{P}_i$. Then, by Theorem 12.7,

$$\prod_{1 \le \ell \le p} \mathbf{P}_\ell \models \varphi_j \quad \text{iff} \quad \mathbf{B}_p \models \Phi_j\big([\![\varphi_{j1}]\!], \ldots, [\![\varphi_{jm_j}]\!]\big)$$

$$\prod_{1 \le \ell \le q} \mathbf{Q}_\ell \models \varphi_j \quad \text{iff} \quad \mathbf{B}_q \models \Phi_j\big([\![\varphi_{j1}]\!], \ldots, [\![\varphi_{jm_j}]\!]\big).$$

In both cases $[\![\varphi_{js}]\!]$ is 1 iff $\mathcal{P}_i \models \varphi_{js}$, and otherwise it is 0. Thus the Boolean sentences in the two cases are the same sentence Ψ, and

$$\prod_{1 \le \ell \le p} \mathbf{P}_\ell \models \varphi_j \quad \text{iff} \quad \mathbf{B}_p \models \Psi$$

$$\prod_{1 \le \ell \le q} \mathbf{Q}_\ell \models \varphi_j \quad \text{iff} \quad \mathbf{B}_q \models \Psi.$$

If $p = q$ then it is clear that the two products $\prod \mathbf{P}_\ell$ and $\prod \mathbf{Q}_\ell$ must both satisfy φ_j, or both not satisfy φ_j, since $\mathbf{B}_p = \mathbf{B}_q$. If $p, q \ge n$ then, by Corollary 12.15, both or neither of $\mathbf{B}_p$ and $\mathbf{B}_q$ satisfy Ψ since $n \ge n_\Psi$, so both or neither of $\prod \mathbf{P}_\ell$ and $\prod \mathbf{Q}_\ell$ satisfy φ_j.

Secondly, given $\mathbf{S}_i, \mathbf{S}'_i \in \mathcal{P}_i^{\gamma_i}$, apply the Feferman-Vaught theorem again to show that either both or neither of $\mathbf{S}_1 \times \cdots \times \mathbf{S}_k$ and $\mathbf{S}'_1 \times \cdots \times \mathbf{S}'_k$ satisfy φ since

$$\prod_{1 \le \ell \le k} \mathbf{S}_\ell \models \varphi \quad \text{iff} \quad \mathbf{B}_k \models \Phi\big([\![\varphi_1]\!], \ldots, [\![\varphi_k]\!]\big)$$

$$\prod_{1 \le \ell \le k} \mathbf{S}'_\ell \models \varphi \quad \text{iff} \quad \mathbf{B}_k \models \Phi\big([\![\varphi_1]\!], \ldots, [\![\varphi_k]\!]\big),$$

noting that $[\![\varphi_j]\!]$ is the same in either of the two products, by the previous paragraph.

Since the union of the $\mathcal{P}_1^{\gamma_1} \cdots \mathcal{P}_k^{\gamma_k}$, for the γ_i ranging over the exponents $0, 1, \ldots, n-1, (\ge n)$, is $\mathcal{K}$, it follows that $\mathcal{K}_\varphi$ is the union of the $\mathcal{P}_1^{\gamma_1} \cdots \mathcal{P}_k^{\gamma_k}$ which satisfy φ. □

Theorem 12.18. *Let $\mathcal{K}$ be an adequate class of finite* L*-structures, for a finite language* L*. If all partition sets of the multiplicative number system $\mathcal{A}_\mathcal{K}$ have local [global] asymptotic density in $\mathcal{A}_\mathcal{K}$ then, for φ a first-order* L*-sentence, the class $\mathcal{K}_\varphi$ has local [global] asymptotic density in $\mathcal{K}$.*

12.5. Applications

Given an adequate class $\mathcal{K}$ let $a(n), p(n)$ be the local counting functions of $\mathcal{A}_\mathcal{K}$, the multiplicative number system of isomorphism classes from $\mathcal{K}$, and let $A(x), P(x)$ be the global counting functions.

Proposition 12.19. *Let $\mathcal{K}$ be an adequate class of algebras. Then the following results concerning first-order limit laws hold:*

(a) $A(x) \in \mathsf{RV}_0$ *implies $\mathcal{K}$ has a global first-order 0–1 law.*

(b) $A(x) \in \mathsf{RV}_\alpha$, *with $0 < \alpha < \infty$, and $A(n/k)/A(n) \le C\alpha^{-k}$, for any k, n satisfying $K \le k \le n$, with C and K both positive, imply $\mathcal{K}$ has a global first-order limit law.*
(This holds if $A(n) \sim cn^\alpha$, where c and α are positive.)

(c) $p(n) = an^\mu + \mathrm{O}(n^\eta)$, *with $a > 0$, $\mu > -1$, and $\eta < \mu$, implies $\mathcal{K}$ has a global first-order limit law.*

(d) *If $P(x)$ is polylog bounded then $A(x) \in \mathsf{RV}_0$, hence $\mathcal{K}$ has a global first-order 0–1 law.*

PROOF. These follow from Theorem 12.18 and: Theorem 10.2 for (a), Theorem 11.15 for (b), Theorem 11.20 (with the previous item) for (c), and [**6**] for (d). □

Such results have been called *spectrally determined* logical limit laws since they depend only on the growth rate of the counting functions (or, spectra) of $\mathcal{K}$, given that $\mathcal{K}$ is adequate. For a given adequate $\mathcal{K}$, only a portion of the partition classes of $\mathcal{K}$ can be defined by first-order sentences—after all, there are continuum many partition classes if $\mathcal{P}$ is infinite, but only a countable number of sentences when the language is countable. However, it turns out that it is necessary that *all* partition classes have asymptotic density in order to deduce a logical limit law from the counting functions alone.

Proposition 12.20. *Suppose $\mathcal{K}$ is adequate, but not all partition classes have asymptotic density. Then one can construct another adequate class $\widehat{\mathcal{K}}$, with the same counting functions as $\mathcal{K}$, by expanding the language of $\mathcal{K}$ by two binary operations, and finitely many unary operations, with the property that there is a first-order sentence φ such that $\widehat{\mathcal{K}}_\varphi$ does not have asymptotic density in $\widehat{\mathcal{K}}$.*

PROOF. Suppose

$$\mathcal{B} \;=\; \mathcal{P}_1^{\gamma_1}\cdots\mathcal{P}_k^{\gamma_k}$$

is a partition class of $\mathcal{K}$ that does not have asymptotic density.

Let $\mathcal{Q} \subseteq \mathcal{P}$ consist of one representative from each isomorphism class of indecomposables. Linearly order each indecomposable in $\mathcal{Q}$, and then define two binary operations $\vee, \wedge$ on these indecomposables by

$$\begin{aligned} x \vee y &= \max(x, y) \\ x \wedge y &= \min(x, y). \end{aligned}$$

Also, add new unary functions $f_1, \dots, f_k$ to each member of $\mathcal{Q}$ such that f_i is constant on members of $\mathcal{Q} \cap \mathcal{P}_i$, and the identity function on members of $\mathcal{Q} \smallsetminus \mathcal{P}_i$. Now close this expansion of $\mathcal{Q}$ under finite direct product, defining the new operations coordinatewise, and then close this class under isomorphism to obtain an expansion $\widehat{\mathcal{K}}$ of $\mathcal{K}$, with the new functions $\vee, \wedge, f_1, \ldots, f_k$, but with the same count functions.

The operations $\vee, \wedge$ are *lattice* operations on the members of $\widehat{\mathcal{K}}$, and one has a partial order defined on each member of $\widehat{\mathcal{K}}$ by

$$x \leq y \quad \text{iff} \quad x \wedge y = x.$$

For the original expansions of the members of $\mathcal{Q}$ this partial ordering is the original linear ordering. Now define the following formulas:

Chain(x)	the elements below x form a chain
MaxChain(x)	x is a maximal element satisfying Chain(x)
Color$_i(x)$	$\exists y \Big[(x \leq y) \wedge \text{MaxChain}(y) \wedge \forall z\big((z \leq y) \to f_i(z) \wedge y \approx f_i(y) \wedge y\big)\Big].$

Each γ_i (in the expression for $\mathcal{B}$) is in one of the forms $(\leq m_i)$, m_i, $(\geq m_i)$, so construct a first-order sentence φ that says:

- there are m_i, but not $m_i + 1$, pairwise incomparable elements that satisfy Chain(x) and Color$_i(x)$, for $\gamma_i = m_i$;
- there are at most m_i pairwise incomparable elements that satisfy Chain(x) and Color$_i(x)$, for $\gamma_i = (\leq m_i)$;
- there are m_i pairwise incomparable elements that satisfy Chain(x) and Color$_i(x)$, for $\gamma_i = (\geq m_i)$.

Then such a sentence is satisfied by precisely the members of $\widehat{\mathcal{B}}$, the expansion of $\mathcal{B}$, so $\widehat{\mathcal{K}}_\varphi$ does not have asymptotic density.

□

Here are some examples of adequate classes with a first-order limit law—the discrete examples are listed first.

(a) $\mathcal{K}$ is the class of finite *Boolean algebras*. In this case there is only one indecomposable in $\widetilde{\mathcal{P}}$, namely the isomorphism type of the two-element Boolean algebra **2**. Thus $\mathcal{A}_\mathcal{K}$ is discrete, and $\mathcal{K}$ has a local first-order 0–1 law.

(b) $\mathcal{K}$ is the class of finite *Post n-algebras*. In this case there is only one indecomposable in $\widetilde{\mathcal{P}}$, namely the isomorphism type of the n-element Post n-algebra $\mathbf{P}_n$. Thus $\mathcal{A}_\mathcal{K}$ is discrete, and $\mathcal{K}$ has a local first-order 0–1 law.

(c) $\mathcal{K}$ is the class of finite *vector spaces* over a finite field $\mathbf{F}$. There is only one indecomposable in $\widetilde{\mathcal{P}}$, namely the isomorphism type of the one-dimensional vector space over $\mathbf{F}$. Thus $\mathcal{A}_\mathcal{K}$ is discrete, and has a local first-order 0–1 law.

(d) $\mathcal{K}$ is the class of finite *monadic algebras*. Monadic algebras are Boolean algebras augmented by a unary operation that plays the role of a quantifier. There is one indecomposable in $\widetilde{\mathcal{P}}$ with norm 2^n, for each $n \geq 1$, and no other indecomposables. Thus $\mathcal{A}_\mathcal{K}$ is discrete, and applying the results of Part 1 it follows that $\mathcal{K}$ has a local first-order 0–1 law.

(e) $\mathcal{K}$ is the class of finite $\mathbf{GF}[q,t]$ *modules*. Then $\mathcal{A}_{\mathcal{K}}$ is discrete and $p(n)$ is uniformly bounded. Thus $\mathcal{K}$ has a local first-order 0–1 law. (See Knopfmacher [**30**].)

(f) $\mathcal{K}$ is the class of finite *semisimple* $\mathbf{GF}[q,t]$ *algebras*. Then $\mathcal{A}_{\mathcal{K}}$ is discrete and $p(n)$ is uniformly bounded. Thus $\mathcal{K}$ has a local first-order 0–1 law. (See Knopfmacher [**30**].)

(g) $\mathcal{K}$ is the class of finite $\mathbf{D}_q$ *modules*, where $\mathbf{D}_q$ is the ring of integral functions in an algebraic function field in one variable over $\mathbf{GA}(q)$. Then $\mathcal{A}_{\mathcal{K}}$ is discrete and $p(n)$ is uniformly bounded. Thus $\mathcal{K}$ has a local first-order 0–1 law. (See Knopfmacher [**30**].)

(h) $\mathcal{K}$ is the class of finite *semisimple* $\mathbf{D}_q$ *algebras*. Then $\mathcal{A}_{\mathcal{K}}$ is discrete and $p(n)$ is uniformly bounded. Thus $\mathcal{K}$ has a local first-order 0–1 law. (See Knopfmacher [**30**].)

(i) $\mathcal{K}$ is the class of finite $(x^n \approx x)$*-rings*. In this case there are finitely many indecomposables in $\widetilde{\mathcal{P}}$, namely the isomorphism types of the finite fields in $\mathcal{K}$. Indeed, $\mathbf{GF}(p^k)$ is in $\mathcal{K}$ iff $p^k - 1\big|n-1$. $\mathcal{A}_{\mathcal{K}}$ is not discrete unless there is only one prime p such that $p-1\big|n-1$. $\mathcal{K}$ has a global first-order 0–1 law.

(j) $\mathcal{K}$ is the class of finite *sets*. These are algebras with no operations. There is one indecomposable of size p, for p a prime number, and no other indecomposables. Indeed, $\mathcal{A}_{\mathcal{K}}$ is isomorphic to the positive integers under multiplication. Since $A(n) = n$ it follows that $\mathcal{K}$ has a global first-order limit law.

(k) $\mathcal{K}$ is the class of finite *Abelian groups*. There is one indecomposable with norm n, for each n which is a positive power of a prime number, and no other indecomposables. By the Erdös and Szekeres result of §9.12.4, $A(x) \sim cx$, where $c = \prod_{n\geq 2}\zeta(n)$. Thus $\mathcal{K}$ has a global first-order limit law.

(l) $\mathcal{K}$ is the class of finite *lattices* that can be decomposed into a product of chains. Then there is exactly one indecomposable of each size n, so $P(n) = n$. Thus $\mathcal{K}$ has a global first-order limit law.

(m) $\mathcal{K}$ is the class of finite $\mathbf{Z_K}$*-modules* where $\mathbf{Z_K}$ is the ring of integers of the algebraic number field $\mathbf{K}$. Since $A(x) \sim cx$, $\mathcal{K}$ has a global first-order limit law. (See Knopfmacher [**29**].)

(n) $\mathcal{K}$ is the class of finite *semisimple rings*. Since $A(x) \sim cx$, $\mathcal{K}$ has a global first-order limit law. (See Knopfmacher [**29**].)

(o) $\mathcal{K}$ is the class of finite *semisimple* $\mathbf{Z}_k$ *algebras*. Since $A(x) \sim cx$, $\mathcal{K}$ has a global first-order limit law. (See Knopfmacher [**29**].)

(p) $\mathcal{K}$ is the class of finite *Heyting algebras* in the variety generated by the 3-element Heyting algebra $\mathbf{H}_3$. There is one indecomposable p_n of size $2^n + 1$, for each $n \geq 0$, and no other indecomposables. Thus $P(n) = \mathrm{O}(\log n)$, so $\mathcal{K}$ has a global first-order 0–1 law.

12.6. Finite dimensional structures

The same results can be obtained for classes of finite dimensional structures provided the definition of "adequate class" is changed by replacing the word "finite" by "finite dimensional", assuming

$$\dim(\mathbf{S}_1 \times \mathbf{S}_2) \;=\; \dim(\mathbf{S}_1) + \dim(\mathbf{S}_2),$$

and the norm function is defined by

$$\|\mathbf{S}\| \;=\; \lambda^{\dim \mathbf{S}},$$

where $\lambda > 1$ is a fixed positive integer. The description of $\mathcal{K}_\varphi$ as a disjoint union of partition classes does not depend on the structures being finite. This applies to the following to give first-order limit laws:

(a) $\mathcal{K}$ is the class of finite dimensional *vector spaces* over a given field $\mathbf{F}$.
(b) $\mathcal{K}$ is the class of finite dimensional *semisimple* $\mathbf{F}$*-algebras*, for a given field $\mathbf{F}$. (See Knopfmacher [**29**].)
(c) $\mathcal{K}$ is the class of finite dimensional *compact Lie groups.* (See Knopfmacher [**29**].)
(d) $\mathcal{K}$ is the class of finite dimensional *semisimple Lie algebras.* (See Knopfmacher [**29**].)

12.7. The main problem

The following problem is surely the main question brought up by the incredible parallels between the local additive results of Part 1 and the global multiplicative results of Part 2.

Problem 12.21. Find general principles to explain why local additive results lift to global multiplicative results.

APPENDIX A

Formal Power Series

A detailed and elementary treatment of the foundations of formal power series is given. To motivate the development, first the basic theory of functions is reviewed.

Real functions

A function is defined to be what is commonly called the *graph of a function*, that is, a function is to be considered as a collection of ordered pairs of real numbers.[1] This set theoretic approach may seem unnecessarily cumbersome at first, but it leads to the possibility of presenting simply, precisely formulated results. A number of key concepts are collected together in the following.

Definition A.1.

- A *real function* f is a subset of $\mathbb{R} \times \mathbb{R}$, the set of ordered pairs of real numbers, with the *single valued* property:

$$\left.\begin{array}{l}(a,b) \in f\\(a,c) \in f\end{array}\right\} \Rightarrow b = c.$$

- $\mathfrak{F}$ is the set of all real functions.
- For $f \in \mathfrak{F}$ the *domain* of f, $\mathrm{Dom}(f)$, and the *range* of f, $\mathrm{Rg}(f)$, are given by

$$\begin{aligned}\mathrm{Dom}(f) &= \big\{a \in \mathbb{R} : (\exists b \in \mathbb{R})\big((a,b) \in f\big)\big\}\\ \mathrm{Rg}(f) &= \big\{b \in \mathbb{R} : (\exists a \in \mathbb{R})\big((a,b) \in f\big)\big\}.\end{aligned}$$

  $\mathrm{Dom}(f)$ is also called *the domain of definition* of f.
- For $a \in \mathrm{Dom}(f)$, let $f(a)$ denote the unique $b \in \mathbb{R}$ such that $(a,b) \in f$. $f(a)$ is the *value of f at a*.
- Given $c \in \mathbb{R}$, $\mathbf{c}$ is the function on $\mathbb{R}$ whose value is always c.
- x denotes the identity function on $\mathbb{R}$.
- $\emptyset$ denotes the *empty* function, the function with no ordered pairs in it.

For $f \in \mathfrak{F}$, one has the trivial observation

$$f = \{(a, f(a)) : a \in \mathrm{Dom}(f)\}.$$

[1]The focus here is on real valued functions with their domain in the reals. This discussion can be applied to other situations, for example, complex valued functions defined on a subset of the complex numbers.

In calculus courses a function is described by an expression, like $\sqrt{x}$. The convention is to give the function the largest domain for which the expression is defined, so $\mathrm{Dom}(\sqrt{x}) = [0, \infty)$. In the more precise notation

$$\surd \;=\; \big\{(a,b) : a,b \geq 0,\ a = b^2\big\}.$$

Since functions are sets of order pairs, the usual notation $f \subseteq g$ is used to signify that every pair in f is also in g.

Definition A.2. If $f \subseteq g$ then f is a *restriction* of g and g is an *extension* of f.

For example the function defined by the power series $\sum_n x^n$ is a restriction of the function (defined by) $(1-x)^{-1}$.

Definition A.3. Suppose $f \in \mathfrak{F}$ and $S \subseteq \mathbb{R}$. Then $f|_S$, the *restriction of f to S*, is given by

$$f|_S \;=\; f \cap (S \times \mathbb{R}),$$

that is, $f|_S \;=\; \{(a,b) \in f : a \in S\}$.

Clearly $f|_S \subseteq f$, so the restriction of f to S is indeed a restriction of f. Furthermore

$$\mathrm{Dom}(f|_S) \;=\; \mathrm{Dom}(f) \cap S,$$

that is, a is in the domain of $f|_S$ iff it is in the domain of f as well as in S.

Thus the function defined by the power series $\sum_n x^n$ is $(1-x)^{-1}\big|_{(-1,1)}$, and not the function $(1-x)^{-1}$ with the larger domain of definition $\mathbb{R} \setminus \{1\}$.

Operations on real functions. Next the common operations used with real functions are considered.

negative: For $f \in \mathfrak{F}$

$$-f \;=\; \{\big(a, -f(a)\big) : a \in \mathrm{Dom}(f)\}$$

so $(-f)(a) \;=\; -f(a)$, for $a \in \mathrm{Dom}(f)$.

scalar multiplication: For $c \in \mathbb{R}$ and $f \in \mathfrak{F}$

$$cf \;=\; \big\{\big(a, c \cdot f(a)\big) : a \in \mathrm{Dom}(f)\big\}$$

so $(cf)(a) \;=\; c \cdot f(a)$, for $a \in \mathrm{Dom}(f)$.

addition: For $f, g \in \mathfrak{F}$

$$f + g \;=\; \big\{\big(a, f(a) + g(a)\big) : a \in \mathrm{Dom}(f) \cap \mathrm{Dom}(g)\big\}$$

so $(f+g)(a) \;=\; f(a) + g(a)$, for $a \in \mathrm{Dom}(f) \cap \mathrm{Dom}(g)$.

The empty function comes in handy here—if f and g have disjoint domains, then $f + g$ is just the empty function. A similar remark applies to the next two operations.

subtraction: Reduce this to addition by defining

$$f - g \;=\; f + (-g)$$

Here is another example of how this notation can be more precise than the usual, namely

$$\sqrt{} - \sqrt{} \;=\; \mathbf{0}\big|_{[0,\infty)}$$

and not $\sqrt{} - \sqrt{} = \mathbf{0}$.

multiplication: For $f, g \in \mathfrak{F}$

$$f \cdot g \;=\; \big\{\big(a, f(a)\cdot g(a)\big) : a \in \mathrm{Dom}(f) \cap \mathrm{Dom}(g)\big\}$$

so $(f \cdot g)(a) \;=\; f(a)\cdot g(a)$, for $a \in \mathrm{Dom}(f) \cap \mathrm{Dom}(g)$.

composition: For $f, g \in \mathfrak{F}$

$$f \circ g \;=\; \big\{\big(a, f(g(a))\big) : a \in \mathrm{Dom}(g),\ g(a) \in \mathrm{Dom}(f)\big\}$$

so $(f \circ g)(a) \;=\; f\big(g(a)\big)$, for $a \in \mathrm{Dom}(g) \cap g^{-1}\big(\mathrm{Dom}(f)\big)$.

Composition is often thought of as *substitution*, for example, if $f(x) = g(x)$ then $f(\sin x) = g(\sin x)$ by substitution. But the latter can also be written as $f \circ \sin = g \circ \sin$.

differentiation: Given $f \in \mathfrak{F}$, define the derivative f' to be the function whose domain is the set of all $a \in \mathrm{Dom}(f)$ for which one can find a neighborhood[2] N_a of a with $N_a \subseteq \mathrm{Dom}(f)$, that is, f is defined on N_a, and the following limit exists:

$$\lim_{\substack{u \to 0 \\ u \in N_a}} \frac{f(a+u) - f(a)}{u}.$$

If a is in the domain of f', then $f'(a)$ is this limit.

Remark A.4. The operations $+, \cdot, \circ, '$ and scalar multiplication are *always defined*, no matter which functions they are applied to.

Identities of real functions. Now the basic properties of these operations on functions are listed. (It took many years during the 19th and early 20th centuries for mathematicians to explicitly list exactly which rules they were using.)

Proposition A.5. *For $a, b \in \mathbb{R}$ and $f, g, h \in \mathfrak{F}$:*

scalar multiplication:

$$\begin{aligned} a\mathbf{0} &= \mathbf{0} & a\mathbf{1} &= \mathbf{a} \\ (-1)f &= -f & a(-f) &= (-a)f \;=\; -(af) \\ a(bf) &= (ab)f & (a+b)f &= (af) + (bf) \\ a(f+g) &= (af) + (ag) & a(f\cdot g) &= (af)\cdot g \;=\; f\cdot(ag) \end{aligned}$$

---

[2]A *neighborhood* of a is just an interval (c, d) with $c < a < d$.

addition, multiplication, and composition:

$$f+\mathbf{0} = f \qquad f\cdot\mathbf{1} = f$$

$$f+g = g+f \qquad f\cdot g = g\cdot f$$

$$(f+g)+h = f+(g+h) \qquad (f\cdot g)\cdot h = f\cdot(g\cdot h)$$

$$f\cdot(g+h) = (f\cdot g) + (f\cdot h) \qquad (f+g)\circ h = (f\circ h)+(g\circ h)$$

$$(f\cdot g)\circ h = (f\circ h)\cdot(g\circ h) \qquad (f\circ g)\circ h = f\circ(g\circ h)$$

differentiation:

$$(af)' \supseteq af' \qquad (f+g)' \supseteq f'+g'$$

$$(f\cdot g)' \supseteq (f'\cdot g)+(f\cdot g') \qquad (f\circ g)' \supseteq g'\cdot(f'\circ g)$$

$$\left(e^f\right)' = f'\cdot e^f$$

In this list one sees the famous product rule, and the chain rule of differential calculus. The last item is just a special case of the chain rule, but one that we want to emphasize.

Polynomials

If one regards $s(0)+\cdots+s(n)x^n$ merely as a formal expression, and not as a function, one is dealing with *formal polynomials.* There is an obvious correspondence between such polynomials and their sequences of coefficients $\big(s(0),\ldots,s(n)\big)$. So the sequence of coefficients will be taken as the definition of a formal polynomial. To make the description of addition and multiplication of polynomials simple to express, and to be able to consider polynomials as special cases of power series, a polynomial is viewed as having an infinite sequence of coefficients (that are eventually 0).

Throughout this appendix all polynomials are to be thought of as formal polynomials. A new notation will be introduced to discuss the functions they define. Consequently, from this point on the adjective 'formal' is dropped when talking about polynomials, but feel free to insert it everywhere you see the word polynomial.

Definition A.6. A *polynomial* is an infinite sequence $\big(s(n)\big)$ of real numbers such that the $s(n)$ are eventually 0. The $s(n)$ are the *coefficients* of the polynomial.

By tradition the polynomial $\mathbf{S} = \big(s(n)\big)$ is written in longer form as $\mathbf{S}(x) = s(0)+s(1)x+\cdots+s(k)x^k$, when $s(m)=0$ for $m>k$, and also as $\mathbf{S}(x) = \sum_n s(n)x^n$. At this point x is merely a symbol, as are $+$ and $\sum$. It is only when real numbers are substituted for x that one is dealing with functions.

Functions defined by polynomials.

Definition A.7. Given a polynomial $\mathbf{S}$ let $f_{\mathbf{S}}$ be the real valued function defined on the reals by $f_{\mathbf{S}}(a) = \sum_n s(n)a^n$. $f_{\mathbf{S}}$ is called a *polynomial function.*

Knowing how functions defined by polynomials behave under the standard operations is crucial to the development.

$f^{(n)}$ means the *nth derivative* of f.

Proposition A.8. *Let* $\mathbf{S}, \mathbf{T}$ *be polynomials and let* $a, c \in \mathbb{R}$. *Then one has*

$$\begin{aligned}
(cf_{\mathbf{S}})(a) &= \sum_n cs(n)a^n \\
(f_{\mathbf{S}} + f_{\mathbf{T}})(a) &= \sum_n \big(s(n) + t(n)\big)a^n \\
(f_{\mathbf{S}} \cdot f_{\mathbf{T}})(a) &= \sum_n \Big(\sum_k s(k) \cdot t(n-k)\Big)a^n \\
(f_{\mathbf{S}} \circ f_{\mathbf{T}})(a) &= \sum_n \Big(\sum_k s(k) \sum_{\ell_1+\cdots+\ell_k=n} t(\ell_1)\cdots t(\ell_k)\Big)a^n \\
f_{\mathbf{S}}'(a) &= \sum_{n\geq 1} ns(n)a^{n-1} \\
s(n) &= f_{\mathbf{S}}^{(n)}(0)/n! \qquad \textit{for } n \geq 0 \\
f_{\mathbf{S}} = f_{\mathbf{T}} &\Leftrightarrow \mathbf{S} = \mathbf{T}.
\end{aligned}$$

PROOF. Only basic facts about finite sums and products are needed. For $cf_{\mathbf{S}}$:

$$(cf_{\mathbf{S}})(a) = cf_{\mathbf{S}}(a) = c\sum_n s(n)a^n = \sum_n cs(n)a^n.$$

For $f_{\mathbf{S}} + f_{\mathbf{T}}$:

$$\begin{aligned}
(f_{\mathbf{S}} + f_{\mathbf{T}})(a) = f_{\mathbf{S}}(a) + f_{\mathbf{T}}(a) &= \sum_n s(n)a^n + \sum_n t(n)a^n \\
&= \sum_n \big(s(n) + t(n)\big)a^n.
\end{aligned}$$

For $f_{\mathbf{S}} \cdot f_{\mathbf{T}}$:

$$\begin{aligned}
(f_{\mathbf{S}} \cdot f_{\mathbf{T}})(a) &= f_{\mathbf{S}}(a) \cdot f_{\mathbf{T}}(a) = \sum_n s(n)a^n \cdot \sum_n t(n)a^n \\
&= \sum_{i,j} \big(s(i) \cdot t(j)\big)a^{i+j} = \sum_n \Big(\sum_{i+j=n} s(i) \cdot t(j)a^{i+j}\Big) \\
&= \sum_n \Big(\sum_{i+j=n} s(i) \cdot t(j)\Big)a^n.
\end{aligned}$$

For $f_{\mathbf{S}} \circ f_{\mathbf{T}}$:

$$\begin{aligned}(f_{\mathbf{S}} \circ f_{\mathbf{T}})(a) = f_{\mathbf{S}}\big(f_{\mathbf{T}}(a)\big) &= \sum_k s(k)\Big(\sum_m t(m)a^m\Big)^k \\ &= \sum_k s(k)\Big(\sum_{\ell_1,\dots,\ell_k \geq 0} t(\ell_1)\cdots t(\ell_k)a^{\ell_1+\cdots+\ell_k}\Big) \\ &= \sum_k s(k)\sum_n\Big(\sum_{\ell_1+\cdots+\ell_k=n} t(\ell_1)\cdots t(\ell_k)a^n\Big) \\ &= \sum_k\sum_n s(k)\Big(\sum_{\ell_1+\cdots+\ell_k=n} t(\ell_1)\cdots t(\ell_k)\Big)a^n \\ &= \sum_n\Big(\sum_k s(k)\sum_{\ell_1+\cdots+\ell_k=n} t(\ell_1)\cdots t(\ell_k)\Big)a^n.\end{aligned}$$

For $f_{\mathbf{S}}'$, apply the rules for differentiation of sums and products, along with $x' = 1$, to obtain the usual termwise differentiation formula. This leads to the formula for $s(n)$, and to the proof that $f_{\mathbf{S}} = f_{\mathbf{T}}$ iff $\mathbf{S} = \mathbf{T}$. □

Operations on polynomials. Now use Proposition A.8 to suggest appropriate definitions for the operations on polynomials.

Definition A.9. Let $\mathbf{S}, \mathbf{T}$ be polynomials and let $c \in \mathbb{R}$.

coefficients: $\big[x^n\big]\mathbf{S}$ is $s(n)$.
equality: $\mathbf{S} = \mathbf{T}$ holds precisely when the corresponding coefficients are equal, that is, for all $n \geq 0$, $\big[x^n\big]\mathbf{S} = \big[x^n\big]\mathbf{T}$.
constant polynomials:

$$\big[x^n\big]\mathbf{c} = \begin{cases} c & \text{if } n = 0 \\ 0 & \text{otherwise.}\end{cases}$$

scalar multiplication: $\big[x^n\big]\big(c\mathbf{S}\big) = cs(n)$.
addition: $\big[x^n\big]\big(\mathbf{S}+\mathbf{T}\big) = s(n) + t(n)$.
multiplication: Use the Cauchy product:

$$\big[x^n\big]\big(\mathbf{S} * \mathbf{T}\big) = \sum_{k \leq n} s(k) \cdot t(n-k).$$

composition:

$$\big[x^n\big]\big(\mathbf{S} \circ \mathbf{T}\big) = \sum_k s(k) \sum_{\ell_1+\cdots+\ell_k=n} t(\ell_1)\cdots t(\ell_k).$$

differentiation: $\big[x^n\big]\mathbf{S}' = (n+1)s(n+1)$.

Since the operations on polynomials have been defined to harmonize with Proposition A.8, immediately one has the following.

Identities of polynomial functions.

Proposition A.10. *Let* $\mathbf{S}, \mathbf{T}$ *be polynomials and let* $c \in \mathbb{R}$. *Then one has*

$$\begin{aligned} f_{c\mathbf{S}} &= cf_{\mathbf{S}} \\ f_{\mathbf{S}+\mathbf{T}} &= f_{\mathbf{S}} + f_{\mathbf{T}} \\ f_{\mathbf{S}*\mathbf{T}} &= f_{\mathbf{S}} \cdot f_{\mathbf{T}} \\ f_{\mathbf{S}\circ\mathbf{T}} &= f_{\mathbf{S}} \circ f_{\mathbf{T}} \\ f_{\mathbf{S}}{}' &= f_{\mathbf{S}'}. \end{aligned}$$

With this, and the fact that $f_{\mathbf{S}} = f_{\mathbf{T}} \Leftrightarrow \mathbf{S} = \mathbf{T}$, the entire Proposition A.5 can be lifted to the setting of polynomials.[3] The results on differentiation are improved to equality as the functions defined by polynomials have domain $\mathbb{R}$, and they are everywhere differentiable.

Proposition A.11. *For* $a, b \in \mathbb{R}$ *and polynomials* $\mathbf{R}, \mathbf{S}, \mathbf{T}$ *the following hold:*

scalar multiplication:

$$\begin{aligned} a\mathbf{0} &= \mathbf{0} & a\mathbf{1} &= \mathbf{a} \\ (-1)\mathbf{R} &= -\mathbf{R} & a(-\mathbf{R}) &= (-a)\mathbf{R} = -(a\mathbf{R}) \\ a(b\mathbf{R}) &= (ab)\mathbf{R} & (a+b)\mathbf{R} &= (a\mathbf{R}) + (b\mathbf{R}) \\ a(\mathbf{R}+\mathbf{S}) &= (a\mathbf{R}) + (a\mathbf{S}) & a(\mathbf{R}\cdot\mathbf{S}) &= (a\mathbf{R})\cdot\mathbf{S} = \mathbf{R}\cdot(a\mathbf{S}) \end{aligned}$$

addition, multiplication, and composition:

$$\begin{aligned} \mathbf{R}+\mathbf{0} &= \mathbf{R} & \mathbf{R}\cdot\mathbf{1} &= \mathbf{R} \\ \mathbf{R}+\mathbf{S} &= \mathbf{S}+\mathbf{R} & \mathbf{R}\cdot\mathbf{S} &= \mathbf{S}\cdot\mathbf{R} \\ (\mathbf{R}+\mathbf{S})+\mathbf{T} &= \mathbf{R}+(\mathbf{S}+\mathbf{T}) & (\mathbf{R}\cdot\mathbf{S})\cdot\mathbf{T} &= \mathbf{R}\cdot(\mathbf{S}\cdot\mathbf{T}) \\ \mathbf{R}\cdot(\mathbf{S}+\mathbf{T}) &= (\mathbf{R}\cdot\mathbf{S}) + (\mathbf{R}\cdot\mathbf{T}) & (\mathbf{R}+\mathbf{S})\circ\mathbf{T} &= (\mathbf{R}\circ\mathbf{T}) + (\mathbf{S}\circ\mathbf{T}) \\ (\mathbf{R}\cdot\mathbf{S})\circ\mathbf{T} &= (\mathbf{R}\circ\mathbf{T})\cdot(\mathbf{S}\circ\mathbf{T}) & (\mathbf{R}\circ\mathbf{S})\circ\mathbf{T} &= \mathbf{R}\circ(\mathbf{S}\circ\mathbf{T}) \end{aligned}$$

differentiation:

$$\begin{aligned} (a\mathbf{R})' &= a\mathbf{R}' & (\mathbf{R}+\mathbf{S})' &= \mathbf{R}'+\mathbf{S}' \\ (\mathbf{R}\cdot\mathbf{S})' &= (\mathbf{R}'\cdot\mathbf{S}) + (\mathbf{R}\cdot\mathbf{S}') & (\mathbf{R}\circ\mathbf{S})' &= \mathbf{S}'\cdot(\mathbf{R}'\circ\mathbf{S}). \end{aligned}$$

---

[3] An algebraist would say that the algebraic structure $\{\text{Poly}, +, *, \circ, ', (c)_{c\in\mathbb{R}}, 0, 1\}$, where Poly is the set of polynomials over $\mathbb{R}$, is *isomorphic* to the corresponding algebraic structure of polynomial functions under the mapping $\mathbf{S} \mapsto f_{\mathbf{S}}$.

Power series

As with polynomials, if one regards $\sum a_n x^n$ merely as a formal expression, and not as a function, then one is dealing with *formal power series.* Many power series have radius of convergence 0, and one cannot learn much from a function that is only defined at 0. However, as formal expressions they contain information, namely one can read off the sequence $s(0), s(1), \ldots$, and this sequence may very well be of interest. Of course, if all that is wanted is the sequence $\big(s(n)\big)$ then the formal power series $\sum s(n)x^n$ would seem a rather clumsy way of capturing this information.

However, it turns out that identities between formal power series can capture valuable information about sequences, for example in Appendix C the formal identity

$$\mathbf{A}(x) \;=\; e^{\mathbf{H}(x)}$$

plays a pivotal role. By appropriately defining operations on formal power series, this identity holds even when the radius of convergence of $\mathbf{A}(x)$ is 0, and then one can differentiate and obtain another formal identity

$$\mathbf{A}'(x) \;=\; \mathbf{H}'(x) * \mathbf{A}(x).$$

Thus new information is obtained about the coefficients. This is the significance of formal power series for us.

A precise definition of a formal power series is given by focusing on the information contained in the sequence of coefficients. One might say that one writes $\big(s(n)\big)$ and thinks $\sum s(n)x^n$. Throughout this appendix all power series are to be thought of as formal power series, just as all polynomials are formal polynomials. Notation will be introduced to discuss the functions that power series define. Consequently, from this point on the adjective 'formal' is dropped, but feel free to insert it anywhere you see the words power series.

Definition A.12. A *power series* is a sequence $\big(s(n)\big)$ of real numbers. The $s(n)$ are the *coefficients* of the power series.

By tradition the power series $\mathbf{S} = \big(s(n)\big)$ is written as $\mathbf{S}(x) = \sum_n s(n)x^n$. Of course it will also be written as $\mathbf{S}(x) = s(0) + s(1)x + s(2)x^2 + \cdots$. At this point x is merely a symbol, as are $\sum$ and $+$. It is only when a real number is substituted for x that one is talking about infinite series.

Functions defined by power series.

Definition A.13. The *domain of convergence* $\mathrm{Dom}(\mathbf{S})$ of a power series $\mathbf{S}$ is the set of reals a such that the series $\sum_n s(n)a^n$ converges.

Given a power series $\mathbf{S}$ let $f_{\mathbf{S}}$ be the real valued function defined on $\mathrm{Dom}(\mathbf{S})$ by $f_{\mathbf{S}}(a) = \sum_n s(n)a^n$. $f_{\mathbf{S}}$ is called a *power series function.*

Definition A.14. Given a power series $\mathbf{S}$ the *radius of convergence* $\rho_{\mathbf{S}}$ is given by

$$\rho_{\mathbf{S}} \;=\; 1/\limsup_{n\to\infty} |s(n)|^{1/n}.$$

($\rho_{\mathbf{S}}$ is allowed to be ∞.)

The well known result of Cauchy-Hadamard is:

$$(-\rho_{\mathbf{S}}, \rho_{\mathbf{S}}) \;\subseteq\; \mathrm{Dom}(\mathbf{S}) \;\subseteq\; [-\rho_{\mathbf{S}}, \rho_{\mathbf{S}}].$$

The following is a modification of Proposition A.8 to give the corresponding facts for power series.

Proposition A.15. *Let $\mathbf{S}, \mathbf{T}$ be power series and let $a, c \in \mathbb{R}$. Then one has*

$$\begin{aligned}
(cf_{\mathbf{S}})(a) &= \sum_n cs(n)a^n \qquad \textit{for } a \in \mathrm{Dom}(\mathbf{S})\\
(f_{\mathbf{S}} + f_{\mathbf{T}})(a) &= \sum_n \big(s(n)+t(n)\big)a^n \qquad \textit{for } a \in \mathrm{Dom}(\mathbf{S}) \cap \mathrm{Dom}(\mathbf{T})\\
(f_{\mathbf{S}} \cdot f_{\mathbf{T}})(a) &= \sum_n \Big(\sum_k s(k)\cdot t(n-k)\Big)a^n \qquad \textit{for } |a| < \min(\rho_{\mathbf{S}}, \rho_{\mathbf{T}})\\
(f_{\mathbf{S}} \circ f_{\mathbf{T}})(a) &= \sum_n \Big(\sum_k s(k) \sum_{\ell_1,\dots,\ell_k \ge 0} t(\ell_1)\cdots t(\ell_k)\Big)a^n\\
&\qquad\qquad \textit{for } |a| < \rho_{\mathbf{T}} \textit{ and } |f_{\mathbf{T}}(a)| < \rho_{\mathbf{S}}\\
f_{\mathbf{S}}{}'(a) &= \sum_{n\ge 1} ns(n)a^{n-1} \qquad \textit{if } |a| < \rho_{\mathbf{S}}\\
s(n) &= f_{\mathbf{S}}^{(n)}(0)/n! \qquad \textit{if } \rho_{\mathbf{S}} > 0\\
\mathbf{S} = \mathbf{T} &\Leftrightarrow f_{\mathbf{S}} = f_{\mathbf{T}} \textit{ on a neighborhood of } 0 \quad (\textit{provided } \rho_{\mathbf{S}}, \rho_{\mathbf{T}} > 0).
\end{aligned}$$

PROOF. These are standard results about infinite series. In the item on addition, note that one can choose any $a \in \mathrm{Dom}(f_{\mathbf{S}}) \cap \mathrm{Dom}(f_{\mathbf{T}})$, even if a is an endpoint of the radius of convergence for one or both functions. However, for the multiplication item, the proof involves a consideration of the series of absolute values of the terms, and to be sure that this converges one is forced inside the interval of convergence. The same comment applies to the composition and differentiation of series. □

Operations on power series. Now Proposition A.15 is used to suggest appropriate definitions for operations on power series.

Definition A.16. Let $\mathbf{S}, \mathbf{T}$ be power series and let $c \in \mathbb{R}$.

coefficients: $[x^n]\mathbf{S}(x)$ is $s(n)$.
equality: $\mathbf{S} = \mathbf{T}$ holds iff $[x^n]\mathbf{S} = [x^n]\mathbf{T}$, for $n \ge 0$.

constant power series:

$$[x^n]\mathbf{c} \;=\; \begin{cases} c & \text{if } n = 0 \\ 0 & \text{otherwise.} \end{cases}$$

scalar multiplication: $[x^n]\big(c\mathbf{S}\big) \;=\; cs(n).$
addition: $[x^n]\big(\mathbf{S}+\mathbf{T}\big) \;=\; s(n)+t(n).$
multiplication: Use the Cauchy product:

$$[x^n]\big(\mathbf{S}*\mathbf{T}\big) \;=\; \sum_{k\le n} s(k)\cdot t(n-k).$$

composition: $\mathbf{S}\circ\mathbf{T}$ will only be defined when $t(0)=0$:[4]

$$[x^n]\big(\mathbf{S}\circ\mathbf{T}\big) \;=\; \sum_{k\le n} s(k) \sum_{\ell_1+\ldots+\ell_k=n} t(\ell_1)\cdots t(\ell_k).$$

exponentiation: This is an instance of composition, and is defined only for $\mathbf{T}$ with $t(0)=0$:

$$[x^n]e^{\mathbf{T}} \;=\; \begin{cases} 1 & \text{for } n=0 \\ \displaystyle\sum_{k\le n} [x^n]\frac{1}{k!}\mathbf{T}^k & \text{for } n\ge 1. \end{cases}$$

differentiation: $[x^n]\mathbf{S}' \;=\; (n+1)s(n+1).$

Connection between power series and functions. Operations on power series were defined to mimic what happens with power series functions, so one has the following translation of Proposition A.15.

Proposition A.17. *Let* $\mathbf{S},\mathbf{T}$ *be power series and let* $a,c\in\mathbb{R}$. *Then one has*

$$\begin{array}{lcll} f_{c\mathbf{S}}(a) &=& cf_{\mathbf{S}}(a) & \textit{for } |a|<\rho_{\mathbf{S}} \\ f_{\mathbf{S}+\mathbf{T}}(a) &=& f_{\mathbf{S}}(a)+f_{\mathbf{T}}(a) & \textit{for } |a|<\min(\rho_{\mathbf{S}},\rho_{\mathbf{T}}) \\ f_{\mathbf{S}*\mathbf{T}}(a) &=& f_{\mathbf{S}}(a)\cdot f_{\mathbf{T}}(a) & \textit{for } |a|<\min(\rho_{\mathbf{S}},\rho_{\mathbf{T}}) \\ f_{\mathbf{S}\circ\mathbf{T}}(a) &=& (f_{\mathbf{S}}\circ f_{\mathbf{T}})(a) & \textit{for } |a|<\rho_{\mathbf{T}},\ |f_{\mathbf{T}}(a)|<\rho_{\mathbf{S}} \\ {f_{\mathbf{S}}}'(a) &=& f_{\mathbf{S}'}(a) & \textit{for } |a|<\rho_{\mathbf{S}}. \end{array}$$

Truncation of power series. Since the function defined by a power series may easily have radius of convergence 0, one cannot proceed as with polynomials and simply say that all the identities of functions lift immediately. The key to lifting the results to power series is the observation that the fundamental operations are such that the definition of the nth coefficient in an application of an operation depends only on the first n coefficients of the given series, with the exception of differentiation which requires the $n+1$st coefficient.

[4]The restriction $t(0)=0$ guarantees that the definition of the nth coefficient of the composition $\mathbf{S}\circ\mathbf{T}$ involves only finitely many of the coefficients of $\mathbf{S}$ and $\mathbf{T}$. [Some condition is needed to guarantee that $\sum_k s(k) \sum_{\ell_1,\ldots,\ell_k\ge 0} t(\ell_1)\cdots t(\ell_k)$ converges!]

Definition A.18. The mth *truncation* $\mathbf{S}|_m$ of a power series $\mathbf{S}$ is the polynomial defined by

$$[x^n]\mathbf{S}|_m = \begin{cases} s(n) & \text{if } n \le m \\ 0 & \text{if } n > m. \end{cases}$$

Thus $\mathbf{S}|_m(x) = s(0) + \cdots + s(m)x^m$.

Lemma A.19. *Given power series* $\mathbf{S}, \mathbf{T}$, *a constant* $c \in \mathbb{R}$, *and* $n \ge 0$*:*

$$\begin{aligned}
[x^n](c\mathbf{S}) &= [x^n](c\mathbf{S}|_m) && \textit{for } m \ge n \\
[x^n](\mathbf{S}+\mathbf{T}) &= [x^n](\mathbf{S}|_m + \mathbf{T}|_m) && \textit{for } m \ge n \\
[x^n](\mathbf{S}\cdot\mathbf{T}) &= [x^n](\mathbf{S}|_m \cdot \mathbf{T}|_m) && \textit{for } m \ge n \\
[x^n](\mathbf{S}\circ\mathbf{T}) &= [x^n](\mathbf{S}|_m \circ \mathbf{T}|_m) && \textit{for } m \ge n \\
[x^n]\mathbf{S}' &= [x^n](\mathbf{S}|_m{}') && \textit{for } m \ge n+1.
\end{aligned}$$

Identities of power series.

Proposition A.20. *For* $a, b \in \mathbb{R}$ *and power series* $\mathbf{R}, \mathbf{S}, \mathbf{T}$ *the following hold:*

scalar multiplication:

$$\begin{aligned}
a\mathbf{0} &= \mathbf{0} & a\mathbf{1} &= \mathbf{a} \\
(-1)\mathbf{R} &= -\mathbf{R} & a(-\mathbf{R}) &= (-a)\mathbf{R} = -(a\mathbf{R}) \\
a(b\mathbf{R}) &= (ab)\mathbf{R} & (a+b)\mathbf{R} &= (a\mathbf{R}) + (b\mathbf{R}) \\
a(\mathbf{R}+\mathbf{S}) &= (a\mathbf{R}) + (a\mathbf{S}) & a(\mathbf{R}\cdot\mathbf{S}) &= (a\mathbf{R})\cdot\mathbf{S} = \mathbf{R}\cdot(a\mathbf{S})
\end{aligned}$$

addition, multiplication, and composition:

$$\begin{aligned}
\mathbf{R}+\mathbf{0} &= \mathbf{R} & \mathbf{R}\cdot\mathbf{1} &= \mathbf{R} \\
\mathbf{R}+\mathbf{S} &= \mathbf{S}+\mathbf{R} & \mathbf{R}\cdot\mathbf{S} &= \mathbf{S}\cdot\mathbf{R} \\
(\mathbf{R}+\mathbf{S})+\mathbf{T} &= \mathbf{R}+(\mathbf{S}+\mathbf{T}) & (\mathbf{R}\cdot\mathbf{S})\cdot\mathbf{T} &= \mathbf{R}\cdot(\mathbf{S}\cdot\mathbf{T}) \\
\mathbf{R}\cdot(\mathbf{S}+\mathbf{T}) &= (\mathbf{R}\cdot\mathbf{S}) + (\mathbf{R}\cdot\mathbf{T}) & (\mathbf{R}+\mathbf{S})\circ\mathbf{T} &= (\mathbf{R}\circ\mathbf{T}) + (\mathbf{S}\circ\mathbf{T}) \\
(\mathbf{R}\cdot\mathbf{S})\circ\mathbf{T} &= (\mathbf{R}\circ\mathbf{T})\cdot(\mathbf{S}\circ\mathbf{T}) & (\mathbf{R}\circ\mathbf{S})\circ\mathbf{T} &= \mathbf{R}\circ(\mathbf{S}\circ\mathbf{T})
\end{aligned}$$

differentiation:

$$\begin{aligned}
(a\mathbf{R})' &= a\mathbf{R}' & (\mathbf{R}+\mathbf{S})' &= \mathbf{R}'+\mathbf{S}' \\
(\mathbf{R}\cdot\mathbf{S})' &= (\mathbf{R}'\cdot\mathbf{S}) + (\mathbf{R}\cdot\mathbf{S}') & (\mathbf{R}\circ\mathbf{S})' &= \mathbf{S}'\cdot(\mathbf{R}'\circ\mathbf{S}) \\
\left(e^{\mathbf{R}}\right)' &= \mathbf{R}'\cdot e^{\mathbf{R}}.
\end{aligned}$$

PROOF. To verify each identity one only has to show that the corresponding coefficients of the two sides are equal. This is straightforward with scalar multiplication and addition.

The commutativity of multiplication is easy to see from the definition. For the associativity let us employ truncations,[5] and use the fact that multiplication of polynomials is associative:

$$\begin{aligned}[x^n]\big((\mathbf{R}*\mathbf{S})*\mathbf{T}\big) &= [x^n]\big((\mathbf{R}|_n*\mathbf{S}|_n)*\mathbf{T}|_n\big) = [x^n]\big(\mathbf{R}|_n*(\mathbf{S}|_n*\mathbf{T}|_n)\big)\\ &= [x^n]\big(\mathbf{R}*(\mathbf{S}*\mathbf{T})\big).\end{aligned}$$

This technique of reduction to polynomials can be used for all the identities under the heading of 'addition, multiplication, and composition'. This can also be used for differentiation. The differentiation of scalar multiplication and of addition are straightforward. For the product rule first note that

$$\begin{aligned}\mathbf{S}|_{n+1}{}' &= \mathbf{S}'|_n\\ [x^n]\mathbf{S}' &= [x^n]\mathbf{S}|_{n+1}{}'.\end{aligned}$$

Then

$$\begin{aligned}[x^n](\mathbf{R}*\mathbf{S})' &= [x^n](\mathbf{R}*\mathbf{S})|_{n+1}{}'\\ &= [x^n]\big(\mathbf{R}|_{n+1}*\mathbf{S}|_{n+1}\big)'\\ &= [x^n]\big(\mathbf{R}|_{n+1}{}'*\mathbf{S}|_{n+1}+\mathbf{R}|_{n+1}*\mathbf{S}|_{n+1}{}'\big)\\ &= [x^n]\big(\mathbf{R}'|_n*\mathbf{S}|_{n+1}+\mathbf{R}|_{n+1}*\mathbf{S}'|_n\big)\\ &= [x^n]\big(\mathbf{R}'|_n*\mathbf{S}|_n+\mathbf{R}|_n*\mathbf{S}'|_n\big)\\ &= [x^n]\big(\mathbf{R}'*\mathbf{S}+\mathbf{R}*\mathbf{S}'\big).\end{aligned}$$

For the chain rule:

$$\begin{aligned}[x^n](\mathbf{R}\circ\mathbf{S})' &= [x^n](\mathbf{R}\circ\mathbf{S})|_{n+1}{}'\\ &= [x^n]\big(\mathbf{R}|_{n+1}\circ\mathbf{S}|_{n+1}\big)'\\ &= [x^n]\big(\mathbf{S}|_{n+1}{}'*(\mathbf{R}|_{n+1}{}'\circ\mathbf{S}|_{n+1})\big)\\ &= [x^n]\big(\mathbf{S}'|_n*(\mathbf{R}'|_n\circ\mathbf{S}|_{n+1})\big)\\ &= [x^n]\big(\mathbf{S}'|_n*(\mathbf{R}'|_n\circ\mathbf{S}|_n)\big)\\ &= [x^n]\big(\mathbf{S}'*(\mathbf{R}'\circ\mathbf{S})\big).\end{aligned}$$

Thus $\big(e^{\mathbf{R}}\big)' = \mathbf{R}'*e^{\mathbf{R}}$. □

An identity for number systems

For any additive number system $\mathcal{A}$, even with $\rho = 0$, the alternate form of the partition identity, $\mathbf{A}(x) = e^{\mathbf{H}(x)}$, still holds as an identity between power series. This will be used later (see Theorems C.3 and C.6).

[5] In these arguments using truncations the reader may find that the individual steps do not follow immediately from what has been written before, but they are not difficult to verify.

Proposition A.21. *For $\mathcal{A}$ an additive number system define* $\mathbf{H}(x) = \sum h(n)x^n$, *where* $h(0) = 0$ *and, for* $n \geq 1$, $h(n) = \sum_{ij=n} p(i)/j$. *Then*

$$\mathbf{A}(x) \;=\; e^{\mathbf{H}(x)}.$$

PROOF. Clearly both sides have the same constant term 1. So let $m \geq 1$ be given, and let $\mathcal{B} = (\mathsf{B}, \mathsf{Q}, +, 0, \|\ \|)$ be the subsystem of $\mathcal{A}$ generated by the set Q of indecomposables of size at most m. Let $\mathbf{H}_{\mathsf{B}}(x) = \sum h_{\mathsf{B}}(n)x^n$, where $h_{\mathsf{B}}(0) = 0$ and, for $n \geq 1$, $h_{\mathsf{B}}(n) = \sum_{ij=n} q(i)/j$, $q(n)$ being the counting function for Q.

Let $\mathbf{S}(x)$ be the power series $e^{\mathbf{H}_{\mathsf{B}}(x)}$. Since $\mathcal{B}$ is finitely generated the radius of convergence of $\mathbf{B}(x)$ is 1, so from §2.5 one has, for $x \in [0, 1)$,

$$f_{\mathbf{B}}(x) \;=\; f_{\mathbf{S}}(x).$$

This gives the equality of the power series $\mathbf{B}(x)$ and $\mathbf{S}(x)$, that is,

$$\mathbf{B}(x) \;=\; e^{\mathbf{H}_{\mathsf{B}}(x)}.$$

Now $a(1), \ldots, a(m)$ are completely determined by Q, and hence must agree with the corresponding $b(1), \ldots, b(m)$, that is, $b(j) = a(j)$, for $j \leq m$. Since $p(j) = q(j)$, for $1 \leq j \leq m$, it follows that $h_{\mathsf{B}}(j) = h(j)$, for $1 \leq j \leq m$. From this, for the given m, one has

$$\begin{aligned} a(m) \;=\; b(m) \;=\; \big[x^m\big]e^{\mathbf{H}_{\mathsf{B}}(x)} \;&=\; \big[x^m\big]e^{\mathbf{H}_{\mathsf{B}}|_m(x)} \\ &=\; \big[x^m\big]e^{\mathbf{H}|_m(x)} \;=\; \big[x^m\big]e^{\mathbf{H}(x)}. \end{aligned}$$

This proves the proposition. □

Division of power series

Proposition A.22. *If* $\mathbf{R}(x), \mathbf{S}(x)$ *are power series with* $s(0) \neq 0$ *then there is a unique power series* $\mathbf{T}(x)$ *such that* $\mathbf{R}(x) = \mathbf{S}(x) * \mathbf{T}(x)$. *One writes* $\mathbf{R}(x)/\mathbf{S}(x)$ *for* $\mathbf{T}(x)$, *and calls it the* quotient of $\mathbf{R}(x)$ by $\mathbf{S}(x)$.

PROOF. Given $\mathbf{R}(x) = \sum_n r(n)x^n$ and $\mathbf{S}(x) = \sum_n s(n)x^n$ with $s(0) \neq 0$, the goal is to show that there is a unique $\mathbf{T}(x) = \sum_n t(n)x^n$ such that $\mathbf{R}(x) = \mathbf{S}(x) * \mathbf{T}(x)$. This is equivalent to proving that there is a unique solution for $t(0), t(1), \ldots$ to the system of equations

$$r(n) = \sum_{k \leq n} s(k) \cdot t(n-k),$$

or, in more detail,

$$\begin{aligned} r(0) \;&=\; s(0)t(0) \\ r(1) \;&=\; s(0)t(1) + s(1) \cdot t(0) \\ r(2) \;&=\; s(0)t(2) + s(1) \cdot t(1) + s(2) \cdot t(0) \\ &\textit{etc.} \end{aligned}$$

Since $s(0) \neq 0$ it is clear that indeed one can recursively solve this system for the $t(n)$, and there is only one solution. □

The composition inverse of a power series

Inversion (under composition) can be carried out over a wide range of power series.

Proposition A.23.

(a) *If* $\mathbf{S}(x)$ *is such that* $s(0) = 0$ *and* $s(1) \neq 0$ *then there is a unique* $\mathbf{R}(x)$ *such that* $(\mathbf{R} \circ \mathbf{S})(x) = x$. *For this* $\mathbf{R}(x)$ *one has* $r(0) = 0$ *and* $r(1) = 1/s(1)$. $\mathbf{R}(x)$ *is the* left inverse *of* $\mathbf{S}(x)$ *under composition.*

(b) *Given* $\mathbf{S}(x)$ *with* $s(0) = 0$ *and* $s(1) \neq 0$ *there is a unique power series* $\mathbf{T}(x) = \sum_n t(n)x^n$ *with* $t(0) = 0$ *such that* $(\mathbf{S} \circ \mathbf{T})(x) = x$. $\mathbf{T}(x)$ *is the* right inverse *of* $\mathbf{S}(x)$.

PROOF. (a) This is equivalent to showing that there is a unique solution for the coefficients $r(0), r(1), \ldots$ in the infinite system of equations

$$\big[x^n\big](\mathbf{R} \circ \mathbf{S}) \;=\; \begin{cases} 1 & \text{if } n = 1 \\ 0 & \text{otherwise,} \end{cases}$$

or, in more detail,

$$\begin{aligned}
&r(0) &&= 0\\
&r(1) \cdot s(1) &&= 1\\
&r(2) \cdot s(1)^2 + r(1) \cdot s(2) &&= 0\\
& &&\;\vdots\\
&r(n) \cdot s(1)^n + q_n\big(r(1), \ldots, r(n-1), s(1), \ldots, s(n)\big) &&= 0\\
& &&\;\vdots
\end{aligned}$$

for suitable polynomials q_n. Since $s(1) \neq 0$, clearly this system is uniquely solvable for the $r(n)$, starting with $r(0) = 0$, $r(1) = 1/s(1)$.

(b) Since $t(0) = 0$ this is equivalent to showing that there is a unique solution $t(1), t(2), \ldots$ to the infinite system of equations

$$\big[x^n\big](\mathbf{S} \circ \mathbf{T}) \;=\; \begin{cases} 1 & \text{if } n = 1 \\ 0 & \text{otherwise,} \end{cases}$$

or, in more detail,

$$\begin{aligned}
&s(1) \cdot t(1) &&= 1\\
&s(1) \cdot t(2) + s(2) \cdot t(1)^2 &&= 0\\
&s(1) \cdot t(3) + 2s(2) \cdot t(1) \cdot t(2) + s(3) \cdot t(1)^3 &&= 0\\
& &&\;\vdots\\
&s(1) \cdot t(n) + q_n\big(s(1), \ldots, s(n), t(1), \ldots, t(n-1)\big) &&= 0\\
& &&\;\vdots
\end{aligned}$$

for suitable polynomials q_n. Since $s(1) \neq 0$, clearly this system is uniquely solvable for the $t(n)$, starting with $t(1) = 1/s(1)$. □

A group under composition

Having established a left inverse $\mathbf{R}(x)$ and a right inverse $\mathbf{T}(x)$ of any power series $\mathbf{S}(x)$ satisfying $s(0) = 0$ and $s(1) \neq 0$, now one would like to show that these two inverses are the same. The simplest proof is to show that the collection of power series $\mathbf{S}(x)$ that satisfy $s(0) = 0$ and $s(1) \neq 0$ forms a group under composition.

Definition A.24. Let

$$\mathcal{G} = \{\mathbf{S}(x) : s(0) = 0, \ s(1) \neq 0\}.$$

Recall that a group is a collection of elements with a binary operation that is associative, there is a left and right identity element with respect to this operation, and every element has a two-sided (both left and right) inverse.

Proposition A.25. *$\mathcal{G}$ is a group under composition.*

PROOF. It is easy to see that $\mathcal{G}$ is closed under the binary operation of composition. Composition is an associative operation on $\mathcal{G}$, by Proposition A.20, and x is a two-sided identity, that is, $x \circ \mathbf{S} = \mathbf{S}$ and $\mathbf{S} \circ x = \mathbf{S}$. For $\mathbf{S} \in \mathcal{G}$, the left inverse of $\mathbf{S}$ in $\mathcal{G}$ exists and is unique, by Proposition A.23 (a), so let $\mathbf{S}^{-1}$ be this left inverse. Now a simple algebraic argument shows that $\mathbf{S}^{-1}$ is also a right inverse of $\mathbf{S}$, namely:

$$\begin{aligned}
\mathbf{S} \circ \mathbf{S}^{-1} &= x \circ \left(\mathbf{S} \circ \mathbf{S}^{-1}\right) \\
&= \left(\left(\mathbf{S}^{-1}\right)^{-1} \circ \mathbf{S}^{-1}\right) \circ \left(\mathbf{S} \circ \mathbf{S}^{-1}\right) \\
&= \left(\mathbf{S}^{-1}\right)^{-1} \circ \mathbf{S}^{-1} \circ \mathbf{S} \circ \mathbf{S}^{-1} \\
&= \left(\mathbf{S}^{-1}\right)^{-1} \circ x \circ \mathbf{S}^{-1} \\
&= \left(\mathbf{S}^{-1}\right)^{-1} \circ \mathbf{S}^{-1} \\
&= x.
\end{aligned}$$

□

Since the left inverse of a power series in $\mathcal{G}$ is a right inverse, and since the right inverse is also unique, one simply refers to *the* inverse of $\mathbf{S} \in \mathcal{G}$.

Now the goal is to prove[6] that if $\mathbf{S} \in \mathcal{G}$ has a positive radius of convergence then so does $\mathbf{S}^{-1}$.

A special case. First the special case where $s(1) = 1$ is considered:

$$\mathbf{S}(x) = x + s(2)x^2 + s(3)x^3 + \cdots$$

Lemma A.26. *Suppose that $\mathbf{S}(x) = x + \sum_{n \geq 2} s(n)x^n$ and $\mathbf{R}(x) = \sum_n r(n)x^n$ is the inverse of $\mathbf{S}$ in $\mathcal{G}$. If $\rho_{\mathbf{S}} > 0$ then $\rho_{\mathbf{R}} > 0$.*

[6]This proof follows Knopp [**34**], pp. 184–188, which he says is due to Cauchy. Two other proofs can be found in Hille [**27**], an intermediate level proof on pp. 86–90, where one uses successive approximations; and a more advanced level proof on p. 265, where one gets an explicit form for the right inverse using an integral in the complex plane.

PROOF. $\mathbf{R}(x)$, as a left inverse, is specified by the system of equations

$$\begin{array}{lcl} r(0) & = & 0 \\ s(1)\cdot r(1) & = & 1 \\ s(1)^2\cdot r(2)+s(2)\cdot r(1) & = & 0 \\ s(1)^3\cdot r(3)+2s(1)\cdot s(2)\cdot r(2)+s(3)\cdot r(1) & = & 0 \\ & \vdots & \\ s(1)^n\cdot r(n)+q_n\big(s(1),\dots,s(n),r(1),\dots,r(n-1)\big) & = & 0 \\ & \vdots & \end{array}$$

Let $\lambda = 1 + 1/\rho_{\mathbf{S}}$. Then, for some K,

$$|s(n)| \ \le\ K\lambda^n, \qquad \text{for } n \ge 0.$$

Now define the power series $\mathbf{U}(x)$ by

$$\mathbf{U}(x) \ =\ x - \sum_{n\ge 2}(K\lambda^n)x^n.$$

$\mathbf{U}(x)$ is clearly in $\mathcal{G}$, so let $\mathbf{V}(x) = \sum_n v(n)x^n$ be its inverse. As a left inverse one has

$$\begin{array}{lcl} v(0) & = & 0 \\ v(1) & = & 1 \\ v(2)-(K\lambda^2)v(1) & = & 0 \\ v(3)-2(K\lambda^2)v(2)-(K\lambda^3)v(1)^3 & = & 0 \\ & \vdots & \end{array}$$

so the $v(n)$ are positive, and indeed, $v(n) \ge |r(n)|$. Thus

$$\rho_{\mathbf{V}} > 0 \ \Rightarrow\ \rho_{\mathbf{R}} > 0. \tag{A.26.1}$$

There is a compact description of the function $f_{\mathbf{U}}(x)$ since there is a geometric series in the definition:

$$f_{\mathbf{U}}(x) \ =\ x - K\lambda^2x^2\Big(\frac{1}{1-\lambda x}\Big), \qquad \text{for } |x| < 1/\lambda.$$

A routine calculation shows that $f_{\mathbf{U}}{}'(0) = 1$, so $f_{\mathbf{U}}(x)$ is a one-one function in some neighborhood of 0. There is a simple formula for the inverse function, namely

$$\begin{aligned} y = f_{\mathbf{U}}(x) \ &\Rightarrow\ y - \lambda yx \ =\ x - \lambda(\lambda K+1)x^2 \\ &\Rightarrow\ \lambda(\lambda K+1)x^2 \ -\ (1+\lambda y)x\ +\ y\ =\ 0 \\ &\Rightarrow\ x \ =\ \frac{(1+\lambda y)\pm\sqrt{(1+\lambda y)^2-4\lambda(\lambda K+1)y}}{2\lambda(\lambda K+1)}. \end{aligned}$$

So if

$$g(x) \ =\ \frac{(1+\lambda x)-\sqrt{(1+\lambda x)^2-4\lambda(\lambda K+1)x}}{2\lambda(\lambda K+1)}$$

then $g(0) = 0$, g is defined on a neighborhood of 0 since the discriminant is a continuous function of x and takes on the value 1 at $x = 0$, and

$$f_{\mathbf{U}}\big(g(x)\big) \;=\; x$$

on some neighborhood of 0.

Let

$$g_0(x) = \sqrt{(1+\lambda x)^2 - 4\lambda(\lambda K + 1)x}.$$

Then $g_0(x)$ can be written in the form $\sqrt{1-q(x)}$, where $q(x)$ is a polynomial in x with $f_q(0) = 0$. Since $\sqrt{1-x}$ has a power series expansion on a neighborhood of 0, by the result on composition of functions defined by power series in Proposition A.17, $g_0(x) = \big(\sqrt{1-x}\big) \circ q(x)$ also has a power series expansion on a neighborhood of 0. Thus $g(x)$ has a power series expansion on some neighborhood of 0.

Let $\mathbf{W}(x) = \sum_n w(n)x^n$ be the power series expansion of $g(x)$. Then $\rho_{\mathbf{W}} > 0$, $w(0) = 0$, and $g(x) = f_{\mathbf{W}}(x)$ on some neighborhood of 0. From $f_{\mathbf{U}}\big(g(x)\big) = x$ on some neighborhood of 0 follows $f_{\mathbf{U}}\big(f_{\mathbf{W}}(x)\big) = x$ on a neighborhood of 0, and thus $\big(f_{\mathbf{U}} \circ f_{\mathbf{W}}\big)(x) = x$ on a neighborhood of 0. Now $\big(f_{\mathbf{U}} \circ f_{\mathbf{W}}\big)(x) = f_{\mathbf{U}\circ\mathbf{W}}(x)$ in that neighborhood, so $f_{\mathbf{U}\circ\mathbf{W}}(x) = x$ in a neighborhood of 0. This implies that $(\mathbf{U}\circ\mathbf{W})(x) = x$, so $\mathbf{W}(x)$ is the inverse of $\mathbf{U}(x)$.

But $\mathbf{V}(x)$ was defined to be that inverse. Consequently $\mathbf{V}(x) = \mathbf{W}(x)$, and thus $\rho_{\mathbf{V}} > 0$. From (A.26.1) it follows that $\rho_{\mathbf{R}} > 0$. □

Removing the condition $s(1) = 1$**.** Now the goal is to lift these results to any power series in $\mathcal{G}$.

Proposition A.27. *Suppose* $\mathbf{S} \in \mathcal{G}$*, that is,* $s(0) = 0$ *and* $s(1) \neq 0$*. If* $\rho_{\mathbf{S}} > 0$*, that is, if* $\mathbf{S}$ *has positive radius of convergence, then the composition inverse of* $\mathbf{S}$ *also has a positive radius of convergence.*

PROOF. Let $\mathbf{U}(x) = \sum_n u(n)x^n$ be the power series defined by

$$\mathbf{U}(x) \;=\; \mathbf{S}(x) \circ \Big(\frac{1}{s(1)}x\Big) \;=\; \sum_n \frac{s(n)}{s(1)^n}x^n.$$

Then $\rho_{\mathbf{U}} = s(1)\rho_{\mathbf{S}} > 0$ and $u(0) = 0$, $u(1) = 1$. Thus, by Lemma A.26, it follows that the inverse $\mathbf{V}(x) = \sum_n v(n)x^n$ of $\mathbf{U}(x)$ has a positive radius of convergence. Of course $v(0) = 0$ and $v(1) = 1$.

Define $\mathbf{T}(x)$ to be the power series $\big(1/s(1)\big)\mathbf{V}(x)$, which can be written as $\big(x/s(1)\big) \circ \mathbf{V}(x)$. Then $\rho_{\mathbf{T}} = \rho_{\mathbf{V}} > 0$, and

$$\begin{aligned}\mathbf{S}(x) \circ \mathbf{T}(x) \;&=\; \mathbf{S}(x) \circ \Big(\frac{1}{s(1)}x\Big) \circ \mathbf{V}(x)\\ &=\; \mathbf{U}(x) \circ \mathbf{V}(x)\\ &=\; x.\end{aligned}$$

This means that $\mathbf{T}$ is the inverse of $\mathbf{S}$, and it has a positive radius of convergence. □

This result will be used to prove Theorem C.3 of Bell, which implies that one is fully justified in avoiding additive number systems with radius of convergence zero when searching for systems with all partition sets having asymptotic density.

APPENDIX B

Refined Counting

The additive case

This topic concerns the number of elements in A of norm n that can be expressed as a sum of m indecomposables.

Definition B.1. The *refined (local) counting function* $a(n,m)$ of A is defined on $\mathbb{N} \times \mathbb{N}$ by

$$a(n,m) \;=\; \big|\{\mathsf{a} \in m\mathsf{P} : \|\mathsf{a}\| = n\}\big|.$$

Corresponding to the function $a(n,m)$ there is the *two variable generating series*

$$\mathbf{A}(x,y) \;=\; \sum_{m,n \geq 0} a(n,m) x^n y^m.$$

This can also be expressed by

$$\mathbf{A}(x,y) \;=\; \sum_{m \geq 0} \sum_{\mathsf{a} \in m\mathsf{P}} x^{\|\mathsf{a}\|} y^m.$$

The *two variable fundamental identity* of $\mathcal{A}$ is

$$\text{(B.1.1)} \qquad \sum_{m,n \geq 0} a(n,m) x^n y^m \;=\; \prod_{n \geq 1} (1 - x^n y)^{-p(n)}.$$

The next section shows that this identity describes a method to compute the values of $a(n,m)$ from $p(n)$, and the following section gives a proof that it holds as an identity between two functions when $0 \leq x < \rho$ and $0 \leq y < 1$.

Meaning of the two variable fundamental identity. The meaning of the two variable fundamental identity (B.1.1), initially, is just that it is a shorthand device to remember how to compute $a(n,m)$ from $p(n)$. To this end think of each factor

$$(1 - x^n y)^{-p(n)}$$

in the product part of the fundamental identity as a power series to the power $p(n)$, namely

$$(1 - x^n y)^{-p(n)} \;=\; \big(1 + x^n y + x^{2n} y^2 + \cdots\big)^{p(n)}.$$

What follows explains why $a(n,m)$ is the coefficient of $x^n y^m$ in the expansion of

$$\left(1+x+x^2+\cdots\right)^{p(1)} \cdots \left(1+x^k+x^{2k}y^2+\cdots\right)^{p(k)},$$

for *any* choice of $k \geq n$. Actually, it is more transparent to say that $a(n,m)$ is the coefficient of x^n in the expansion of

$$\prod_{\|\mathsf{p}\| \leq k} \left(1+x^{\|\mathsf{p}\|}y+x^{2\|\mathsf{p}\|}y^2+\cdots\right),$$

for any choice of $k \geq n$.

To see that this is true proceed as follows. Let $\mathsf{p}_0, \mathsf{p}_1, \ldots$ be a listing of the indecomposable elements of $\mathcal{A}$. As before, one has a natural one-to-one correspondence between the elements of A and the sequences (m_i) of nonnegative integers with only finitely many of the m_i nonzero, namely (m_i) corresponds to the element a such that m_i is the coefficient of p_i in the decomposition of a into indecomposables. Thus the sequences $\left(x^{m_i\|\mathsf{p}_i\|}y^{m_i}\right)$ of monomials, where almost all monomials are 1, are also naturally in a one-to-one correspondence with A.

Now $a(n,m)$ is the number of sequences (m_i) of nonnegative integers such that $n = \sum m_i\|\mathsf{p}_i\|$ and $m = \sum m_i$. This is because each element of $m\mathsf{P}$ with norm n can be expressed as a sum $\sum m_i\mathsf{p}_i$ with $\sum m_i\|\mathsf{p}_i\| = n$ and $\sum m_i = m$, and distinct sequences (m_i) with $\sum m_i = m$ give rise to distinct elements $\sum m_i\mathsf{p}_i$ of $m\mathsf{P}$. Note that

$$n = \sum m_i\|\mathsf{p}_i\| \quad \text{and} \quad m = \sum m_i \quad \text{iff} \quad x^n y^m = \prod x^{m_i\|\mathsf{p}_i\|}y^{m_i}.$$

Thus $a(n,m)$ is the number of sequences (m_i) of nonnegative integers such that $x^n y^m = \prod x^{m_i\|\mathsf{p}_i\|}y^{m_i}$. Now consider the product

$$\text{(B.1.2)} \quad \left(1+x^{\|\mathsf{p}_0\|}y+x^{2\|\mathsf{p}_0\|}y^2+\cdots\right) \cdots \left(1+x^{\|\mathsf{p}_k\|}y+x^{2\|\mathsf{p}_k\|}y^2+\cdots\right),$$

where k is sufficiently large to ensure that all indecomposables p with norm at most n appear in the list $\mathsf{p}_0, \ldots, \mathsf{p}_k$. Then each sequence (m_i) with $x^n y^m = \prod x^{m_i\|\mathsf{p}_i\|}y^{m_i}$ corresponds to a choice of one monomial from each of the sums $1+x^{\|\mathsf{p}_i\|}y+x^{2\|\mathsf{p}_i\|}y^2+\cdots$ above, and thus $a(n,m)$ is the coefficient of $x^n y^m$ in the expansion of (B.1.2).

Validity of the two variable fundamental identity. Let $\mathcal{A}$ be an additive number system with radius of convergence ρ.

Proposition B.2. *The two variable fundamental identity*

$$\sum_{m,n\geq 0} a(n,m)x^n y^m \;=\; \prod_{n\geq 1}(1-x^n y)^{-p(n)}$$

of $\mathcal{A}$ *holds for* $0 \leq x, y < 1$, *with convergence if* $0 \leq x < \rho$.

PROOF. This follows from the observation that, for $x, y \geq 0$, and, for any $\ell \geq 1$,

$$\sum_{0 \leq m,n \leq \ell} a(n,m)x^n y^m \;\leq\; \prod_{1 \leq j \leq \ell} \Big(\sum_{k \geq 0} x^{kj} y^k\Big)^{p(j)} \;\leq\; \sum_{m,n \geq 0} a(n,m)x^n y^m.$$

Thus, for $0 \leq x, y < 1$,

$$\sum_{0 \leq m,n \leq \ell} a(n,m)x^n y^m \;\leq\; \prod_{1 \leq j \leq \ell} (1 - x^j y)^{-p(j)} \;\leq\; \sum_{m,n \geq 0} a(n,m)x^n y^m,$$

so,

$$\sum_{m,n \geq 0} a(n,m)x^n y^m \;=\; \prod_{n \geq 1} (1 - x^n y)^{-p(n)}.$$

If $0 \leq x < \rho$ and $0 \leq y < 1$ then

$$\begin{aligned}
\sum_{m,n} a(n,m)x^n y^m \;&\leq\; \sum_{m,n} a(n,m)x^n \\
&=\; \sum_{n} \Big(\sum_{m} a(n,m)\Big) x^n \\
&=\; \sum_{n} a(n)x^n \\
&=\; \mathbf{A}(x) \;<\; \infty.
\end{aligned}$$

□

Alternate version of the two variable fundamental identity. Just as with the fundamental identity, there is a useful alternate form involving exponentiation.

Proposition B.3. *Let $\mathcal{A}$ be an additive number system with radius of convergence $\rho > 0$. Then the identity*

$$\mathbf{A}(x,y) \;=\; \exp\Big(\sum_{m \geq 1} \mathbf{P}(x^m) y^m / m\Big)$$

holds when $0 \leq x < \rho$ and $0 \leq y < 1$.

PROOF. For $0 \leq x < \rho$ and $0 \leq y < 1$,

$$\begin{aligned}
(1 - x^n y)^{-p(n)} \;&=\; \exp\Big(\log\big((1 - x^n y)^{-p(n)}\big)\Big) \\
&=\; \exp\big(-p(n)\log(1 - x^n y)\big) \\
&=\; \exp\Big(p(n) \sum_{m \geq 1} x^{mn} y^m / m\Big),
\end{aligned}$$

thus,

$$\prod_{1 \leq n \leq M} (1 - x^n y)^{-p(n)} \;=\; \prod_{1 \leq n \leq M} \exp\Big(p(n) \sum_{m \geq 1} x^{mn} y^m / m\Big).$$

So, from Proposition B.2,

$$\begin{aligned}\mathbf{A}(x,y) &= \prod_{n\ge 1}(1-x^n y)^{-p(n)}\\ &= \lim_{M\to\infty}\prod_{1\le n\le M}\exp\Big(\sum_{m\ge 1}p(n)x^{mn}y^m/m\Big)\\ &= \lim_{M\to\infty}\exp\Big(\sum_{1\le n\le M}\sum_{m\ge 1}p(n)x^{mn}y^m/m\Big)\\ &= \exp\Big(\sum_{n\ge 1}\sum_{m\ge 1}p(n)x^{mn}y^m/m\Big)\\ &= \exp\Big(\sum_{m\ge 1}\sum_{n\ge 1}p(n)x^{mn}y^m/m\Big)\\ &= \exp\Big(\sum_{m\ge 1}\mathbf{P}(x^m)y^m/m\Big).\end{aligned}$$

□

The generating series of $m\mathsf{P}$. The generating series of $m\mathsf{P}$ can be expressed as a polynomial in $\mathbf{P}(x),\dots,\mathbf{P}(x^m)$ with nonnegative rational coefficients. The case $m=0$ is trivial as $[0\mathbf{P}](x)=1$, so a proof is only needed when $m>0$.

Proposition B.4. *Given a positive integer m, for any $x\ge 0$,*

$$\text{(B.4.1)}\qquad [m\mathbf{P}](x) = \sum_{j=1}^{m}\frac{1}{j!}\sum_{\substack{m_1+\cdots+m_j=m\\ m_i\ge 1}}\frac{1}{m_1\cdots m_j}\mathbf{P}(x^{m_1})\cdots\mathbf{P}(x^{m_j}),$$

and the two sides converge iff $\mathbf{P}(x)$ converges.

PROOF. Since

$$[m\mathbf{P}](x) = \sum_{n\ge 0}a(n,m)x^n,$$

it follows that

$$\text{(B.4.2)}\qquad \mathbf{A}(x,y) = \sum_{m\ge 0}[m\mathbf{P}](x)y^m,$$

so, from Proposition B.3,

$$\begin{aligned}\sum_{m\ge 0}[m\mathbf{P}](x)y^m &= \exp\Big(\sum_{m\ge 1}\mathbf{P}(x^m)y^m/m\Big)\\ \text{(B.4.3)}\qquad &= \sum_{j\ge 0}\frac{1}{j!}\Big(\sum_{m\ge 1}\mathbf{P}(x^m)y^m/m\Big)^j.\end{aligned}$$

This means that one obtains $[m\mathbf{P}](x)$ by collecting the terms from (B.4.3) that contribute to the coefficient of y^m, and this gives equation (B.4.1), at least for $0 \le x < \rho$.

To see that (B.4.1) holds when $x \ge 0$, with convergence precisely when $\mathbf{P}(x)$ converges, note that $\mathbf{P}(x), \ldots, \mathbf{P}(x^m)$ are power series with nonnegative integer coefficients, so, when $x \ge 0$, if they converge at x then they converge absolutely. Furthermore, if $\mathbf{P}(x)$ converges at $x \ge 0$ then so do each of $\mathbf{P}(x^2), \ldots, \mathbf{P}(x^m)$. After noting that one of the summands in the right side of (B.4.1) is

$$\frac{1}{m!}\mathbf{P}(x)^m$$

(this comes from $j = m$), it follows that, for $x \ge 0$, the right side of (B.4.1) converges precisely when $\mathbf{P}(x)$ converges. By Lemma 1.19, there is a power series $\mathbf{R}(x)$ such that the right side of (B.4.1) equals $\mathbf{R}(x)$ when $x \ge 0$, so

$$[m\mathbf{P}](x) \;=\; \mathbf{R}(x)$$

on $[0, \rho)$. Since $[m\mathbf{P}](x)$ and $\mathbf{R}(x)$ are both power series, they must be the same power series. Thus (B.4.1) holds when $x \ge 0$, with the two sides converging precisely when $\mathbf{P}(x)$ converges. $\square$

Here are two examples to illustrate (B.4.1):

$$\begin{aligned} [2\mathbf{P}](x) &= \tfrac{1}{2}\mathbf{P}(x)^2 + \tfrac{1}{2}\mathbf{P}(x^2) \\ [3\mathbf{P}](x) &= \tfrac{1}{6}\mathbf{P}(x)^3 + \tfrac{1}{2}\mathbf{P}(x)\cdot\mathbf{P}(x^2) + \tfrac{1}{3}\mathbf{P}(x^3). \end{aligned}$$

The reader may have noticed that the sums of the coefficients in each of the two examples is 1. This is no accident. For from (B.4.1) there is a polynomial $\mathbf{r}(y_1, \ldots, y_m)$ with rational coefficients such that

$$[m\mathbf{P}](x) \;=\; \mathbf{r}\big(\mathbf{P}(x), \ldots, \mathbf{P}(x^m)\big).$$

The sum of the coefficients is $\mathbf{r}(1, \ldots, 1)$. From (B.4.3)

$$[m\mathbf{P}](x) \;=\; [y^m]\sum_{j\ge 0}\frac{1}{j!}\Big(\sum_{k\ge 1}\mathbf{P}(x^k)y^k/k\Big)^j.$$

Thus

$$\begin{aligned} \mathbf{r}(1, \ldots, 1) &= [y^m]\sum_{j\ge 0}\frac{1}{j!}\Big(\sum_{k\ge 1} y^k/k\Big)^j \\ &= [y^m]\exp\Big(\sum_{k\ge 1} y^k/k\Big) \\ &= [y^m]e^{-\log(1-y)} \\ &= [y^m]\frac{1}{1-y} \;=\; 1, \end{aligned}$$

so $\mathbf{r}(1, \ldots, 1) = 1$.

The multiplicative case

This section concerns the number of elements in A of norm n that can be expressed as a product of m indecomposables.

Definition B.5. The *refined (local) counting function* $a(n,m)$ of A, defined on $\mathbb{P} \times \mathbb{N}$, is given by

$$a(n,m) \;=\; \big|\{\mathsf{a} \in \mathsf{P}^m : \|\mathsf{a}\| = n\}\big|.$$

Corresponding to the function $a(n,m)$ there is the *two variable generating series*

$$\mathbf{A}(x,y) \;=\; \sum_{\substack{m \geq 0 \\ n \geq 1}} a(n,m) n^{-x} y^m.$$

This can also be expressed by

$$\mathbf{A}(x,y) \;=\; \sum_{m \geq 0} \sum_{\mathsf{a} \in \mathsf{P}^m} \|\mathsf{a}\|^{-x} y^m.$$

The *two variable fundamental identity* of $\mathcal{A}$ is

$$\text{(B.5.1)} \qquad \sum_{\substack{m \geq 0 \\ n \geq 1}} a(n,m) n^{-x} y^m \;=\; \prod_{n \geq 2} (1 - n^{-x} y)^{-p(n)}.$$

The next section shows that this identity describes a method to compute the values of $a(n,m)$ from $p(n)$, and the following section gives a proof that it holds as an identity between two functions when $x > \alpha$ and $0 \leq y < 1$.

Meaning of the two variable fundamental identity. The meaning of the two variable fundamental identity (B.5.1), initially, is just that it is a shorthand device to remember how to compute $a(n,m)$ from $p(n)$. (This is a 'formal' aspect of the identity—in the next section convergence is examined.) To this end think of each factor

$$(1 - n^{-x} y)^{-p(n)}$$

in the product part of the fundamental identity as a geometric series to the power $p(n)$, namely,

$$(1 - n^{-x} y)^{-p(n)} \;=\; \big(1 + n^{-x} y + n^{-2x} y^2 + \cdots\big)^{p(n)}.$$

What follows explains why $a(n,m)$ is the coefficient of $n^{-x} y^m$ in the expansion of

$$\big(1 + 2^{-x} y + 2^{-2x} y^2 \cdots\big)^{p(2)} \;\cdots\; \big(1 + k^{-x} y + k^{-2x} y^2 + \cdots\big)^{p(k)},$$

for *any* choice of $k \geq n$. Actually it is more transparent to say that $a(n,m)$ is the coefficient of $n^{-x}y^m$ in the expansion of

$$\prod_{\|\mathsf{p}\| \leq k} \left(1 + \|\mathsf{p}\|^{-x}y + \|\mathsf{p}\|^{-2x}y^2 + \cdots\right),$$

for any choice of $k \geq n$.

To see that this is true proceed as follows. Let $\mathsf{p}_0, \mathsf{p}_1, \ldots$ be a listing of the indecomposable elements of $\mathcal{A}$. As before, one has a natural one-to-one correspondence between the elements of A and the sequences (m_i) of nonnegative integers with only finitely many of the m_i nonzero, namely, (m_i) corresponds to the element a such that m_i is the exponent of p_i in the factorization of a into indecomposables. Thus the sequences $\left(\|\mathsf{p}_i\|^{-m_i x}y^{m_i}\right)$ of monomials, where almost all monomials are 1, are also naturally in a one-to-one correspondence with A.

Now $a(n,m)$ is the number of sequences (m_i) of nonnegative integers such that $n = \prod \|\mathsf{p}_i\|^{m_i}$ and $m = \sum m_i$. This is because each element of P^m with norm n can be expressed as a product $\prod \mathsf{p}_i^{m_i}$ with $\prod \|\mathsf{p}_i\|^{m_i} = n$ and $\sum m_i = m$, and distinct sequences (m_i) with $\sum m_i = m$ give rise to distinct elements $\prod \mathsf{p}_i^{m_i}$ of P^m. Note that

$$n = \prod \|\mathsf{p}_i\|^{m_i} \quad \text{and} \quad m = \sum m_i \quad \text{iff} \quad n^{-x}y^m = \prod \|\mathsf{p}_i\|^{-m_i x}y^{m_i}.$$

Thus $a(n,m)$ is the number of sequences (m_i) of nonnegative integers such that $n^{-x}y^m = \prod \|\mathsf{p}_i\|^{-m_i x}y^{m_i}$. Now consider the product

(B.5.2)
$$\left(1 + \|\mathsf{p}_0\|^{-x}y + \|\mathsf{p}_0\|^{-2x}y^2 + \cdots\right) \cdots \left(1 + \|\mathsf{p}_k\|^{-x}y + \|\mathsf{p}_k\|^{-2x}y^2 + \cdots\right),$$

where k is sufficiently large to ensure that all indecomposables p with norm at most n appear in the list $\mathsf{p}_0, \ldots, \mathsf{p}_k$. Then each sequence (m_i) with $n^{-x}y^m = \prod \|\mathsf{p}_i\|^{-m_i x}y^{m_i}$ corresponds to a choice of one monomial from each of the sums $1 + \|\mathsf{p}_i\|^{-x}y + \|\mathsf{p}_i\|^{-2x}y^2 + \cdots$ above, and thus $a(n,m)$ is the coefficient of $n^{-x}y^m$ in the expansion of (B.5.2).

Validity of the two variable fundamental identity. Let $\mathcal{A}$ be a multiplicative number system with abscissa of convergence α.

Proposition B.6. *The two variable fundamental identity*

$$\sum_{\substack{m \geq 0 \\ n \geq 1}} a(n,m)n^{-x}y^m = \prod_{n \geq 2} (1 - n^{-x}y)^{-p(n)}$$

of $\mathcal{A}$ holds for $x \geq 0$ and $0 \leq y < 1$, with convergence if $x > \alpha$.

PROOF. This follows from the observation that, for $x \geq 0$ and $0 \leq y < 1$, and for any $\ell \geq 1$,

$$\sum_{\substack{0\leq m\leq \ell\\ 1\leq n\leq \ell}} a(n,m)n^{-x}y^m \;\leq\; \prod_{2\leq k\leq \ell}\Big(\sum_{j\geq 0} k^{-jx}y^j\Big)^{p(k)} \;\leq\; \sum_{\substack{m\geq 0\\ n\geq 1}} a(n,m)n^{-x}y^m,$$

and thus,

$$\sum_{\substack{0\leq m\leq \ell\\ 1\leq n\leq \ell}} a(n,m)n^{-x}y^m \;\leq\; \prod_{2\leq k\leq \ell}(1-k^{-x}y)^{-p(k)} \;\leq\; \sum_{\substack{m\geq 0\\ n\geq 1}} a(n,m)n^{-x}y^m,$$

so,

$$\sum_{\substack{m\geq 0\\ n\geq 1}} a(n,m)n^{-x}y^m \;=\; \prod_{n\geq 2}(1-n^{-x}y)^{-p(n)}.$$

If $x > \alpha$ and $0 \leq y < 1$ then

$$\begin{aligned}\sum_{m,n} a(n,m)n^{-x}y^m \;&\leq\; \sum_{m,n} a(n,m)n^{-x}\\ &=\; \sum_{n}\Big(\sum_{m} a(n,m)\Big)n^{-x}\\ &=\; \sum_{n} a(n)n^{-x}\\ &=\; \mathbf{A}(x) \;<\; \infty.\end{aligned}$$

□

Alternate version of the two variable fundamental identity. Just as with the fundamental identity, there is a useful alternate form involving exponentiation.

Proposition B.7. *Let $\mathcal{A}$ be a multiplicative number system with abscissa of convergence $\alpha < \infty$. Then the identity*

$$\mathbf{A}(x,y) \;=\; \exp\Big(\sum_{m\geq 1}\mathbf{P}(mx)y^m/m\Big)$$

holds when $x > \alpha$ and $0 \leq y < 1$.

PROOF. For $x > \alpha$ and $0 \leq y < 1$,

$$\begin{aligned}(1-n^{-x}y)^{-p(n)} \;&=\; \exp\Big(\log\big((1-n^{-x}y)^{-p(n)}\big)\Big)\\ &=\; \exp\big(-p(n)\log(1-n^{-x}y)\big)\\ &=\; \exp\Big(p(n)\sum_{m\geq 1} n^{-mx}y^m/m\Big),\end{aligned}$$

thus,

$$\prod_{1\le n\le M}\left(1-n^{-x}y\right)^{-p(n)} = \prod_{1\le n\le M}\exp\Big(p(n)\sum_{m\ge 1}n^{-mx}y^m/m\Big).$$

So, from Proposition B.6,

$$\begin{aligned}
\mathbf{A}(x,y) &= \prod_{n\ge 2}\left(1-n^{-x}y\right)^{-p(n)}\\
&= \lim_{M\to\infty}\prod_{2\le n\le M}\exp\Big(\sum_{m\ge 1}p(n)n^{-mx}y^m/m\Big)\\
&= \lim_{M\to\infty}\exp\Big(\sum_{2\le n\le M}\sum_{m\ge 1}p(n)n^{-mx}y^m/m\Big)\\
&= \exp\Big(\sum_{n\ge 2}\sum_{m\ge 1}p(n)n^{-mx}y^m/m\Big)\\
&= \exp\Big(\sum_{m\ge 1}\sum_{n\ge 2}p(n)n^{-mx}y^m/m\Big)\\
&= \exp\Big(\sum_{m\ge 1}\mathbf{P}(mx)y^m/m\Big).
\end{aligned}$$

□

The generating series of P^m. Now it will be proved that the generating series of P^m can be expressed as a polynomial in $\mathbf{P}(x),\ldots,\mathbf{P}(mx)$ with nonnegative rational coefficients. The case $m=0$ is trivial as one has $\mathbf{P}^0(x)=1$, so a proof is only needed when $m>0$.

Proposition B.8. *Given a positive integer m, for any $x\ge 0$,*

$$\text{(B.8.1)}\qquad \mathbf{P}^m(x) = \sum_{j=1}^{m}\frac{1}{j!}\sum_{\substack{m_1+\cdots+m_j=m\\ m_i\ge 1}}\frac{1}{m_1\cdots m_j}\mathbf{P}(m_1x)\cdots\mathbf{P}(m_jx),$$

and the two sides converge iff $\mathbf{P}(x)$ converges.

Proof. Since $\mathbf{P}^m(x)=\sum_{n\ge 2}a(n,m)n^{-x}$, it follows that

$$\text{(B.8.2)}\qquad \mathbf{A}(x,y) = \sum_{m\ge 0}\mathbf{P}^m(x)y^m,$$

so, from Proposition B.7,

$$\begin{aligned}
\sum_{m\ge 0}\mathbf{P}^m(x)y^m &= \exp\Big(\sum_{m\ge 1}\mathbf{P}(mx)y^m/m\Big)\\
\text{(B.8.3)}\qquad &= \sum_{j\ge 0}\frac{1}{j!}\Big(\sum_{m\ge 1}\mathbf{P}(mx)y^m/m\Big)^j.
\end{aligned}$$

This means that one has $\mathbf{P}^m(x)$ by collecting the terms from (B.8.3) that contribute to the coefficient of y^m, and this gives equation (B.8.1), at least for $x > \alpha$.

To see that (B.8.1) holds when $x \geq 0$, with convergence precisely when $\mathbf{P}(x)$ converges, note that $\mathbf{P}(x), \ldots, \mathbf{P}(mx)$ are Dirichlet series with non-negative integer coefficients, so when $x \geq 0$, if they converge at x then they converge absolutely. Furthermore, if $\mathbf{P}(x)$ converges at $x \geq 0$ then so do each of $\mathbf{P}(2x), \ldots, \mathbf{P}(mx)$. After noting that one of the summands in the right side of (B.8.1) is

$$\frac{1}{m!}\mathbf{P}(x)^m$$

(this comes from $j = m$), it follows that, for $x \geq 0$, the right side of (B.8.1) converges precisely when $\mathbf{P}(x)$ converges. By Lemma 7.15 there is a Dirichlet series $\mathbf{R}(x)$ such that the right side of (B.8.1) equals $\mathbf{R}(x)$ when $x \geq 0$, so

$$\mathbf{P}^m(x) \;=\; \mathbf{R}(x)$$

on (α, ∞). Since $\mathbf{P}^m(x)$ and $\mathbf{R}(x)$ are both Dirichlet series, they must be the same Dirichlet series. Thus (B.8.1) holds when $x \geq 0$, with the two sides converging precisely when $\mathbf{P}(x)$ converges. □

Here are two examples to illustrate (B.8.1):

$$\begin{aligned}\mathbf{P}^2(x) &= \tfrac{1}{2}\mathbf{P}(x)^2 + \tfrac{1}{2}\mathbf{P}(2x)\\ \mathbf{P}^3(x) &= \tfrac{1}{6}\mathbf{P}(x)^3 + \tfrac{1}{2}\mathbf{P}(x)\cdot\mathbf{P}(2x) + \tfrac{1}{3}\mathbf{P}(3x).\end{aligned}$$

In the additive case one has a polynomial $\mathbf{r}\big(\mathbf{P}(x), \ldots, \mathbf{P}(x^m)\big)$ to express $[m\mathbf{P}](x)$, and now one has, with the same polynomial,

$$\mathbf{P}^m(x) \;=\; \mathbf{r}\big(\mathbf{P}(x), \ldots, \mathbf{P}(mx)\big),$$

since the expression (B.8.1) looks just like the corresponding expression (B.4.1) for the additive case.

APPENDIX C

Consequences of $\delta(\mathsf{P}) = 0$

The major theme in this study of density is to find necessary as well as sufficient conditions that all partition sets have asymptotic density. From Proposition 3.28, the condition $\delta(\mathsf{P}) = 0$ is necessary. Bell [**5**] and Bell, Bender, Cameron and Richmond [**7**] show that this has strong consequences, namely $\rho > 0$ and $\mathbf{A}(\rho) = \infty$.

Note: The Standard Assumption about $\mathcal{A}$ is not in effect in this appendix.

Bell's Positive Radius Theorem: $\delta(\mathsf{P}) = 0$ implies $\rho > 0$

Knowledge of compositions of power series can be useful in showing that a radius of convergence is positive.

Lemma C.1. *Suppose*

$$\mathbf{R}(x) \ = \ \sum_{n=1}^{\infty} r(n)x^n \qquad \text{and} \qquad \mathbf{T}(x) \ = \ \sum_{n=1}^{\infty} t(n)x^n$$

are two power series such that $t(1) \neq 0$, $\rho_{\mathbf{T}} > 0$ *and* $\rho_{\mathbf{T}\circ\mathbf{R}} > 0$. *Then* $\rho_{\mathbf{R}} > 0$.

Proof. Since $t(1) \neq 0$, the inverse $\mathbf{T}^{-1}(x)$ of the power series $\mathbf{T}(x)$ exists. Since $\rho_{\mathbf{T}} > 0$, by Proposition A.27 one has $\rho_{\mathbf{T}^{-1}} > 0$. Since $\mathbf{R} = \mathbf{T}^{-1} \circ \mathbf{T} \circ \mathbf{R}$ it follows that $\rho_{\mathbf{R}} > 0$. □

Here is the key tool.

Proposition C.2. *Suppose*

$$\mathbf{S}(x) \ = \ \sum_{n=1}^{\infty} s(n)x^n \qquad \text{and} \qquad \mathbf{T}(x) \ = \ \sum_{n=1}^{\infty} t(n)x^n$$

are two power series, with $t(1) \neq 0$, $s(n) \geq 0$, *for* $n \geq 1$, *and* $\rho_{\mathbf{T}} > 0$, *such that*

$$2s(n) \ \leq \ [x^n]e^{\mathbf{S}(x)} \ + \ t(n).$$

Then $\rho_{\mathbf{S}} > 0$.

Proof. Let $\mathbf{R}(x) = \sum_{n=1}^{\infty} r(n)x^n$ be the power series that satisfies the equation

$$2\mathbf{R}(x) \ = \ e^{\mathbf{R}(x)} \ + \ \mathbf{T}(x) \ - \ 1, \tag{C.2.1}$$

that is, $\mathbf{R}(x)$ is defined by the system of equations

$$\begin{aligned} r(0) &= 0 \\ 2r(n) &= [x^n]e^{\mathbf{R}(x)} + t(n) - 1 \qquad (n \geq 1). \end{aligned}$$

(This system indeed has exactly one solution.) Since

$$r(1) = t(1) \neq 0,$$

it follows that $\mathbf{R}(x)$ has an inverse power series $\mathbf{R}^{-1}(x)$. Now, substituting $x = \mathbf{R}^{-1}(u)$ into the equation (C.2.1) gives

$$\mathbf{T}\big(\mathbf{R}^{-1}(u)\big) = -e^u + 2u + 1.$$

Thus $\rho_{\mathbf{R}^{-1}} > 0$ by Lemma C.1. Then, by Lemma A.27, $\rho_{\mathbf{R}} > 0$.

Now it is proved, by induction on n, that $0 \leq s(n) \leq r(n)$, for $n \geq 1$. For $n = 1$,

$$2s(1) \leq [x]e^{\mathbf{S}(x)} + t(1) = s(1) + t(1),$$

so $s(1) \leq t(1) = r(1)$.

Hence the claim is true when $n = 1$. Suppose it is true for values less than n. Then

$$\begin{aligned} 2s(n) &\leq [x^n]e^{\mathbf{S}(x)} + t(n) \\ &= [x^n]e^{s(1)x+s(2)x^2+\cdots+s(n)x^n} + t(n) \\ &\leq [x^n]e^{r(1)x+r(2)x^2+\cdots+r(n)x^n+(s(n)-r(n))x^n} + t(n), \end{aligned}$$

since $s(k) \leq r(k)$, for $k < n$. Thus

$$\begin{aligned} 2s(n) &\leq [x^n]\Big(e^{r(1)x+\cdots+r(n)x^n} \cdot e^{(s(n)-r(n))x^n}\Big) + t(n) \\ &= [x^n]\Big(e^{\mathbf{R}(x)} \cdot \big(1 + \big(s(n) - r(n)\big)x^n\big)\Big) + t(n) \\ &= [x^n]e^{\mathbf{R}(x)} + s(n) - r(n) + t(n) \\ &= \big(2r(n) - t(n)\big) + s(n) - r(n) + t(n) \\ &= r(n) + s(n). \end{aligned}$$

Hence $0 \leq s(n) \leq r(n)$, for $n \geq 1$. Since $\rho_{\mathbf{R}} > 0$, it follows that $\rho_{\mathbf{S}} > 0$. This completes the proof. □

Theorem C.3 (Bell [**5**]). *Let $\mathcal{A}$ be an additive number system such that $\delta(\mathsf{P}) = 0$. Then $\rho > 0$.*

Proof. Let $\widetilde{\mathcal{A}} = (\widetilde{\mathsf{A}}, \widetilde{\mathsf{P}}, +, 0, \|\ \|)$ be an additive number system with counting functions $\widetilde{a}(n), \widetilde{p}(n)$, where $\widetilde{p}(n) = p(n) + 1$. Then, since $\mathbf{P}(x)$ and $\widetilde{\mathbf{P}}(x)$ have the same radius of convergence, it follows from Corollary 2.25 (b) that $\rho = \widetilde{\rho}$. Also, $\delta(\widetilde{\mathsf{P}}) = 0$ since

$$\frac{\widetilde{p}(n)}{\widetilde{a}(n)} = \frac{p(n)+1}{\widetilde{a}(n)} \leq \frac{p(n)+1}{a(n)} \to 0.$$

So assume, without loss of generality, that $\mathcal{A}$ is such that $p(n) > 0$, for $n \geq 1$ (by replacing $\mathcal{A}$ by $\widetilde{\mathcal{A}}$).

Suppose $\rho = 0$. Since $\delta(\mathsf{P}) = 0$ there exists an integer $N \geq 1$ such that

$$\frac{a(n)}{p(n)} > 3, \qquad \text{for } n > N. \tag{C.3.1}$$

Define the two sets

$$S_1 = \left\{n > N : \frac{a(n)}{h(n)} \geq 2\right\} \tag{C.3.2}$$

and

$$S_2 = \left\{n > N : \frac{a(n)}{h(n)} < 2\right\}. \tag{C.3.3}$$

From $h(n) = \sum_{ij=n} p(i)/j$, by Möbius inversion one has

$$p(n) = \sum_{j|n} \frac{\mu(j)h(n/j)}{j}.$$

Suppose $n \in S_2$. By (C.3.1) one has $\dfrac{p(n)}{h(n)} < \dfrac{2}{3}$. Thus

$$\sum_{j|n} \frac{\mu(j)h(n/j)}{j} = p(n) < \frac{2h(n)}{3}. \tag{C.3.4}$$

But

$$\begin{aligned} \sum_{j|n} \frac{\mu(j)h(n/j)}{j} &= h(n) + \sum_{\substack{j|n \\ j\neq 1}} \frac{\mu(j)h(n/j)}{j} \\ &\geq h(n) - \sum_{\substack{j|n \\ j\neq 1}} \frac{h(n/j)}{j}. \end{aligned}$$

Combining this result with (C.3.4) gives

$$\begin{aligned} 2h(n)/3 &> \sum_{j|n} \frac{\mu(j)h(n/j)}{j} \\ &\geq h(n) - \sum_{\substack{j|n \\ j\neq 1}} \frac{h(n/j)}{j}, \end{aligned}$$

so,

$$\sum_{\substack{j|n \\ j\neq 1}} \frac{h(n/j)}{j} > \frac{h(n)}{3}.$$

Choose a divisor $j \neq 1$ of n such that $h(n/j)/j$ is maximum among the values $h(n/2)/2, \ldots, h(n/n)/n$. Then

$$d(n)\frac{h(n/j)}{j} > \frac{h(n)}{3},$$

where $d(n)$ counts the number of divisors of n, so

$$\frac{h(n/j)}{j} > \frac{h(n)/3}{d(n)} \geq \frac{h(n)}{3n}.$$

For this choice of j, and since $n \in S_2$,

$$\begin{aligned} 2h(n) > a(n) &= [x^n]e^{\mathbf{H}(x)} \\ &\geq h(n) + \frac{h(n/j)^j}{j!} \\ &\geq h(n) + \frac{\big(jh(n)\big)^j}{(3n)^j j!} \\ &\geq h(n) + \frac{h(n)^j}{(3n)^j}, \end{aligned}$$

so

$$h(n) < (3n)^{j/(j-1)} \leq 9n^2.$$

Thus, for $n \in S_2$, one has $h(n) < 9n^2$. So, there is a $c > 0$ such that $h(n) < cn^2$, for $n \in S_2 \cup \{1, 2, \ldots, N\}$. Thus $h(n) < cn^2$ if $n \notin S_1$. Let

$$\mathbf{T}(x) = 2c \sum_{j \notin S_1} j^2 x^j.$$

By the Cauchy-Hadamard formula for the radius of convergence, $\mathbf{T}(x)$ has a radius of convergence equal to 1 if S_2 is infinite, and ∞ if S_2 is finite.

For $n \notin S_1$,

$$\begin{aligned} [x^n]\big(2\mathbf{H}(x) - \mathbf{T}(x)\big) &= 2\big(h(n) - cn^2\big) \\ &< 0 \\ &\leq a(n) \\ &= [x^n]e^{\mathbf{H}(x)}. \end{aligned}$$

For $n \in S_1$,

$$[x^n]\big(2\mathbf{H}(x) - \mathbf{T}(x)\big) = 2h(n) \leq a(n) = [x^n]e^{\mathbf{H}(x)}.$$

Hence, for any $n \geq 1$,

$$2h(n) \leq [x^n]e^{\mathbf{H}(x)} + t(n).$$

Now $t(1) = 2c > 0$, so by Proposition C.2, $\rho_{\mathbf{H}} > 0$. Since $\rho = \rho_{\mathbf{H}}$ one has a contradiction. $\square$

Remark C.4. Jason Bell actually proved the stronger result (for reduced $\mathcal{A}$):

$$\rho = 0 \qquad \Rightarrow \qquad \limsup_{n\to\infty} \frac{p(n)}{a(n)} \;=\; 1.$$

This was announced in October, 1999. Richard Warlimont, who had also been looking at this problem (motivated in part by an earlier draft of this book), found a completely different proof in November, and by December he could prove the analog for multiplicative number systems [**55**].

Showing $\mathbf{A}(\rho) = \infty$

Now it is proved that the condition that all partition sets of $\mathcal{A}$ have asymptotic density leads to $\mathbf{A}(\rho) = \infty$. The following technical result is needed.

Lemma C.5 (Stam). *Let $\mathbf{T}(x)$ be a power series with nonnegative coefficients, $t(0) = 0$, and $\rho = \rho_{\mathbf{T}} > 0$. Let*

$$\sum \frac{1}{n!}\mathfrak{q}_n(y)x^n \;=\; e^{y\mathbf{T}(x)}.$$

If $\mathbf{T}(\rho)$ converges, then

$$\limsup_{n\to\infty} \frac{\mathfrak{q}_n(y)}{\mathfrak{q}_n(1)} \;>\; 0, \qquad \textit{for } y > 0.$$

PROOF. See Theorem 3 of Stam [**53**]. □

The following result has stronger hypotheses than needed (see Remark C.9 below), but this version is quite sufficient for our purposes.

Theorem C.6 (Bell, Bender, Cameron and Richmond [**7**]). *Suppose that $\mathcal{A} \in \mathsf{RT}_\rho$ and $\delta(\mathsf{P}) = 0$. Then $\mathbf{A}(\rho) = \infty$.*

PROOF. By Theorem C.3, $\rho > 0$. For the case that $\rho = 1$ the result is trivial. So assume $0 < \rho < 1$. Since $\delta(\mathsf{P}) = 0$ one has $p(n)/a(n) \to 0$. Then from Proposition 3.46 follows

$$\frac{p(n)}{s(n)} \to 0,$$

where

$$\mathbf{S}(x) \;=\; \sum s(n)x^n \;=\; e^{\mathbf{P}(x)},$$

the power series defined by formal exponentiation.

Assume $\mathbf{A}(\rho)$ converges. Then $\mathbf{P}(\rho)$ converges, so apply Lemma C.5, with $\mathbf{T}(x) = 2\mathbf{P}(x)$ and $y = 1/2$, to obtain

$$\text{(C.6.1)} \qquad \limsup_{n\to\infty} \frac{\mathfrak{q}_n(1/2)}{\mathfrak{q}_n(1)} \;>\; 0.$$

Now

$$\begin{aligned}\sum \frac{1}{n!}\mathbf{q}_n(1/2)x^n &= e^{\mathbf{P}(x)} &= \mathbf{S}(x)\\ \sum \frac{1}{n!}\mathbf{q}_n(1)x^n &= e^{2\mathbf{P}(x)} &= \mathbf{S}(x)^2\end{aligned}$$

so

$$\begin{aligned}\frac{1}{n!}\mathbf{q}_n(1/2) &= s(n)\\ \frac{1}{n!}\mathbf{q}_n(1) &= \sum_{k=0}^{n} s(k)s(n-k).\end{aligned}$$

Then, from $\mathbf{S}'(x) = \mathbf{P}'(x) \cdot \mathbf{S}(x)$ follows

$$\begin{aligned}\frac{1}{n!}\mathbf{q}_n(1/2) &= s(n)\\ &= \frac{1}{n}\sum_{k=0}^{n} kp(k)s(n-k)\\ &= \sum_{k=0}^{n} \frac{kp(k)}{ns(k)} s(k)s(n-k)\\ &< \sum_{k<n^{1/2}} n^{-1/2}s(k)s(n-k)\\ &\qquad + \sum_{k\geq n^{1/2}} \frac{p(k)}{s(k)} s(k)s(n-k)\\ &\leq \mathrm{o}(1)\sum_{k=0}^{n} s(k)s(n-k)\\ &= \mathrm{o}(1)\,\frac{1}{n!}\mathbf{q}_n(1),\end{aligned}$$

contradicting (C.6.1). Thus $\mathbf{A}(\rho)$ diverges. □

Corollary C.7. *Suppose $\mathcal{A}$ is such that all partition sets have [global] asymptotic density. Then the [global] asymptotic density of a partition set B is the Dirichlet density $\partial(\mathsf{B})$.*

PROOF. It suffices to prove this when $\mathcal{A}$ is reduced. The hypothesis forces $\delta(\mathsf{P}) = 0$ by Proposition 3.28 (the proof of this proposition does not require that one assume $\rho > 0$). Then $\rho > 0$ follows from Theorem C.3, so $\mathcal{A} \in \mathsf{RT}_\rho$ by Corollary 3.30. Thus $\mathbf{A}(\rho) = \infty$ by Theorem C.6. This means Theorem 3.13 applies. □

Corollary C.8. *Suppose that $\mathcal{A} \in \mathsf{RT}_\rho$ and $\delta(\mathsf{P}) = 0$. Then either P is a finite set or $\mathbf{P}(\rho) = \infty$.*

PROOF. By Corollary 2.25,

$$\mathbf{A}(x) < \infty \quad \text{iff} \quad \mathbf{P}(x) < \infty,$$

for $0 \le x < 1$. If $\rho = 1$ then $\mathbf{P}(\rho) < \infty$ holds iff P is a finite set. If $\rho < 1$ apply Theorem C.6. $\square$

Remark C.9. Actually Bell, et al [**7**] prove

$$\delta(\mathsf{P}) \;=\; 0 \qquad \Rightarrow \qquad \mathbf{A}(\rho) \;=\; \infty,$$

without assuming $\mathcal{A}$ has any special properties other than being an additive number system.

APPENDIX D

On the Monotonicity of $a(n)$ When $p(n) \leq 1$

When $\mathcal{A} \in \mathsf{RT}_\rho$ with $\rho < 1$ then it is easy to prove that $a(n)$ is eventually strictly increasing. One only has to apply Schur's Tauberian Theorem to $(1 - x) \cdot \mathbf{A}(x)$ to see that

$$a(n+1) - a(n) \ \sim \ (1 - \rho)a(n).$$

When $\rho = 1$ it need not be the case that $a(n)$ is eventually strictly increasing. In 1956 Bateman and Erdös [**4**] investigated this question under the hypothesis $p(n) \leq 1$, that is, there is at most one indecomposable of each size. They found necessary and sufficient conditions on the support of $p(n)$ for $a(n)$ to be monotone. Recall the definition of $f^{-}(n)$ from Chapter 4.

Definition D.1. For $f : \mathbb{N} \to \mathbb{N}$ let

$$\begin{aligned} f^{-}(n) &= f(n) - f(n-1) \qquad \text{for } n > 0 \\ f^{-}(0) &= f(0). \end{aligned}$$

f^{--} means $\left(f^{-}\right)^{-}$.

Definition D.2. An additive number system $\mathcal{A}$ with counting functions $a(n), p(n)$ and with $r = \sum p(n)$ has the property BE[1] if:

(a) $r \geq 2$, and
(b) $\gcd(\{\|\mathsf{q}\| : \mathsf{q} \in \mathsf{P},\ \mathsf{q} \neq \mathsf{p}\}) \ = \ 1$, for any $\mathsf{p} \in \mathsf{P}$.

Thus $\mathcal{A}$ has BE means that if one deletes any indecomposable from P then the subsystem generated by the remaining indecomposables is (still) reduced.

First consider the case of finitely many indecomposables.

Theorem D.3. *Suppose $\mathcal{A}$ is an additive number system of finite rank r with property BE. Let $\mathsf{supp}\, p(n) = \{d_1, \ldots, d_k\}$ and let $p(d_i) = m_i$. Then*

$$a^{-}(n) \ \sim \ \frac{1}{(r-2)! \prod d_i{}^{m_i}} \cdot n^{r-2}.$$

Consequently $a^{-}(n)$ is eventually positive and

$$\lim_{n \to \infty} \frac{a^{--}(n)}{a^{-}(n)} \ = \ 0.$$

[1] Bateman and Erdös referred to this property as P_1—here the initials BE of Bateman and Erdös offer an appropriately mnemonic name.

PROOF. By Lemma 4.11

$$\begin{aligned}\sum_{n\geq 0} a^{-}(n)x^n &= (1-x)\cdot\sum_{n\geq 0} a(n)x^n \\ &= (1-x)\cdot\prod_{i=1}^{k}(1-x^{d_i})^{-m_i} \\ &= \frac{C}{(1-x)^{r-1}} + \sum_{\substack{j<r-1\\ \theta^{\ell}=1}} \frac{C_{j,\theta}}{(1-\theta x)^j},\end{aligned}$$

using the partial fraction decomposition technique from the proof of Theorem 2.48. The reason for the bound $r-1$ on the j's is that property BE guarantees that a given factor $1-\theta x$, $\theta\neq 1$, can be a factor in at most $r-2$ of the $1-x^{d_i}$ appearing in $\prod_{i=1}^{k}(1-x^{d_i})^{-m_i}$.

Now repeat the proof of Theorem 2.48 to derive the asymptotic expression for $a^{-}(n)$. The asymptotics show that $a^{-}(n)$ is eventually positive and that $\dfrac{a^{-}(n-1)}{a^{-}(n)}\to 1$ as n goes to infinity. Thus $\lim_{n\to\infty}\dfrac{a^{--}(n)}{a^{-}(n)} = 0.$ □

Corollary D.4. *Suppose $\mathcal{A}$ is an additive number system of finite rank r with counting functions $a(n), p(n)$. Then $a(n)$ is eventually strictly increasing iff* BE *holds.*

PROOF. If BE holds then appeal to Theorem D.3.

For the converse, suppose that $a(n)$ is eventually strictly increasing. Let $\mathsf{supp}\,p(n) = \{d_1,\dots,d_k\}$ and let $p(d_i) = m_i$. In the following δ_{ij} is the Kronecker delta function, namely $\delta_{ij}=1$ if $i=j$, otherwise $\delta_{ij}=0$. Then, for $j\in\{d_1,\dots,d_k\}$,

$$\begin{aligned}(1-x^j)\cdot\sum_{n\geq 0} a^{-}(n)x^n &= (1-x^j)\cdot(1-x)\cdot\prod_{i=1}^{k}(1-x^{d_i})^{-m_i} \\ &= (1-x)\cdot\prod_{i=1}^{k}(1-x^{d_i})^{-m_i+\delta_{ij}}\end{aligned}$$

so

$$(1+x+x^2+\dots+x^{j-1})\cdot\sum_{n\geq 0} a^{-}(n)x^n = \prod_{i=1}^{k}(1-x^{d_i})^{-m_i+\delta_{ij}}.$$

Since $a^{-}(n)$ is eventually positive it follows that in the expansion of the left side of the last equation the coefficients are eventually positive. Since $p(j)\geq 1$ this equation is the partition identity of a finitely generated additive number system, the subsystem of $\mathcal{A}$ generated by P with one indecomposable of norm j removed. It follows from Lemma 2.42 that $\gcd\big(\{d_i : m_i-\delta_{ij}>0\}\big)=1$. Thus $\mathcal{A}$ has the property BE. □

Theorem D.5. *Suppose $\mathcal{A}$ is an additive number system with counting functions $a(n), p(n)$ and $p(n) \leq 1$. Then $a(n)$ is eventually strictly increasing iff* BE *holds.*

PROOF. Suppose that $a(n)$ is eventually strictly increasing. The argument to show that BE holds is a repeat of the argument from the last proof, but the product may be infinite. If $p(j) = 1$ then

$$(1 - x^j) \cdot \sum_{n \geq 0} a^-(n) x^n \;=\; (1 - x) \cdot \prod_{n \geq 1} (1 - x^n)^{-p(n) + \delta_{nj}},$$

so

$$(1 + x + x^2 + \cdots + x^{j-1}) \cdot \sum_{n \geq 0} a^-(n) x^n \;=\; \prod_{n \geq 1} (1 - x^n)^{-p(n) + \delta_{nj}}.$$

Since $a^-(n)$ is eventually positive it follows that in the expansion of the left side the coefficients are eventually positive. Since this equation is the partition identity of an additive number system it follows by Lemma 2.42 that $\gcd\big(\{n : p(n) - \delta_{nj} > 0\}\big) = 1$. Thus $\mathcal{A}$ has the property BE.

Suppose that BE holds. The case that $r = \sum_n p(n)$ is finite is covered by Corollary D.4. So now suppose r is infinite.

Choose $p_1(n)$ with finite support such that $0 \leq p_1(n) \leq p(n)$ and $p_1(n)$ has the property BE. Let $p(n) = p_1(n) + p_2(n)$. Let $a_1(n), a_2(n)$ be defined by the additive partition identities:

$$\sum_{n \geq 0} a_i(n) x^n \;=\; \prod_{n \geq 1} (1 - x^n)^{-p_i(n)}, \qquad \text{for } i = 1, 2.$$

Then, from Theorem D.3,

$$\text{(D.5.1)} \qquad \frac{a_1^-(n)}{1 + \big|a_1^{--}(n)\big|} \;\to\; \infty.$$

Let t be a positive integer. By (D.5.1) there exists a positive integer M such that

$$\text{(D.5.2)} \qquad \frac{a_1^-(n)}{1 + |a_1^{--}(n)|} \;\geq\; \begin{cases} -M & \text{for all } n \\ t + 1 & \text{for } n \geq M. \end{cases}$$

Let

$$\text{(D.5.3)} \qquad c \;=\; \max_{0 \leq n \leq M-1} \Big(1 + \big|a_1^{--}(n)\big|\Big).$$

From Lemma 4.9 $\dfrac{a_2(n)}{A_2(n)} \to 0$, as $\dfrac{a_2^\star(n-1)}{a_2^\star(n)} \to 1$ implies $\dfrac{A_2(n-1)}{A_2(n)} \to 1$.
Thus there is a positive integer N such that

$$\text{(D.5.4)} \qquad \frac{\displaystyle\sum_{m=n-M+1}^{n} a_2(m)}{A_2(n)} \;\leq\; \frac{1}{(M + t + 1)c}, \qquad \text{for } n \geq N.$$

For $n \geq \max(M, N)$,

$$\begin{aligned}
a^-(n) &= \sum_{m=0}^{n} a_1^-(n-m) \cdot a_2(m) \qquad \text{by Lemma 4.12} \\
&\geq \sum_{m=0}^{n-M} a_1^-(n-m) \cdot a_2(m) \;-\; \sum_{m=n-M+1}^{n} \left|a_1^-(n-m)\right| \cdot a_2(m) \\
&\geq (t+1) \cdot \sum_{m=0}^{n-M} \left(1 + \left|a_1^{--}(n-m)\right|\right) \cdot a_2(m) \\
&\qquad - M \cdot \sum_{m=n-M+1}^{n} \left(1 + \left|a_1^{--}(n-m)\right|\right) \cdot a_2(m) \quad \text{by (D.5.2)} \\
&= (t+1) \cdot \sum_{m=0}^{n} \left(1 + \left|a_1^{--}(n-m)\right|\right) \cdot a_2(m) \\
&\qquad - (M+t+1) \cdot \sum_{m=n-M+1}^{n} \left(1 + \left|a_1^{--}(n-m)\right|\right) \cdot a_2(m) \\
&\geq (t+1) \cdot \sum_{m=0}^{n} \left(1 + \left|a_1^{--}(n-m)\right|\right) \cdot a_2(m) \\
&\qquad - (M+t+1) \cdot \sum_{m=n-M+1}^{n} c a_2(m) \qquad \text{by (D.5.3)} \\
&= (t+1) \cdot \sum_{m=0}^{n} \left(1 + \left|a_1^{--}(n-m)\right|\right) \cdot a_2(m) \\
&\qquad - (M+t+1) \cdot c \cdot \frac{\sum_{m=n-M+1}^{n} a_2(m)}{A_2(n)} \cdot A_2(n) \\
&\geq \left[(t+1) \cdot \sum_{m=0}^{n} \left(1 + \left|a_1^{--}(n-m)\right|\right) \cdot a_2(m)\right] - A_2(n) \qquad \text{(D.5.4)} \\
&\geq t \cdot \sum_{m=0}^{n} \left(1 + \left|a_1^{--}(n-m)\right|\right) \cdot a_2(m) \\
&\geq t \cdot \sum_{m=0}^{n} a_2(m) \\
&= t \cdot A_2(n)\,.
\end{aligned}$$

Thus, for $n \geq \max(M, N)$, $\dfrac{a^-(n)}{A_2(n)} \geq t$, and thus $a(n) - a(n-1)$ is eventually positive, so $a(n)$ is eventually strictly increasing. □

APPENDIX E

Results of Woods

Woods was interested in modifying the work of Compton so that it would apply to additive number systems that are derived from (partial or total) unary functions with k additional unary predicates. The following Tauberian theorem of Woods [**57**] is a modification of a result of Meir and Moon [**37**].

Woods' Tauberian Theorem

Theorem E.1 (Woods, 1997). *Let* $\mathbf{S}(x), \mathbf{T}(x)$ *be two power series such that, for some* ρ *with* $0 < \rho \leq 1$, $\mathbf{T}(x) \in \mathsf{RT}_\rho$. *Let* $\mathbf{R}(x) = \mathbf{S}(x) * \mathbf{T}(x)$. *If, for some* $C > 0$ *and* $\mu < 1$,

$$s(n) = \mathrm{O}(\rho^{-n}/n) \tag{E.1.1}$$

$$t(n) \sim C\rho^{-n}/n^{\mu} \tag{E.1.2}$$

and $\mathbf{S}(\rho)$ *converges absolutely then* $\lim_{n\to\infty} \dfrac{r(n)}{t(n)}$ *exists and*

$$\lim_{n\to\infty} \frac{r(n)}{t(n)} = \mathbf{S}(\rho) = \lim_{x\to\rho} \frac{\mathbf{R}(x)}{\mathbf{T}(x)}.$$

Proof. The case that $\mu \leq 0$ is covered by Theorem 5.1, in view of Lemma 5.2, with $t_0(n) = Cn^{-\mu}$. So, for the rest of the proof, assume

$$0 < \mu < 1.$$

Assume (by the change of variable $x \leftarrow \rho x$) that $\rho = 1$. Then

$$\left|\mathbf{S}(1) - \frac{r(n)}{t(n)}\right| \leq \underbrace{\left|\mathbf{S}(1) - \sum_{k\leq n} s(k)\right|}_{\mathrm{I}} + \underbrace{\left|\sum_{k\leq n} s(k) - \frac{r(n)}{t(n)}\right|}_{\mathrm{II}},$$

and term I goes to 0 as $n \to \infty$. So one only needs to show that term II vanishes as $n \to \infty$. For this,

$$\left|\sum_{k\le n} s(k) - \frac{r(n)}{t(n)}\right| \le \underbrace{\sum_{k\le\sqrt{n}} \left|s(k)\cdot\left(1-\frac{t(n-k)}{t(n)}\right)\right|}_{\text{III}}$$

$$+ \underbrace{\sum_{\sqrt{n}<k\le n} |s(k)|}_{\text{IV}} + \underbrace{\sum_{\sqrt{n}<k\le n} |s(k)|\frac{t(n-k)}{t(n)}}_{\text{V}}.$$

Term IV goes to 0 as $n \to \infty$ since $\sum_k |s(k)|$ is convergent. Thus it only remains to show that terms III and V also go to 0.

For term III, let k_n be a value of k in $[0, \sqrt{n}]$ for which $\left|1 - \frac{t(n-k)}{t(n)}\right|$ achieves its maximum value. Then

$$\sum_{k\le\sqrt{n}} \left|s(k)\cdot\left(1-\frac{t(n-k)}{t(n)}\right)\right| \le \sum_{k\le\sqrt{n}} \left|s(k)\cdot\left(1-\frac{t(n-k_n)}{t(n)}\right)\right|$$

$$\le \left(\sum_{k\ge 0} |s(k)|\right)\cdot\left|\left(1-\frac{t(n-k_n)}{t(n)}\right)\right|.$$

Since $k_n \le \sqrt{n}$ it follows that $n - k_n \to \infty$ as $n \to \infty$. Since $t(n) \sim Cn^{-\mu}$ it follows that

$$t(n-k_n) \sim C\cdot(n-k_n)^{-\mu}.$$

Now, from $\lim_{n\to\infty} k_n/n = 0$,

$$(n-k_n)^{-\mu} \sim n^{-\mu},$$

and thus

$$t(n-k_n) \sim Cn^{-\mu} \sim t(n).$$

From this, for any $\varepsilon > 0$, for n sufficiently large,

$$\left|1 - \frac{t(n-k_n)}{t(n)}\right| < \varepsilon.$$

This shows that term III goes to 0 as $n \to \infty$.

So now one only needs to show that term V has limit 0 as $n \to \infty$. Let

$$C_n = \sum_{k>\sqrt{n}} |s(k)|,$$

the tail end of the convergent series $\sum_k |s(k)|$. Next define

$$M_n = n\sqrt{C_n} + \sqrt{n}.$$

Then the following hold:

(E.1.3) $$\lim_{n\to\infty} C_n = 0$$

$$\begin{aligned}
&\text{(E.1.4)} & \lim_{n\to\infty} M_n &= \infty \\
&\text{(E.1.5)} & \lim_{n\to\infty} (n - M_n) &= \infty \\
& & \sqrt{n} &< n - M_n \qquad \text{for } n \text{ sufficiently large} \\
&\text{(E.1.6)} & M_n &= \mathrm{o}(n) \\
&\text{(E.1.7)} & nC_n &= \mathrm{o}(M_n).
\end{aligned}$$

Break up term V into two sums as follows:

$$\sum_{\sqrt{n}<k<n} |s(k)| \frac{t(n-k)}{t(n)} = \underbrace{\sum_{\sqrt{n}<k\le n-M_n} |s(k)| \frac{t(n-k)}{t(n)}}_{\text{VI}} + \underbrace{\sum_{n-M_n<k\le n} |s(k)| \frac{t(n-k)}{t(n)}}_{\text{VII}}.$$

It will be proved that each of the terms VI and VII goes to 0 as $n \to \infty$.

Define k_n to be a k in $(\sqrt{n}, n - M_n]$ for which $\dfrac{t(n-k)}{t(n)}$ has maximum value. Then

$$\begin{aligned}
\sum_{\sqrt{n}<k\le n-M_n} |s(k)| \frac{t(n-k)}{t(n)} &\le \frac{t(n-k_n)}{t(n)} \cdot \sum_{\sqrt{n}<k\le n-M_n} |s(k)| \\
\text{(E.1.8)} \qquad &\le C_n \frac{t(n-k_n)}{t(n)}.
\end{aligned}$$

Clearly $n - k_n \ge M_n$ and (E.1.4) imply

$$\text{(E.1.9)} \qquad \lim_{n\to\infty} (n - k_n) = \infty.$$

From (E.1.2), (E.1.4), and (E.1.9),

$$\begin{aligned}
t(n-k_n) &\sim C \cdot (n-k_n)^{-\mu} \\
\text{(E.1.10)} \qquad t(M_n) &\sim C M_n^{-\mu}
\end{aligned}$$

and thus

$$\text{(E.1.11)} \qquad \frac{t(n-k_n)}{t(M_n)} \sim \Big(\frac{M_n}{n-k_n}\Big)^{\mu}.$$

From $k_n \le n - M_n$, one has $\dfrac{M_n}{n-k_n} \le 1$, so $\Big(\dfrac{M_n}{n-k_n}\Big)^{\mu} \le 1$. Combining this with (E.1.11) leads to

$$\text{(E.1.12)} \qquad t(n-k_n) = \mathrm{O}\big(t(M_n)\big),$$

and applying (E.1.12) to (E.1.8) gives

$$\begin{aligned}\sum_{\sqrt{n}<k\le n-M_n} |s(k)|\frac{t(n-k_n)}{t(n)} &= \mathrm{O}\Big(C_n\frac{t(M_n)}{t(n)}\Big)\\ &= \mathrm{O}\Big(C_n\cdot\big(n/M_n\big)^{\mu}\Big) && \text{(E.1.2), (E.1.10)}\\ &= \mathrm{O}\Big(\big(C_n\cdot n/M_n\big)^{\mu}\Big) && \text{(E.1.3), } 0<\mu<1,\end{aligned}$$

and the last line goes to 0 as $n\to\infty$ by (E.1.7). Thus term VI goes to 0 as $n\to\infty$.

For term VII,

$$\sum_{n-M_n<k\le n} |s(k)|\frac{t(n-k)}{t(n)} = \sum_{k<M_n}|s(n-k)|\frac{t(k)}{t(n)}.$$

Now choose k_n to be a $k<M_n$ such that $|s(n-k)|$ is maximal. Then

$$\sum_{k<M_n}|s(n-k)|\frac{t(k)}{t(n)} \le |s(n-k_n)|\sum_{k<M_n}\frac{t(k)}{t(n)}.$$

From (E.1.1) and (E.1.5),

$$s(n-k_n) = \mathrm{O}\Big(\frac{1}{n-k_n}\Big),$$

and, since $k_n=\mathrm{o}(n)$ by (E.1.6),

$$\frac{1}{n-k_n} \sim \frac{1}{n}.$$

Thus

$$s(n-k_n) = \mathrm{O}\big(1/n\big). \tag{E.1.13}$$

For $0<\mu<1$, one has an upper bound on $k^{-\mu}$ given by

$$\frac{1}{k^{\mu}} < \int_{k-1}^{k}\frac{dx}{x^{\mu}}, \qquad \text{for } x>0,$$

and this leads to

$$\begin{aligned}\sum_{1\le k<n}\frac{1}{k^{\mu}} &< 1+\int_1^n\frac{dx}{x^{\mu}}\\ &= 1+\Big(\frac{n^{1-\mu}}{1-\mu}-\frac{1}{1-\mu}\Big)\\ &= \frac{n^{1-\mu}}{1-\mu}-\frac{\mu}{1-\mu},\end{aligned}$$

and thus

$$\text{(E.1.14)} \qquad \sum_{1\le k<n} \frac{1}{k^\mu} = \mathrm{O}\big(n^{1-\mu}\big).$$

Now, from $t(k) \sim k^{-\mu}$,

$$\text{(E.1.15)} \qquad t(k) = \mathrm{O}\big(k^{-\mu}\big)$$

$$\text{(E.1.16)} \qquad \frac{1}{t(k)} = \mathrm{O}\big(k^{\mu}\big).$$

Then

$$\begin{aligned} |s(n-k_n)| \sum_{k<M_n} \frac{t(k)}{t(n)} &= \mathrm{O}\Big(\frac{1}{nt(n)} \sum_{k<M_n} t(k)\Big) && \text{by (E.1.13)} \\ &= \mathrm{O}\Big(\frac{1}{n^{1-\mu}} \sum_{1\le k<M_n} k^{-\mu}\Big) && \text{by (E.1.15), (E.1.16)} \\ &= \mathrm{O}\Big(\big(M_n/n\big)^{1-\mu}\Big) && \text{by (E.1.14).} \end{aligned}$$

From (E.1.6),

$$\lim_{n\to\infty} \big(M_n/n\big)^{1-\mu} = 0,$$

and thus term VII goes to 0 as $n \to \infty$. This finishes the proof. □

Woods' Density Theorem

The following theorem from [**57**] is used to prove monadic second-order limit laws for classes of unary functions with unary predicates.

Theorem E.2 (Woods, 1997). *Suppose $\mathcal{A}$ is an additive number system with*

$$\text{(E.2.1)} \qquad p(n) = \mathrm{O}\big(\beta^n/n\big)$$

$$\text{(E.2.2)} \qquad a(n) \sim C\beta^n/n^\mu,$$

for some $C > 0$, $\beta \ge 1$, and $\mu < 1$. Then all partition sets of $\mathcal{A}$ have asymptotic density which equals the Dirichlet density.

Proof. From $a(n) \sim C\beta^n/n^\mu$ one has $\rho = 1/\beta$, and $\mathcal{A} \in \mathsf{RT}_\rho$. If $\beta = 1$ then Theorem 4.2 applies. So assume $\beta > 1$. Let B be a partition set.

Case 1: $\partial(\mathsf{B}) = 0$.
From Corollary 5.7 one has $\delta(\mathsf{B}) = 0$.

Case 2: $\partial(\mathsf{B}) > 0$.
For this case it will be proved that Theorem E.1 applies with

$$\mathbf{S}(x) = [\overline{\mathbf{B}}/\mathbf{A}](x) \quad \text{and} \quad \mathbf{T}(x) = \mathbf{A}(x).$$

Note that $\mathbf{S}(x)$ is absolutely convergent on $[0,\rho]$ by Lemma 5.11.

Let $\mathsf{B} = \sum_{i=1}^{k} \gamma_i \mathsf{P}_i$, let J be the set of indices j in $\{1, \dots, k\}$ such that $\mathbf{A}_{\mathsf{P}_j}(\rho) < \infty$, and let $\mathsf{Q} = \bigcup_{j \in J} \mathsf{P}_j$. Then, for $0 \le x < \rho$, since $\partial(\overline{\mathsf{B}}) > 0$,

$$\begin{aligned}
[\overline{\mathbf{B}}/\mathbf{A}](x) &= \overline{\mathbf{B}}(x)/\mathbf{A}(x) \\
&= \prod_{j \in J} [\gamma_j \mathbf{P}_j](x)/\mathbf{A}_{\mathbf{P}_j}(x) \\
&= \Big(\prod_{j \in J} [\gamma_j \mathbf{P}_j](x)\Big)/\mathbf{A}_{\mathsf{Q}}(x) \\
&= \Big(\prod_{j \in J} [\gamma_j \mathbf{P}_j](x)\Big) \cdot \exp\Big(-\sum_{m \ge 1} \mathbf{Q}(x^m)/m\Big),
\end{aligned}$$

so it follows that

$$\begin{aligned}
|s(n)| &= \left| [x^n]\bigg(\Big(\prod_{j \in J} [\gamma_j \mathbf{P}_j](x)\Big) \cdot \exp\Big(-\sum_{m \ge 1} \mathbf{Q}(x^m)/m\Big)\bigg) \right| \\
&\le [x^n]\bigg(\Big(\prod_{j \in J} [\gamma_j \mathbf{P}_j](x)\Big) \cdot \exp\Big(\sum_{m \ge 1} \mathbf{Q}(x^m)/m\Big)\bigg) \\
&= [x^n]\bigg(\Big(\prod_{j \in J} [\gamma_j \mathbf{P}_j](x)\Big) \cdot \mathbf{A}_{\mathsf{Q}}(x)\bigg) \\
&\le [x^n]\big(\mathbf{A}_{\mathsf{Q}}(x) \cdot \mathbf{A}_{\mathsf{Q}}(x)\big) \\
&= [x^n] \exp\Big(2 \sum_{m \ge 1} \mathbf{Q}(x^m)/m\Big).
\end{aligned}$$

Let $\mathbf{U}(x) = \sum_{n \ge 0} u(n) x^n$ be the power series expansion of $2 \sum_{m \ge 1} \mathbf{Q}(x^m)/m$. Then $u(n) = \mathrm{O}\big(\rho^{-n}/n\big)$ follows from:

$$\begin{aligned}
u(n) &\le 2 \sum_{k|n} p(n/k)/k \\
&= 2 \sum_{k|n} (k/n) p(k) \\
&= 2p(n) \;+\; 2 \sum_{\substack{k|n \\ 1 \le k \le n/2}} (k/n) p(k) \\
&= \mathrm{O}\Big(\rho^{-n}/n \;+\; \sum_{\substack{k|n \\ 1 \le k \le n/2}} (k/n) \rho^{-k}/k\Big) \\
&= \mathrm{O}\big(\rho^{-n}/n\big).
\end{aligned}$$

Now let $\mathbf{V}(x) = \sum_{n\geq 0} v(n)x^n$ be the power series expansion of $\exp\big(\mathbf{U}(x)\big)$. Since

$$\mathbf{A}_{\mathsf{Q}}(\rho) \;=\; \prod_{j\in J} \mathbf{A}_{\mathsf{P}_j}(\rho) \;<\; \infty,$$

$\mathbf{U}(\rho) < \infty$, so $\mathbf{V}(x) = \exp\big(\mathbf{U}(x)\big)$, convergent at ρ. The claim is that $v(n) \in \mathrm{O}(\rho^{-n}/n)$. Differentiating $\mathbf{V}(x) = \exp\big(\mathbf{U}(x)\big)$ gives

$$\begin{aligned}\mathbf{V}'(x) &= \mathbf{U}'(x)\cdot \exp\big(\mathbf{U}(x)\big)\\ &= \mathbf{U}'(x)\cdot \mathbf{V}(x).\end{aligned}$$

By equating the coefficients of x^{n-1},

$$\begin{aligned}nv(n) &= \sum_{k=1}^{n} ku(k)v(n-k)\\ &= \mathrm{O}\Big(\sum_{k=1}^{n} k\big(\rho^{-k}/k\big)v(n-k)\Big)\\ &= \mathrm{O}\Big(\sum_{k=1}^{n} \rho^{-k}v(n-k)\Big)\\ &= \mathrm{O}\Big(\rho^{-n}\cdot\sum_{k=1}^{n} \rho^{n-k}v(n-k)\Big)\\ &= \mathrm{O}\Big(\rho^{-n}\cdot\sum_{k=0}^{n-1} v(k)\rho^{k}\Big),\end{aligned}$$

so $v(n) = \mathrm{O}\big(\rho^{-n}/n\big)$ since

$$\sum_{k=0}^{n-1} v(k)\rho^k \;\leq\; \mathbf{V}(\rho) \;=\; \exp\big(\mathbf{U}(\rho)\big) \;<\; \infty.$$

Now, since $|s(n)| \leq v(n)$, it follows that $s(n) = \mathrm{O}\big(\rho^{-n}/n\big)$. Thus $\mathbf{S}(x)$ and $\mathbf{T}(x)$ fulfill the hypotheses of Theorem E.1. Since $\overline{\mathbf{B}}(x) = \mathbf{S}(x) * \mathbf{A}(x)$ it follows that $\overline{\mathsf{B}}$ has asymptotic density $\mathbf{S}(\rho)$, which is the Dirichlet density of $\overline{\mathsf{B}}$. Thus the same applies to B. □

Bibliography

[1] Tom M. Apostol, *Introduction to Analytical Number Theory.* Springer Verlag, 1976.

[2] Tom M. Apostol, *Modular Functions and Dirichlet Series.* Springer Verlag, 2nd ed., 1990.

[3] Paul T. Bateman and Harold G. Diamond, *Asymptotic distribution of Beurling's generalized prime numbers.* In: W. LeVeque, ed., *Studies in Number Theory*, Math. Assoc. Amer., Studies in Math. **6**, Prentice-Hall, 1969, 152–210.

[4] P.T. Bateman and P. Erdös, *Monotonicity of partition functions.* Mathematika **3** (1956), 1–14.

[5] Jason P. Bell, *When structures are almost surely connected.* Electron. J. Combin. **7** (2000), R36.

[6] Jason P. Bell, *Sufficient conditions for zero-one laws.* Preprint, 2000.

[7] Jason P. Bell, Edward A. Bender, Peter J. Cameron, L. Bruce Richmond, *Asymptotics for the probability of connectedness and the distribution of number of components.* Electron. J. Combin. **7** (2000), R33.

[8] Jason P. Bell and Stanley N. Burris, *Asymptotics for logical limit laws. When the growth of the components is in an RT class.* Preprint, 2000.

[9] Edward A. Bender, *Asymptotic methods in enumeration.* SIAM Rev. **16** (1974), 485–515.

[10] Arne Beurling, *Analyse de la loi asymptotique de la distribution des nombres premiers généralisés, I.* Acta Math. **68** (1937), 255–291.

[11] David Bigelow and Stanley Burris, *Boolean algebras of factor congruences.* Acta Sci. Math. (Szeged) **54** (1990), 11–20.

[12] Stanley Burris, *Spectrally determined first-order limit laws.* Logic and Random Structures (New Brunswick, NJ, 1995), 33–52, ed. by Ravi Boppana and James Lynch. DIMACS Ser. Discrete Math. Theoret. Comput. Sci. **33**, Amer. Math. Soc., Providence, RI, 1997.

[13] Stanley N. Burris, *Logic for Mathematics and Computer Science.* Prentice Hall, 1998.

[14] Stanley Burris, Kevin Compton, Andrew Odlyzko, Bruce Richmond, *Fine spectra and limit laws II. First-order 0–1 laws.* Canad. J. Math. **49** (1997), 641–652.

[15] Stanley Burris and Paweł Idziak, *A directly representable variety has a discrete first-order law.* Internat. J. Algebra Comput. **6** (1996), 269–276.

[16] Stanley Burris and H.P. Sankappanavar, *A Course in Universal Algebra.* Originally published by Springer, 1981. Now freely distributed as the *Millenium Edition* on the internet at `www.thoralf.uwaterloo.ca`.

[17] Stanley Burris and András Sárközy, *Fine spectra and limit laws I. First-order laws.* Canad. J. Math. **49** (1997), 468–498.

[18] Peter J. Cameron, *On the probability of connectedness.* 15th British Combinatorics Conference (Stirling 1995), Discrete Math. **167/168** (1997), 175–187.

[19] C.C. Chang and H.J. Keisler, *Model Theory.* Studies in Logic and Foundations of Mathematics **73**, North-Holland, 1973.

[20] Kevin J. Compton, *A logical approach to asymptotic combinatorics. I. First order properties.* Adv. in Math. **65** (1987), 65–96.

[21] Kevin J. Compton, *A logical approach to asymptotic combinatorics. II. Monadic second-order properties.* J. Combin. Theory, Ser. A **50** (1989), 110–131.

[22] Kevin J. Compton, *Laws in logic and combinatorics.* Algorithms and Order (Ottawa, ON, 1987), 353–383, ed. by I. Rival. Kluwer Acad. Publ., 1989.

[23] Heinz-Dieter Ebbinghaus and Jörg Flum, *Finite Model Theory.* Perspectives in Mathematical Logic, 2nd ed., Springer-Verlag, 1999.

[24] P. Erdös and G. Szekeres, *Über die Anzahl der Abelschen Gruppen gegebener Ordnung und über ein verwandtes zahlentheoretisches Problem.* Acta Litt. Sci. Reg. Univ. Hungar. Fr.–Jos. Sect. Sci. Math. **7** (1934), 94–103.

[25] J.L. Geluk and L. de Haan, *Regular variation, extensions and Tauberian theorems.* CWI Tract **40**, Centre for Mathematics and Computer Science, 1987.

[26] G.H. Hardy and E.M. Wright, *An Introduction to the Theory of Numbers.* Oxford Press, 5th ed., 1980.

[27] Einar Hille, *Analytic Function Theory, Vol. I.* Introduction to Higher Mathematics, Ginn and Company, 1959.

[28] Paweł M. Idziak and Jerzy Tyszkiewicz, *Monadic second order probabilities in algebra. Directly representable varieties and groups.* Logic and random structures (New Brunswick, NJ, 1995), 79–107, DIMACS Ser. Discrete Math. Theoret. Comput. Sci., **33**. Amer. Math. Soc., Providence, RI, 1997.

[29] John Knopfmacher, *Abstract Analytic Number Theory.* North-Holland Mathematical Library **12**, North-Holland, 1975. (Available as a Dover Reprint.)

[30] John Knopfmacher, *Analytic Arithmetic of Algebraic Function Fields.* Lecture Notes in Pure and Applied Mathematics **50**, Marcel Dekker, Inc. 1979.

[31] Arnold Knopfmacher, John Knopfmacher and Richard Warlimont, *"Factorisatio numerorum" in arithmetical semigroups.* Acta Arith. **61** (1992), 327–336.

[32] Arnold Knopfmacher and John Knopfmacher, *Arithmetical semigroups related to trees and polyhedra.* J. Combin. Th., Ser. A. **86** (1999), 85–102.

[33] John Knopfmacher and Richard Warlimont, *Arithmetical semigroups related to trees and polyhedra, II — Maps on surfaces.* Preprint.

[34] Konrad Knopp, *Theory and Application of Infinite Series.* Dover edition, 1990.

[35] Ralph N. McKenzie, George F. McNulty and Walter F. Taylor, *Algebras, Lattices, and Varieties. I.* The Wadsworth and Brooks/Cole Mathematics Series, Wadsworth and Brooks/Cole, 1987.

[36] Günter Meinardus, *Asymptotische Aussagen über Partitionen.* Math. Zeitschr. Bd. **59** (1954), 388–398.

[37] A. Meir and J.W. Moon, *Some asymptotic results useful in enumeration problems.* Aequationes Math. **33** (1987), 260–268.

[38] Melvyn B. Nathanson, *Elementary Methods in Number Theory.* Graduate Texts in Mathematics **195**, Springer-Verlag, 2000.

[39] A.M. Odlyzko, *Asymptotic enumeration methods.* Handbook of Combinatorics, Vol. 1, 2, 1063–1229, Elsevier, Amsterdam, 1995.

[40] A. Oppenheim, *On an arithmetic function.* J. London Math. Soc. **1** (1926), 105–211.

[41] A. Oppenheim, *On an arithmetic function (II).* J. London Math. Soc. **2** (1927), 123–130.

[42] O. Perron, *Über das infinitäre Verhalten der Koeffizienten einer gewissen Potenzreihe.* Archiv d. Math. u. Phys., Ser. 3 **22** (1914), 329–340.

[43] O. Perron, *Über das Verhalten einer ausgearteten hypergeometrischen Reihe bei unbegrenztem Wachstum eines Parameters.* Journal f. d. reine u. angew. Mathematik, **151** (1921), 63–78.

[44] G. Pólya and G. Szegö, *Aufgaben und Lehrsätze aus der Analysis. I.* Springer-Verlag, 1970.

[45] L. Bruce Richmond, *Asymptotic relations for partitions.* J. Number Theory **7** (1975), 389–405.

[46] L. Bruce Richmond, *A general asymptotic result for partitions.* Canad. J. Math. **27** (1975), 1083–1091.

[47] L. Bruce Richmond, *Asymptotic results for partitions. I. The distribution of certain integers.* J. Number Theory **8** (1976), 372–389.

[48] L. Bruce Richmond, *Asymptotic results for partitions. II. A conjecture of Bateman and Erdös.* J. Number Theory **8** (1976), 390–396.

[49] L. Bruce Richmond, *The moments of partitions. II.* Acta Arith. **28** (1975), 229–243.

[50] L. Bruce Richmond, *Asymptotic relations for partitions.* Trans. Amer. Math. Soc. **219** (1976), 379–385.

[51] K.F. Roth and G. Szekeres, *Some asymptotic formulae in the theory of partitions.* Quart. J. Math. Oxford Ser. 2 **5** (1954), 241–259.

[52] Th. Skolem, *Untersuchung über die Axiome des Klassenkalküls und über Produktations- und Summationsprobleme, welche gewisse Klassen von Aussagen betreffen.* Videnskabsakademiet i Kristiania, Skrifter I, No. 3, 1919, pp. 37. Also in "Selected Works in Logic by Th. Skolem", ed. by Jens Erik Fenstak, Scand. Univ. Books, Universitetsforlaget, Oslo, 1970.

[53] A.J. Stam, *Polynomials of binomial type and compound Poisson processes.* J. Math. Anal. Appl. **130** (1988), 493–508.

[54] E.C. Titchmarsh, *The Theory of Functions.* Oxford University Press, (first ed. 1932) 1997.

[55] Richard Warlimont, *Handwritten notes on* $\Delta(\mathsf{P}) = 0$. January, 2000.

[56] Herbert S. Wilf, *Generatingfunctionology.* 2nd ed., Academic Press, Inc., 1994.

[57] Alan R. Woods, *Counting finite models.* J. Symbolic Logic **62** (1997), 925–949.

Symbol Index

Subject Index

Selected Titles in This Series

(Continued from the front of this publication)

54 **Casper Goffman, Togo Nishiura, and Daniel Waterman,** Homeomorphisms in analysis, 1997

53 **Andreas Kriegl and Peter W. Michor,** The convenient setting of global analysis, 1997

52 **V. A. Kozlov, V. G. Maz′ya, and J. Rossmann,** Elliptic boundary value problems in domains with point singularities, 1997

51 **Jan Malý and William P. Ziemer,** Fine regularity of solutions of elliptic partial differential equations, 1997

50 **Jon Aaronson,** An introduction to infinite ergodic theory, 1997

49 **R. E. Showalter,** Monotone operators in Banach space and nonlinear partial differential equations, 1997

48 **Paul-Jean Cahen and Jean-Luc Chabert,** Integer-valued polynomials, 1997

47 **A. D. Elmendorf, I. Kriz, M. A. Mandell, and J. P. May (with an appendix by M. Cole),** Rings, modules, and algebras in stable homotopy theory, 1997

46 **Stephen Lipscomb,** Symmetric inverse semigroups, 1996

45 **George M. Bergman and Adam O. Hausknecht,** Cogroups and co-rings in categories of associative rings, 1996

44 **J. Amorós, M. Burger, K. Corlette, D. Kotschick, and D. Toledo,** Fundamental groups of compact Kähler manifolds, 1996

43 **James E. Humphreys,** Conjugacy classes in semisimple algebraic groups, 1995

42 **Ralph Freese, Jaroslav Ježek, and J. B. Nation,** Free lattices, 1995

41 **Hal L. Smith,** Monotone dynamical systems: an introduction to the theory of competitive and cooperative systems, 1995

40.4 **Daniel Gorenstein, Richard Lyons, and Ronald Solomon,** The classification of the finite simple groups, number 4, 1999

40.3 **Daniel Gorenstein, Richard Lyons, and Ronald Solomon,** The classification of the finite simple groups, number 3, 1998

40.2 **Daniel Gorenstein, Richard Lyons, and Ronald Solomon,** The classification of the finite simple groups, number 2, 1995

40.1 **Daniel Gorenstein, Richard Lyons, and Ronald Solomon,** The classification of the finite simple groups, number 1, 1994

39 **Sigurdur Helgason,** Geometric analysis on symmetric spaces, 1994

38 **Guy David and Stephen Semmes,** Analysis of and on uniformly rectifiable sets, 1993

37 **Leonard Lewin, Editor,** Structural properties of polylogarithms, 1991

36 **John B. Conway,** The theory of subnormal operators, 1991

35 **Shreeram S. Abhyankar,** Algebraic geometry for scientists and engineers, 1990

34 **Victor Isakov,** Inverse source problems, 1990

33 **Vladimir G. Berkovich,** Spectral theory and analytic geometry over non-Archimedean fields, 1990

32 **Howard Jacobowitz,** An introduction to CR structures, 1990

31 **Paul J. Sally, Jr. and David A. Vogan, Jr., Editors,** Representation theory and harmonic analysis on semisimple Lie groups, 1989

30 **Thomas W. Cusick and Mary E. Flahive,** The Markoff and Lagrange spectra, 1989

29 **Alan L. T. Paterson,** Amenability, 1988

28 **Richard Beals, Percy Deift, and Carlos Tomei,** Direct and inverse scattering on the line, 1988

27 **Nathan J. Fine,** Basic hypergeometric series and applications, 1988

For a complete list of titles in this series, visit the AMS Bookstore at **www.ams.org/bookstore/**.